ENCYCLOPEDIA OF MATHEMATICS AND ITS APPLICATIONS

EDITED BY G.-C. ROTA

Volume 33

Factorization calculus and geometric probability

ENCYCLOPEDIA OF MATHEMATICS AND ITS APPLICATIONS

ENCYCLOPEDIA OF MATHEMATICS AND ITS APPLICATIONS

Factorization calculus and geometric probability

R. V. AMBARTZUMIAN

Institute of Mathematics, Armenian Academy of Sciences

The right of the
University of Cambridge
to print and sell
all manner of books
was granted by
Henry VIII in 1534.
The University has printed
and published continuously
since 1584.

CAMBRIDGE UNIVERSITY PRESS

Cambridge

New York Port Chester Melbourne Sydney

Published by the Press Syndicate of the University of Cambridge
The Pitt Building, Trumpington Street, Cambridge CB2 1RP
40 West 20th Street, New York, NY 10011, USA
10 Stamford Road, Oakleigh, Melbourne 3166, Australia

First published 1990

Printed in Great Britain at the University Press, Cambridge

British Library cataloguing in publication data
Ambartzumian, R. V. (Ruben V.), *1941–*
Factorization calculus and geometric probability.
1. Integral geometry
I. Title   II. Series
516.3'62

Library of Congress cataloguing in publication data
Ambartzumian, R. V.
Factorization calculus and geometric probability/R. V. Ambartzumian.
p.   cm.–(Encyclopedia of mathematics and its applications; v. 33)
ISBN 0 521 34535 9
1. Stochastic geometry.   2. Geometric probabilities.
3.   Factorization (Mathematics)   I. Title.   II. Series.
QA273.5.A45 1990
519.2–dc20   89-7314   CIP

ISBN 0 521 34535 9 hardback

CONTENTS

PREFACE

The subject of this book is closely related to and expands classical integral geometry. In its most advanced areas it merges with those topics in geometrical probability which are now known as stochastic geometry. By the application of a number of powerful yet simple new ideas, the book makes a sophisticated field accessible to readers with just a modest mathematical background.

Traditionally, integral geometry considers only finite sets of geometrical elements (lines, planes etc.) and measures in the spaces of such sets. In the spirit of the Erlangen program, these measures should be invariant with respect to an appropriate group acting in basic space – to ensure that we are still in the domain of geometry. Assume that the basic space is $\mathbb{R}^n$ (as is the case in the most of this book). If the group contains translations of $\mathbb{R}^n$, then the measures in question are necessarily totally infinite and *cannot* be normalized to become probability measures. Yet a step towards *countably infinite* sets of geometrical elements changes the situation: spaces of such sets admit *probability measures* which are invariant and these measures are numerous.

The step from finite sets to countably infinite sets directly transfers an integral geometrician into the domain of probability. The vast field of inquiry that opens up surely deserves attention by virtue of the mathematical elegance of its problems and as a potentially rich source of models for applied sciences.

Probability theory offers a ready apparatus for the study of countably infinite random sets, i.e. random point processes. In stochastic geometry their realizations should lie on manifolds which represent the spaces of the geometrical objects in question.

The book begins with chapters on integral geometry. The preparation of the aforementioned step to geometrical point processes dictated the choice of the content of these early chapters, with the result that they have very little overlap with earlier books on integral geometry, including the fundamental

book by Santalo [2]. Chapter 1–4 can be considered as an introduction
to Haar factorization, a rather neglected subject within integral geometry.
Chapter 5 is devoted to combinatorial integral geometry (a complete account
of this novel theory can be found in the author's earlier book [3]). A variety
of geometrical integrals is presented in chapter 6.

Our preoccupation with factorization of measures is justified on further
development: a version of this tool plays a major role in the study of geometrical random processes with laws which are invariant under the action of
the prescribed group. This is the way we approach the fundamental notion of
Palm distribution which is a key to the solution of many problems given here.
The content of the chapters on geometrical processes ranges from presentation
of basic notions and examples in chapter 7, via Palm distribution theory for
point processes in Euclidean spaces (chapter 8) and Palm distributions for
point processes on groups (chapter 9), to the synthesis of many previous ideas
in chapter 10.

Among the innovations which contributed to a simplified treatment of the
whole subject we mention the following:

(1) The systematic use of the so-called Cavalieri principle which replaces
    many Jacobian calculations by clear geometrical reasoning.
(2) Strong reliance on the theorems of uniqueness of Haar measures on
    groups. In particular this enables us to give simple rigorous proofs of
    uniqueness of the standard invariant measures in the spaces of lines,
    planes etc.
(3) Derivation of Haar measures on groups based on knowledge of Haar
    measures on their subgroups.
(4) Integral geometrical analysis of 'realizations' with later use of the results
    in stochastic contexts.
(5) A new treatment of the problem of probabilistic description of 'typical
    elements' generated by geometrical processes by reduction to calculation
    of the intensities of partial (thinned) point processes.
(6) Rigorous application of the results of combinatorial integral geometry
    to problems involving geometrical processes. (The first attempt in this
    direction was made in [3].)

Apart from making things simpler these new ideas yield new approaches to a
number of long-standing problems: random shapes, J. J. Sylvester's *Vierpunktproblem*, translation-invariance with probability 1, zonoids and others.
Doubtless zonoids belong to the basic notions of translatory integral geometry and this theme occurs in many sections of the book.

Readers will surely notice that we often treat the planar and spatial cases
separately, or even restrict ourselves to just the planar case, even if a single

treatment in general $\mathbb{R}^n$ may seem possible. In some cases this is because of the author's desire to make things clearer but in others the spatial case simply remains unexplored.

Thus with some justification the book can claim it is a research monograph in a new field. However, as regards level and style it resembles a textbook. Perhaps this duality complies with the purposes proclaimed by the Editor for this Encyclopedia.

The final stages of preparation of the book took place during the blockade of Armenia which was the result of the struggle to end the separation of Karabach from Armenia. The problem of conveying material between the publisher and the author could not have been solved under the circumstances without the vital shuttle missions of Professor C. Mutafian from Paris. It is the author's duty to acknowledge both this and the absolutely essential help offered with unfading enthusiasm by V. K. Oganian at all phases of the work.

1

Cavalieri principle and other prerequisites

The aim of this chapter is to present some basic mathematical tools on which many constructions in the subsequent chapters depend.

Thus we will often refer to what we call the 'Cavalieri principle'. We try to revive this old familiar name because of the surprising frequency with which the transformations Cavalieri considered about 350 years ago occur in integral geometry.

No less useful will be the principles which we call 'Lebesgue factorization' and 'Haar factorization'. The first is a rather simple corollary of a well-known fact that in $\mathbb{R}^n$ there is only one (up to a constant factor) locally-finite measure which is invariant with respect to shifts of $\mathbb{R}^n$, namely the Lebesgue measure. Haar factorization is a similar corollary of a much more general theorem of uniqueness of Haar measures on topological groups. We use the two devices in the construction of Haar measures on groups starting from Haar measures on subgroups.

Integral geometry binds together such notions as metrics, convexity and measures, and these interconnections remain significant throughout the book; §§1.7 and 1.8 are introductory to this topic.

1.1 The Cavalieri principle

The classical Cavalieri principle in two dimensions can be formulated as follows.

Let D_1 and D_2 be two domains in a plane (see fig. 1.1.1).

If for each value of y the length of the chords X_1 and X_2 coincide, then the areas of D_1 and D_2 are equal.

The proof of this beautiful geometrical proposition follows from the representation of the area of D_i, $i = 1, 2$, by the integral

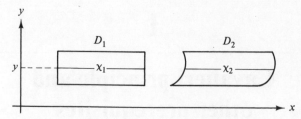

Figure 1.1.1 X_i is the intersection of D_i with the horizontal line on the level y

$$\int X_i \, dy.$$

Pairs of domains having the above property arise whenever we consider transformations of the plane of the type

$$x_1 = x + a(y)$$
$$y_1 = y,$$

which are clearly area-preserving. To these transformations the following interpretation can be given.

We consider the plane as composed of 'rigid' horizontal lines. The transformation

$$(x, y) \rightarrow (x_1, y_1)$$

rigidly shifts each horizontal line along itself (although the shifts can be different for different lines). The domain D_2 in the above example can be considered to be the image of D_1 under some transformation of this type. We call this a *Cavalieri transformation*. Measures in $\mathbb{R}^2$ which remain invariant with respect to Cavalieri transformations are numerous. For instance, measures given by the densities of the form

$$f(y) \, dx \, dy$$

are all invariant as shown by the identity

$$\iint_{D_2} f(y) \, dx \, dy = \int f(y) X_1(y) \, dy = \iint_{D_1} f(y) \, dx \, dy, \qquad (1.1.1)$$

where D_1 and D_2 are as in fig. 1.1.1. Similarly (Fubini theorem) it can be shown that *any* product measure

$$m \times L_1,$$

where L_1 is the Lebesgue measure on the Ox axis and m is *any* measure on the Oy axis, is invariant with respect to Cavalieri transformations of $\mathbb{R}^2$.

In the sequel similar transformations of other product spaces will occur. The typical situation here will be as follows.

Let the basic space $\mathbb{X}$ be a product of two spaces

$$\mathbb{X} = \mathbb{Y} \times \mathbb{R}^k, \quad (x \in \mathbb{X}, y \in \mathbb{Y}, \mathscr{P} \in \mathbb{R}^k)$$

(the second factor is k-dimensional Euclidean).

Assume that a transformation of the space $\mathbb{X}$ has two properties:

(a) it sends each generator, i.e. the set

$$y \times \mathbb{R}^k = \{(y, \mathscr{P}) : y \text{ is constant}, \mathscr{P} \text{ changes in } \mathbb{R}^k\},$$

into the same generator;

(b) it preserves the distances between the points of a generator (i.e. the generators are 'rigid').

Such transformations we again call Cavalieri. The *Cavalieri principle* in this situation is as follows.

Every product measure $m \times L_k$, where L_k is the Lebesgue measure on $\mathbb{R}^k$ and m is any measure on $\mathbb{Y}$, is invariant with respect to Cavalieri transformations of $\mathbb{X}$.

In the spaces of importance described in chapter 2 we actually have natural groups of Cavalieri transformations.

1.2 Lebesgue factorization

We consider a product of two spaces

$$\mathbb{X} = \mathbb{Y} \times \mathbb{R}^k,$$

where $\mathbb{R}^k$ is k-dimensional Euclidean space, while the space $\mathbb{Y}$ here remains unspecified (any separable metric space will suffice).

Let $\mathbb{T}_k$ be the group of translations of $\mathbb{R}^k$. We *define* the action of a translation $t \in \mathbb{T}_k$ on the space $\mathbb{X}$ as follows.

For $(y, \mathscr{P})$, where $y \in \mathbb{Y}$, $\mathscr{P} \in \mathbb{R}^k$, we put

$$t(y, \mathscr{P}) = (y, t\mathscr{P})$$

(this means that t is y-preserving).

A measure μ on $\mathbb{X}$ is called *invariant with respect to* $\mathbb{T}_k$ (or simply $\mathbb{T}_k$-invariant) if for every $t \in \mathbb{T}_k$ and $C \in \mathbb{X}$ we have

$$\mu(tC) = \mu(C), \quad \text{where} \quad tC = \{t(y, \mathscr{P}) : (y, \mathscr{P}) \in C\}. \tag{1.2.1}$$

To check (1.2.1) it is enough to consider the product sets, i.e. to take

$$C = A \times B, \quad A \subset \mathbb{Y}, \quad B \subset \mathbb{R}^k,$$

in which case (1.2.1) reduces to

$$\mu(t(A \times B)) = \mu(A \times tB) = \mu(A \times B),$$

where tB denotes the translation of B by t:

$$tB = \{t\mathscr{P} : \mathscr{P} \in B\}.$$

As usual, measures which are finite on compact sets we call locally-finite. We say that a measure m *on* $\mathbb{X}$ has a locally-finite projection on $\mathbb{R}^k$, if for every compact $B \subset \mathbb{R}^k$ we have

$$m(\mathbb{Y} \times \mathbb{B}) < \infty.$$

The Lebesgue factorization principle states that:

Any locally-finite and $\mathbb{T}_k$-invariant measure μ on $\mathbb{X} = \mathbb{Y} \times \mathbb{R}^k$ is necessarily a product measure:

$$\mu = m \times L_k,$$

where L_k is Lebesgue measure on $\mathbb{R}^k$ and m is a locally-finite measure on $\mathbb{Y}$. If additionally m has a locally-finite projection on $\mathbb{R}^k$, then

$$\mu = \lambda \cdot P \times L_k,$$

where $\lambda \geqslant 0$ is a constant and P is a probability measure on $\mathbb{Y}$, *i.e.* $P(\mathbb{Y}) = 1$.

Proof Let us fix a set $A_0 \subset \mathbb{Y}$ which has compact closure and let us regard $\mu(A_0 \times B)$ as a set function depending on B. It follows from the properties of μ that this is a measure on $\mathbb{Y}$ which is translation-invariant and locally-finite. It is known from analysis that any such measure is proportional to Lebesgue measure, i.e.

$$\mu(A_0 \times B) = m(A_0) \cdot L_k(B).$$

So far $m(A_0)$ has been some constant which does not depend on B, but may depend on our choice of A_0.

Now we fix a set $B_0 \subset \mathbb{R}^k$ which has compact closure and consider

$$\mu(A \times B_0) = m(A) \cdot L_k(B_0)$$

as a function of A. Clearly $\mu(A \times B_0)$ is a measure on $\mathbb{Y}$. This implies that m is a locally-finite measure on $\mathbb{Y}$. This proves the first assertion. In the case where μ has a locally-finite projection on $\mathbb{R}^k$, we have

$$\mu(\mathbb{Y} \times B_0) = m(\mathbb{Y}) \cdot L_k(B_0) < \infty;$$

i.e.

$$m(\mathbb{Y}) < \infty.$$

We get the second assertion when we put (assuming $\lambda > 0$)

$$\lambda = m(\mathbb{Y}), \quad P = \lambda^{-1} m.$$

In the factorization table 2.8.1 we give several important examples where Lebesgue factorization is directly applied.

We mention, however, that the factorizations of table 2.9.1 are valid under quite different conditions: in the corresponding spaces the group of shifts no longer transforms product sets into product sets.

Remark on terminology In this book we consider only locally-finite measures. However, in the text we often omit the adjective 'locally-finite'. Thus a 'measure' will always mean a 'locally-finite measure'.

1.3 Haar factorization

The main idea of Lebesgue factorization can be extended to product spaces where one of the space factors is a group.

For a broad class of locally-compact topological groups (an exact account of the theory can be found in [4]) an important theorem is valid which establishes existence and uniqueness (up to a constant factor) of the so-called *left-invariant* and *right-invariant Haar measures*. In general the two measures need not be proportional. When they are, we have the *bi-invariant* Haar measure (which is again defined up to a constant factor).

We will always tacitly assume that our groups belong to the class mentioned above, as do all the concrete groups we consider in this book.

Let $\mathbb{U}$ be a group. A non-zero measure h on $\mathbb{U}$ is called *left-invariant Haar* if

$$h(uA) = h(A) \tag{1.3.1}$$

for arbitrary $u \in \mathbb{U}$ and $A \subset \mathbb{U}$. Here

$$uA = \{uu_1 : u_1 \in A\},$$

where uu_1 denotes group multiplication.

A non-zero measure h is called *right-invariant Haar* if

$$h(Au) = h(A) \tag{1.3.2}$$

for any $A \subset \mathbb{U}$ and $u \in \mathbb{U}$. Here

$$Au = \{u_1 u : u_1 \in A\}.$$

We will use the notation $h_{\mathbb{U}}^{(l)}$, $h_{\mathbb{U}}^{(r)}$ and $h_{\mathbb{U}}$, respectively, for left-, right- and bi-invariant measures on $\mathbb{U}$.

By essentially repeating the proof of the previous section we can extend its result to product spaces

$$\mathbb{X} = \mathbb{Y} \times \mathbb{U},$$

where the factor $\mathbb{U}$ is a group (it replaces $\mathbb{R}^k$), $\mathbb{Y}$ again is a separable metric space.

Any measure μ on $\mathbb{X}$ which is invariant with respect to the transformations

$$u_1(y, u) = (y, u_1 u)$$

necessarily factorizes:

$$\mu = m \times h_{\mathbb{U}}^{(l)}, \tag{1.3.3}$$

where m is some measure on $\mathbb{Y}$.

If we use right-multiplication, i.e.

$$u_1(y, u) = (y, uu_1),$$

then the right-invariant Haar measure $h_{\mathbb{U}}^{(r)}$ will appear in (1.3.3). Below we will refer to these factorizations as 'Haar factorizations'

Remark In chapters 8 and 9 (in the point processes context) we apply the above proposition in the situation in which $\mathbb{Y}$ is the space of 'realizations'.

There we gloss over the question of introducing the metric on such $\mathbb{Y}$. (This work has been carried out in detail in [18].)

In some cases we can apply Haar factorization to find Haar measures explicitly, as well as to obtain the criteria of existence of *bi-invariant Haar measures* (i.e. measures which satisfy both (1.3.1) and (1.3.2)).

Let $\mathbb{X}$ be a (non-commutative) group, and let $\mathbb{U}$ and $\mathbb{V}$ be two subgroups of $\mathbb{X}$. Assume that each $x \in \mathbb{X}$ admits both representations

$$
\begin{aligned}
x &= u_l v_r, \quad u_l \in \mathbb{U}, \quad v_r \in \mathbb{V} \\
x &= v_l u_r, \quad u_r \in \mathbb{U}, \quad v_l \in \mathbb{V}
\end{aligned}
\tag{1.3.4}
$$

and that each of these representations is unique. (The letters 'l' and 'r' in the subscripts stand for 'left' and 'right'.)

According to (1.3.4) the *set-theoretical* product $\mathbb{U} \times \mathbb{V}$ can be mapped on $\mathbb{X}$ in two ways:

$$
\begin{aligned}
f_1 &: (u, v) \to uv \\
f_2 &: (u, v) \to vu
\end{aligned}
\tag{1.3.5}
$$

and these maps are one-to-one. In other words, the product $\mathbb{U} \times \mathbb{V}$ can be used as a model for $\mathbb{X}$ *in two different ways*:

$$
\begin{aligned}
(u_l, v_r) &= f_1^{-1}(x) \\
(u_r, v_l) &= f_2^{-1}(x),
\end{aligned}
$$

where f^{-1} denotes the inverse of f.

Now

$$
\begin{aligned}
f_1^{-1}(ux) &= (uu_l, v_r), \quad u \in \mathbb{U}, \\
f_2^{-1}(vx) &= (u_r, vv_l), \quad v \in \mathbb{V}.
\end{aligned}
$$

Therefore the left-invariant Haar measure $h_{\mathbb{X}}^{(l)}$ necessarily admits two Haar factorizations, namely

$$
\begin{aligned}
h_{\mathbb{X}}^{(l)} &\overset{f_1}{\underline{\underline{\ \ }}} h_{\mathbb{U}}^{(l)} \times m_1 \\
h_{\mathbb{X}}^{(l)} &\overset{f_2}{\underline{\underline{\ \ }}} m_2 \times h_{\mathbb{V}}^{(l)},
\end{aligned}
\tag{1.3.6}
$$

where m_1 and m_2 are some measures on $\mathbb{V}$ and $\mathbb{U}$, respectively. The symbol $\overset{f}{\underline{\underline{\ \ }}}$ denotes the image of the measure under the map f.

There are similar equations for the right-invariant Haar measure on $\mathbb{X}$:

$$
\begin{aligned}
h_{\mathbb{X}}^{(r)} &\overset{f_1}{\underline{\underline{\ \ }}} m_1' \times h_{\mathbb{V}}^{(r)} \\
h_{\mathbb{X}}^{(r)} &\overset{f_2}{\underline{\underline{\ \ }}} h_{\mathbb{U}}^{(r)} \times m_2',
\end{aligned}
\tag{1.3.7}
$$

where m_1' and m_2' are some measures on $\mathbb{U}$ and $\mathbb{V}$, respectively.

In the cases where the Haar measures on the subgroups $\mathbb{U}$ and $\mathbb{V}$ are known, the partial information given by these equations can be used for the purpose of finding Haar measures on $\mathbb{X}$. Some examples are given in chapter 4.

The maps f_1 and f_2 can be used to formulate a necessary and sufficient condition of bi-invariance of $h_{\mathbb{X}}$. By repeated application of Haar factorization we conclude that any measure on $\mathbb{X}$ which is invariant with respect to the transformation

$$x \to uxv$$

is necessarily the image under f_1 of the measure

$$c \cdot h_{\mathbb{U}}^{(l)} \times h_{\mathbb{V}}^{(r)},$$

where c is a constant. Similarly, any measure on $\mathbb{X}$ which is invariant with respect to the transformations

$$x \to vxu$$

is necessarily the image of the measure

$$c \cdot h_{\mathbb{V}}^{(l)} \times h_{\mathbb{U}}^{(r)}$$

(perhaps with a different constant) under f_2. These images can be substantially different. But let us assume that *both images are proportional to a measure h on* $\mathbb{X}$. For any $A \subset \mathbb{X}$ we will have

$$h(u_2 v_1 A u_1 v_2) = h(v_1 A u_1) = h(A).$$

Since both $u_2 v_1$ and $u_1 v_2$ represent general elements from $\mathbb{X}$, this is essentially the condition defining the bi-invariant Haar measure on $\mathbb{X}$. We have come to the following result.

On $\mathbb{U} \times \mathbb{V}$ we consider two measures:

$$h_{\mathbb{U}}^{(l)} \times h_{\mathbb{V}}^{(r)} \quad \text{and} \quad h_{\mathbb{U}}^{(r)} \times h_{\mathbb{V}}^{(l)}.$$

Their respective images under f_1 and f_2 are proportional if and only if there exists a (unique) bi-invariant measure $h_{\mathbb{X}}$ on $\mathbb{X}$, $h_{\mathbb{X}}$ being proportional to the above-mentioned image measures.

The practical application of this criterion can be as follows. Each of the pairs (u_1, v_r) or (u_r, v_1) can serve as coordinates on the group $\mathbb{X}$. We express u_1 and v_r in terms of u_r and v_1 (both pairs of variables correspond to the same x as in (1.3.4)):

$$\begin{aligned} u_1 &= \varphi_1(u_r, v_1), \\ v_r &= \varphi_2(u_r, v_1). \end{aligned} \tag{1.3.8}$$

We can assume that f_1 is trivial, i.e.

$$(u_1, v_r) = x.$$

Then f_2^{-1} is given by (1.3.8). Now application of the above criterion reduces to the usual Jacobian calculation, i.e. to a check that the transformation (1.3.8) maps $h_{\mathbb{U}}^{(r)} \times h_{\mathbb{V}}^{(l)}$ into $c \cdot h_{\mathbb{U}}^{(l)} \times h_{\mathbb{V}}^{(r)}$. In all the cases we consider in this book, bi-invariance of $h_{\mathbb{X}}$ implies that the constant c equals *one*. The typical situation will be as follows.

The elements $u \in \mathbb{U}$ and $v \in \mathbb{V}$ will depend on a finite number of parameters. Therefore c will equal the absolute value of the determinant of a matrix which we briefly denote by

$$c = \begin{vmatrix} \dfrac{\partial \varphi_1}{\partial u} & \dfrac{\partial \varphi_1}{\partial v} \\[2mm] \dfrac{\partial \varphi_2}{\partial u} & \dfrac{\partial \varphi_2}{\partial v} \end{vmatrix}.$$

Since c is a constant it is enough to calculate the value of this Jacobian at the points

$$u = 1_{\mathbb{U}} \quad \text{(the unit element of } \mathbb{U})$$

and

$$v = 1_{\mathbb{V}} \quad \text{(the unit element of } \mathbb{V}).$$

We have

$$\begin{vmatrix} \dfrac{\partial \varphi_1(u, v)}{\partial u} & \dfrac{\partial \varphi_1(u, v)}{\partial v} \\[2mm] \dfrac{\partial \varphi_2(u, v)}{\partial u} & \dfrac{\partial \varphi_2(u, v)}{\partial v} \end{vmatrix}_{\substack{u=1_{\mathbb{U}} \\ v=1_{\mathbb{V}}}} = \begin{vmatrix} \dfrac{\partial \varphi_1(u, 1_{\mathbb{V}})}{\partial u} & \dfrac{\partial \varphi_1(1_{\mathbb{U}}, v)}{\partial v} \\[2mm] \dfrac{\partial \varphi_2(u, 1_{\mathbb{V}})}{\partial u} & \dfrac{\partial \varphi_2(1_{\mathbb{U}}, v)}{\partial v} \end{vmatrix}_{\substack{u=1_{\mathbb{U}} \\ v=1_{\mathbb{V}}}}.$$

Because $\mathbb{U}$ and $\mathbb{V}$ will always be groups of transformations of the same space, from (1.3.4) we find

$$\varphi_1(u, 1_{\mathbb{V}}) \equiv u, \qquad \varphi_1(1_{\mathbb{U}}, v) \equiv 1_{\mathbb{U}},$$
$$\varphi_2(u, 1_{\mathbb{V}}) \equiv 1_{\mathbb{V}}, \qquad \varphi_2(1_{\mathbb{U}}, v) \equiv v.$$

We obtain the determinant of the unit matrix, i.e. $c = 1$.

In the chapters that follow we often apply 'differential' notation for Haar measures according to table 1.3.1. Similar notation for uniquely determined invariant measures are also applied in other spaces; for instance, if $\mathscr{P}$ is the generic notation for a point in $\mathbb{R}^n$ then $d\mathscr{P}$ will denote Lebesgue measure in $\mathbb{R}^n$.

Also we use lower indexation as in (1.3.4) to avoid explicit mention of the maps f_1, f_2. Thus, $d^{(l)}u_1\, d^{(r)}v_r$ will denote a measure on $\mathbb{X}$ which is the image of the product measure $d^{(l)}u\, d^{(r)}v$ on $\mathbb{U} \times \mathbb{V}$ under the map f_1. Similarly $d^{(r)}u_r\, d^{(l)}v_l$ will denote the image of $d^{(r)}u\, d^{(l)}v$ (another measure on $\mathbb{U} \times \mathbb{V}$) under f_2. Our result for bi-invariant measures now becomes

$$dx = d^{(l)}u_1\, d^{(r)}v_r = d^{(r)}u_r\, d^{(l)}v_l. \tag{1.3.9}$$

This corresponds to writing a measure in terms of coordinates.

Table 1.3.1

Group	Element	Bi-invariant Haar	Left-invariant Haar	Right-invariant Haar
$\mathbb{X}$	$x \in \mathbb{X}$	dx	$d^{(l)}x$	$d^{(r)}x$

Remark In some cases (especially in chapter 3) we use notation like dl, dV etc. to denote infinitesimal lengths, volumes etc. The exact meaning of the notation will always be clear from the context.

1.4 Further remarks on measures

I One of the principles of the general theory of Haar measures is that on compact groups Haar measures are both finite and bi-invariant. The uniqueness up to a constant factor of course follows from the general statement quoted in §1.3. The above principle can be useful in concrete situations whenever we can point out a *finite left-invariant* (say) Haar measure h_0 (as we do in the case say, of a rotation group in §3.2). Then we automatically conclude bi-invariance and essential uniqueness of h_0. Now we show that the bi-invariance property of a finite left-invariant h_0 can be demonstrated effortlessly.

Suppose h_0 is a left-invariant Haar measure on a (compact) group $\mathbb{U}$. We take $h_0(\mathbb{U}) = 1$ for convenience. Consider the right-transformed measure $h(A) = h_0(Au)$, $A \subset \mathbb{U}$, for a fixed $u \in \mathbb{U}$. This is still a left-invariant Haar measure and obviously

$$h(\mathbb{U}) = h_0(\mathbb{U}u) = h_0(\mathbb{U}) = 1.$$

We argue by the uniqueness of left-invariant measures that

$$h = h_0.$$

This holds for all $u \in \mathbb{U}$ and so h_0 is also a right-invariant Haar.

II In chapter 4 we will use the Haar measure on the multiplicative group of positive numbers. We now denote this group by $\mathbb{X}$, $x \in \mathbb{X}$.

The map

$$y = \ln x$$

isomorphically transforms $\mathbb{X}$ into $\mathbb{T}_1$. Therefore the Haar measure on $\mathbb{X}$ is necessarily the image of the Haar (Lebesgue) measure on $\mathbb{T}_1$ under the map

$$x = e^y.$$

We have

$$dy = \frac{dx}{x},$$

thus *the measure dx/x is the (unique) bi-invariant Haar measure on $\mathbb{X}$*. We also call it the 'logarithmic measure'.

The following precise result often hides behind the name of 'homothety consideration'; it will be of use in chapter 4.

Let m be a locally-finite measure on $(0, \infty)$ for which always

$$m(hB) = h^k m(B),$$

where hB denotes the image of Borel B under a homothety h (alternatively h is the corresponding rescaling factor). Then necessarily m has a density proportional to $x^{k-1} \, \mathrm{d}x$.

Proof the measure $x^{-k}m(\mathrm{d}x)$ is invariant with respect to homotheties and is therefore proportional to the logarithmic measure.

III In line with our concern with the question of uniqueness lies the following proposition.

Let us assume that a measure m which is defined on a product of two spaces

$$\mathbb{Y} \times \mathbb{Z}$$

has two product representations:

$$m = m_1 \times m'$$

and

$$m = m_2 \times m',$$

where m_1 and m_2 are measures on $\mathbb{Y}$ and m' is a measure on $\mathbb{Z}$. If there is a set $A_0 \subset \mathbb{Z}$ for which

$$0 < m'(A_0) < \infty$$

then the measures m_1 and m_2 are identical.

Proof For any $B \subset \mathbb{Y}$ we have

$$m(B \times A_0) = m_1(B) \cdot m'(A_0) = m_2(B) \cdot m'(A_0).$$

Therefore

$$m_1(B) = m_2(B).$$

We call the above the 'elimination of a measure factor' and use it several times in chapters 2–4.

1.5 Some topological remarks

I A number of spaces of integral geometry belong to the class of so-called fibered spaces which generalize the notion of product spaces. A space $\mathbb{X}$ is referred to as fibered when there is a map

$$\pi : \mathbb{X} \to \mathbb{Y}$$

(the projection of $\mathbb{X}$ into a space $\mathbb{Y}$) such that each fiber

$$\{x : \pi(x) = y\}$$

is homeomorphic to a space $\mathbb{Z}$ (the fiber model) which does not depend on y. Note that in §1.1, where $\mathbb{X} = \mathbb{Y} \times \mathbb{Z}$, the fibers are called *generators*. We now give an example of a fibered space which we use in §2.5.

We take the unit sphere $\mathbb{S}_2$ in $\mathbb{R}^3$. At each point $\omega \in \mathbb{S}_2$ we construct the tangent plane $t(\omega)$. Let $\mathbb{X}_1$ be the set of pairs

$$(\omega, \mathscr{P}) \quad \text{where always} \quad \mathscr{P} \in t(\omega).$$

We endow $\mathbb{X}_1$ with a topology in the following way: by definition, a sequence $\{(\omega_n, \mathscr{P}_n)\}$ converges to a point $(\omega, \mathscr{P})$ if and only if

(1) ω_n converges to ω in the usual topology on $\mathbb{S}_2$;
(2) $\mathscr{P}_n$ converges to $\mathscr{P}$ in the topology of $\mathbb{R}^3$.

The space $\mathbb{X}_1$ thus obtained is called the *tangent bundle* of $\mathbb{S}_2$; it is topologically *different* from the product $\mathbb{S}_2 \times \mathbb{R}^2$ (see [56]). The fact that it is impossible to choose coordinate systems for each tangent plane so that they vary continuously over all of the unit sphere is a simple example of the famous topological phenomenon concerning the non-existence of non-zero continuous tangent vector-fields on spheres.

Our $\mathbb{X}_1$ is a fibered space with

$$\pi(\omega, \mathscr{P}) = \omega$$

and we consider the fibers (tangent planes) as Euclidean replicas of $\mathbb{R}^2$ (i.e. we can consider congruent figures on different tangent planes).

Let us consider the planar Lebesgue measure L_2 on each $t(\omega)$. We assume that L_2 is independent of ω in the sense that congruent domains on different tangent planes have equal L_2-measures. With every measure m on $\mathbb{S}_2$ we now associate a measure μ on $\mathbb{X}_1$ by the formula

$$\mu(A) = \int L_2(A_\omega) m(d\omega), \qquad (1.5.1)$$

where $A_\omega = A \cap t(\omega)$ is the trace of A on $t(\omega)$. We call μ a *composition* of Lebesgue measures on fibers.

We call transformation a of $\mathbb{X}_1$ *Cavalieri* if

(1) a maps a fiber into a fiber;
(2) the image of L_2 on each fiber is again L_2 (on another fiber);
(3) a induces a map $\mathbb{Y} \rightarrow \mathbb{Y}$ of fibers which preserve m.

The non-product topology on $\mathbb{X}_1$ does not restrict the use of a type of Cavalieri principle:

On $\mathbb{X}_1$ any composition of Lebesgue measures is invariant under a Cavalieri transformation.

The proof is almost tautological.

Similar Cavalieri principles also hold for other fibered spaces in this book. We stress that in all cases the measures on fibers we compose are invariant with respect to the choice of coordinates on the fibers. As a result, their composition is uniquely determined by the measure in the space of fibers. We had this advantage in the above example.

II The remark made at the end of I concerns the different spaces of figures we consider in this book (such as spaces of lines, planes etc.). There is a general principle which governs our choice of topologies in these spaces:

They comply with the topology in the space $\mathbb{F}$ of closed sets.

By this we mean the following.

Let $\mathbb{X}$ be the basic space where our figures belong (in the case of figures which are lines, $\mathbb{X} = \mathbb{R}^2$ or $\mathbb{R}^3$). By F we now denote closed sets in $\mathbb{X} : F \in \mathbb{F}$. By definition, a sequence F_n converges in $\mathbb{F}$ if and only if it satisfies the two conditions:

(1) if an open set G hits F (i.e. if $G \cap F \neq \varnothing$) then G hits all the F_n except, at most, a finite number of them;
(2) if a compact K is disjoint of F, it is disjoint of all the F_n except, at most, of a finite number of them.

This convergence notion defines the topology on $\mathbb{F}$ (see [1]).

In many cases our figures can be considered as closed sets in $\mathbb{X}$. Then, each time, the topology on $\mathbb{F}$ induces a topology in the space of figures in question: a set A is declared open if A is an intersection of an open set in $\mathbb{F}$ with the total set of figures.

The compliance means that we will be considering homeomorphic models of the spaces of figures where the topology is induced by $\mathbb{F}$ in the above sense.

III An adequate description of a number of spaces of integral geometry requires the notion of elliptical (projective) space. We denote n-dimensional elliptical space by $\mathbf{E}_n$. A model of this space can be obtained from the unit n-dimensional sphere $\mathbb{S}_n$ by 'gluing together' every two points of $\mathbb{S}_n$ which are symmetrical (antipodal). In other words each pair of antipodal points on $\mathbb{S}_n$ is considered to be a single point of the space $\mathbf{E}_n$. Equivalently, we can take a closed half of $\mathbb{S}_n$ (a closed hemisphere) and 'glue together' the points on the boundary which are opposite to each other. Fig. 1.5.1 illustrates the latter

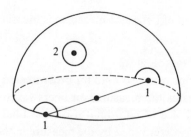

Figure 1.5.1

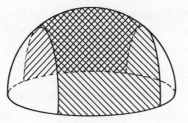

Figure 1.5.2

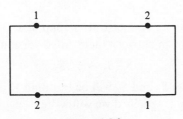

Figure 1.5.3

operation on the example of the construction of $\mathbf{E}_2$ from a two-dimensional hemisphere. The diagram shows the neighborhoods of the points on our model: the neighborhood of point 1 consists of two semicircular parts; the neighbourhood of point 2 is shown as a circle.

Consider that part of our model which is obtained by cutting off the two closed semicircular parts, as shown in fig. 1.5.2. The shaded region is homeomorphic to a rectangle with points on a pair of opposite sides glued together, as shown in fig. 1.5.3. This is the usual construction of an open Möbius band. The topology of the region that remains does not change when the region of $\mathbf{E}_2$ which is removed reduces to a point. Thus the space

$$\mathbf{E}_2 \setminus \{\text{a point}\}$$

is homeomorphic to a Möbius band.

There is a clear one-to-one map between $\mathbf{E}_2$ and spaces of such figures in $\mathbb{R}^3$ as

(1) diameters of $\mathbb{S}_2$;
(2) lines through a point O;
(3) planes through a point O (each plane of this bundle is determined by a line through O normal to the plane).

The topology of $\mathbf{E}_2$ *complies with* the closed sets' trace topology in these spaces. Therefore the spaces (1)–(3) are often described simply as $\mathbf{E}_2$.

$\mathbf{E}_1$ is obtained from $[0, \pi]$ (closed interval) by 'gluing together' its endpoints. Thus $\mathbf{E}_1$ is homeomorphic to a circle. It represents both the space of diameters of $\mathbb{S}_1$ and the space of lines through O in the plane.

1.6 Parametrization maps

Usual 'geographical' coordinates on the sphere $\mathbb{S}_2$ provide the best known example of parametrization. Actually we have a map

$$\mathbb{S}_2 \to \mathbb{S}_1 \times (0, \pi),$$

as shown in fig. 1.6.1. The image (v, Φ) is defined for all points $\omega \in \mathbb{S}_2$ except for the 'poles' N and S.

In a typical situation a parametrization map of a space $\mathbb{X}$ onto a space $\mathbb{Y}$ will be a homeomorphism between their slitted versions

$$\mathbb{X} \backslash S_1 \to \mathbb{Y} \backslash S_2, \tag{1.6.1}$$

where the excluded sets S_1 and S_2 will be less than $\mathbb{X}$ or $\mathbb{Y}$ *in dimensionality*. As soon as such a map is specified we will write

$$\mathbb{X} \approx \mathbb{Y}. \tag{1.6.2}$$

Let m be a measure on $\mathbb{X}$. The image of m under parametrization (1.6.1) will provide an adequate description of m whenever

$$m(S_1) = 0. \tag{1.6.3}$$

Yet in general not every measure m_1 on $\mathbb{Y}$ for which

$$m_1(S_2) = 0 \tag{1.6.4}$$

can be considered to be an image of a measure on $\mathbb{X}$ (recall that in our usage measures are necessarily locally-finite).

Clearly the map converse to (1.6.1) can send a non-compact set $B \subset \mathbb{Y}$ into a subset of a compact set in $\mathbb{X}$. Therefore a measure m_1 on $\mathbb{Y}$ happens to be

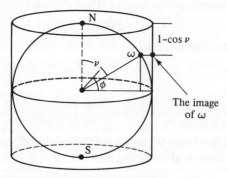

Figure 1.6.1 The circumcylinder is $\mathbb{S}_1 \times (-1, 1)$

an image of a measure on $\mathbb{X}$ whenever an additional condition

$$m_1(B) < \infty \qquad (1.6.5)$$

is satisfied for every $B \subset \mathbb{Y} \setminus S_2$ with the described property.

If both (1.6.4) and (1.6.5) are satisfied, then m_1 in a sense represents a measure on $\mathbb{X}$. For instance, the measure on $S_1 \times (0, \pi)$ given by the density $(v)^{-1}\, dv\, d\Phi$ is locally but not totally finite and fails to represent a measure on S_2. The measure $\sin v\, dv\, d\Phi$ represents the area measure on S_2.

The precautions (1.6.3)–(1.6.5) would be pointless if we could complement the map (1.6.1) by a one-to-one map between S_1 and S_2 with the property that the one-to-one map between $\mathbb{X}$ and $\mathbb{Y}$ that arises sends a compact $C \subset \mathbb{X}$ into a relatively compact set $C' \subset \mathbb{Y}$ and vice versa. Recall that a set is called relatively compact if it can be covered by a compact set.

If such a map between S_1 and S_2 can be established, then *each* measure on $\mathbb{Y}$ represents a measure on $\mathbb{X}$ and vice versa. Such a map turns $\mathbb{Y}$ into a *measure-representing model* of $\mathbb{X}$. Some examples will be given in chapter 2.

1.7 Metrics and convexity

One of the concerns of contemporary integral geometry is the interrelation between the notions of metrics, convexity and measures in the spaces of lines and planes. In this and the next section we outline the simplest facts and leave more detailed discussion of this topic for chapter 5.

Given a bounded convex domain $D \subset \mathbb{R}^2$ we define its breadth function $b(\varphi)$ to be the distance between the pair of parallel support lines of D which are orthogonal to the direction φ (see fig. 1.7.1; by definition, a support line has a point in common with ∂_D but not with the interior of D). In general $b(\varphi)$ *does not* determine a convex D in a unique way; but it does if we additionally assume that D is centrally-symmetric.

After Minkowski [30] we consider *linear continuations* b^* of the breadth

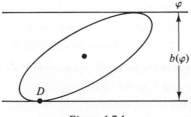

Figure 1.7.1

functions. Given a breadth function $b(\varphi)$ we put

$$b^*(\mathscr{P}) = rb(\varphi),$$

where (r, φ) are the usual polar coordinates of $\mathscr{P} \in \mathbb{R}^2$ with the origin at the symmetry center of D. There is a fundamental proposition [14]:

If $b(\varphi)$ is a breadth function of a centrally-symmetric bounded convex $D \subset \mathbb{R}^2$ which is not a line segment then

$$\rho(\mathscr{P}_1, \mathscr{P}_2) = b^*(\mathscr{P}_2 - \mathscr{P}_1)$$

is a metric in $\mathbb{R}^2$. If D is a line segment then ρ is a pseudometric.

(In this case

$$b(\varphi) = l \cdot |\cos(\varphi - \alpha)|,$$

where l is the length and α is the direction of the segment.)

We recall that a metric in $\mathbb{R}^n$ is a non-negative symmetrical function $\rho(\mathscr{P}_1, \mathscr{P}_2)$, $\mathscr{P}_1, \mathscr{P}_2 \in \mathbb{R}^n$, which satisfies the conditions

(a) $\rho(\mathscr{P}_1, \mathscr{P}_2) = 0$ if and only if $\mathscr{P}_1 = \mathscr{P}_2$;
(b) $\rho(\mathscr{P}_1, \mathscr{P}_3) \leqslant \rho(\mathscr{P}_1, \mathscr{P}_2) + \rho(\mathscr{P}_2, \mathscr{P}_3)$ for every $\mathscr{P}_1, \mathscr{P}_2, \mathscr{P}_3$.

If ρ satisfies (b) but $\rho(\mathscr{P}_1, \mathscr{P}_2) = 0$ does not imply that $\mathscr{P}_1 = \mathscr{P}_2$ then ρ is called a *pseudometric*.

Remarkably the complete Minkowski statement also includes the inversion of the above.

If for a planar metric ρ we have

$$\rho(\mathscr{P}_1, \mathscr{P}_2) = |\mathscr{P}_1, \mathscr{P}_2| \cdot h(\varphi),$$

where $|\mathscr{P}_1, \mathscr{P}_2|$ is the Euclidean distance between $\mathscr{P}_1$ and $\mathscr{P}_2$ and the function h depends only on the direction φ from $\mathscr{P}_1$ to $\mathscr{P}_2$, then $h(\varphi)$ is the breadth function of some centrally-symmetrical convex bounded domain in $\mathbb{R}^2$ which is not a line segment. Under the same conditions any pseudometric ρ necessarily corresponds to a line segment.

Let us turn now to connections of (in general no longer translation-invariant) pseudometrics with measures in the space $\mathbb{G}$ of lines in $\mathbb{R}^2$. We describe this space in §2.2.

Let us denote by $\mathscr{P}_1|\mathscr{P}_2$ the set of lines which separate the points $\mathscr{P}_1$ and $\mathscr{P}_2$; and by $\mathscr{P}_1|\mathscr{P}_2, \mathscr{P}_3$ we denote the set of lines which separate $\mathscr{P}_1$ from $\mathscr{P}_2$ and $\mathscr{P}_3$. We have an identity which can be checked directly:

$$2I_{\mathscr{P}_1|\mathscr{P}_2\mathscr{P}_3}(g) = I_{\mathscr{P}_1|\mathscr{P}_2}(g) + I_{\mathscr{P}_1|\mathscr{P}_3}(g) - I_{\mathscr{P}_2|\mathscr{P}_3}(g),$$

where g denotes a line and I_A is the indicator function of the set A. Integration of the above with respect to any measure m in the space of lines which ascribes zero to any bundle of lines through a point yields

$$2m(\mathscr{P}_1|\mathscr{P}_2, \mathscr{P}_3) = m(\mathscr{P}_1|\mathscr{P}_2) + m(\mathscr{P}_1|\mathscr{P}_3) - m(\mathscr{P}_2|\mathscr{P}_3). \qquad (1.7.1)$$

If we restrict ourselves to measures m on $\mathbb{G}$ whose values on bundles is zero, then it is quite straightforward that *each* function

$$\rho(\mathscr{P}_1, \mathscr{P}_2) = m(\mathscr{P}_1|\mathscr{P}_2) \qquad (1.7.2)$$

is a linearly-additive continuous pseudometric. In particular the triangle inequality property (b) follows from (1.7.1) (where the right-hand side is nonnegative).

Remarkably the following converse statement is also true.

Any pseudometric in $\mathbb{R}^2$ which is linearly-additive and continuous is generated via (1.7.2) by some measure in the space of lines, and this measure is unique.

A complete proof of this statement can be found in [3], where it was derived within the framework of combinatorial ideas (to be outlined in chapter 5). For translation-invariant cases a similar partial conclusion can be drawn using the ideas of §2.11. By Minkowski's proposition, this means that every planar symmetrical bounded convex domain is generated by a translation-invariant measure in the space $\mathbb{G}$.

Which of the above notions virtually generalize to many dimensions, in particular to $\mathbb{R}^3$?

The significance of the breadth functions $b(\omega)$ for a complete description of centrally-symmetrical convex domains survives together with Minkowski's propositions. (In $\mathbb{R}^3$, $b(\omega)$ equals the distance between parallel support planes of a convex D which are orthogonal to the spatial direction ω.) Also the principle (1.7.2) that *measures generate metrics* remains true. (In $\mathbb{R}^3$ we have to interpret $\mathscr{P}_1|\mathscr{P}_2$ as the set of planes separating $\mathscr{P}_1$ from $\mathscr{P}_2$; m becomes a measure in $\mathbb{E}$, the space of planes in $\mathbb{R}^3$.)

Yet in $\mathbb{R}^3$ the situation with the *inversion* of the latter principle changes: in $\mathbb{R}^3$ there exist linearly-additive, continuous metrics which do not admit the representation (1.7.2) with any measure m on $\mathbb{E}$. Accordingly the bounded symmetrical convex domains (bodies) in $\mathbb{R}^3$ split into two subclasses: *zonoids*, i.e. those which correspond to metrics generated by measures in $\mathbb{E}$, and those which do not. (We dwell upon these questions later in §§2.12, 5.10, 6.1 and 6.2.)

Breadth functions are useful in the study of projections of convex bodies on planes (of course the projections are planar convex domains).

Let Ω be the direction normal to the plane on which we project a convex body $D \subset \mathbb{R}^3$, and let $b(\omega)$ be the breadth function of D. By $\langle \Omega \rangle$ we denote the circle of directions orthogonal to Ω (they lie in the plane of projection). The *breadth function $b(\varphi)$ of the projection coincides with the restriction of $b(\omega)$ to the set $\langle \Omega \rangle$*. This is clearly demonstrated by fig. 1.7.2.

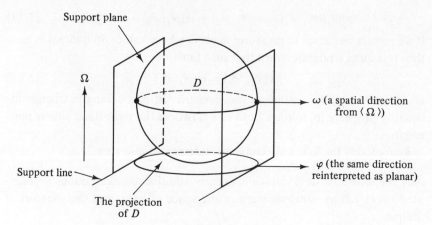

Support plane

Ω

D

ω (a spatial direction from $\langle \Omega \rangle$)

φ (the same direction reinterpreted as planar)

Support line

The projection of D

Figure 1.7.2

1.8 Versions of Crofton's theorem

In books on integral geometry (including [1] and [2]) the following problem is discussed in detail. Given two non-intersecting planar convex domains D_1 and D_2, find the *invariant measure* of the set of lines separating D_1 from D_2. By invariant measure we understand the unique (up to a constant factor) measure in the space of lines in the plane which is invariant with respect to Euclidean motions. We discuss this measure in detail in chapter 3, and it reappears frequently in other chapters.

The solution attributed to Crofton [2] is that the value in question equals 'the least length of a closed string drawn round D_1 and D_2 and crossing over itself at a point O, minus the lengths of the perimeters of D_1 and D_2'. (See fig. 1.8.1.)

Let us consider a version of this result in which D_1 and D_2 are replaced by line segments.

On the plane we have two line segments, δ_1 and δ_2, situated as shown in fig. 1.8.2.

The invariant measure of the lines which hit both δ_1 and δ_2 (or, equivalently, separate s_1 from s_2) equals

$$|d_1| + |d_2| - |s_1| - |s_2|, \tag{1.8.1}$$

where $|d|$ stands for the length of d.

In fact, versions of these simple results for non-invariant measures in the space of lines in the plane lie at the source of the theory of combinatorial integral geometry ([3]).

Although we outline the theory later on (in chapter 5), we will need the following simple fact in chapter 2.

Let us denote by $[\delta]$ the set of lines which hit the segment δ. Except for the lines passing through the endpoints of δ_1 and δ_2 we have (see fig. 1.8.2)

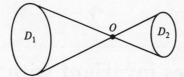

Figure 1.8.1

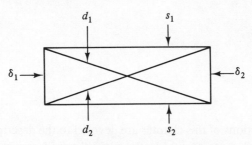

Figure 1.8.2 We denote by d_1 and d_2 the *diagonals* and by s_1 and s_2 the *sides* of the quadrilateral

$$2I_{[\delta_1] \cap [\delta_2]}(g) = I_{[d_1]}(g) + I_{[d_2]}(g) - I_{[s_1]}(g) - I_{[s_2]}(g), \qquad (1.8.2)$$

where g denotes a line and I_A is the indicator function of the set A. (To get the proof it is enough to consider four different positions of g). Integration of (1.8.2) with respect to any measure m in the space of lines (which ascribes zero to any bundle of lines through a point) yields

$$2m([\delta_1] \cap [\delta_2]) = m([d_1]) + m([d_2]) - m([s_1]) - m([s_2]). \qquad (1.8.3)$$

Of course (1.7.1) can be considered as a special case of this relation when δ_1 and δ_2 are situated so as to form two sides of a triangle.

2

Measures invariant with respect to translations

The first few sections of this chapter are devoted to the description of several spaces of integral geometry and to the derivation of factorization results for translation invariant measures on these spaces. As an intermediate step we consider 'slitted' versions of the spaces in question where Lebesgue factorization applies directly. The concept of a 'rose' of a $\mathbb{T}$-invariant measure in the space of lines or planes is introduced, which relates such measures to the theory of convex sets in corresponding $\mathbb{R}^n$. We also apply Lebesgue factorization to problems concerning *random* measures, namely their translation-invariance with probability 1.

2.1 The space $\overline{\mathbb{G}}$ of directed lines on $\mathbb{R}^2$

A directed line g in the Euclidean plane $\mathbb{R}^2$ is a line with an arrow on it. The space $\overline{\mathbb{G}}$ of directed lines can be mapped in a one-to-one way onto the surface of a unit cylinder with the usual cylindrical coordinates (φ, p). Given $g \in \overline{\mathbb{G}}$ we define φ to be the direction of g (φ is a point on the unit circle $\mathbb{S}_1$ or, equivalently, the angle measured as shown in fig. 2.1.1); p is the signed perpendicular distance from the origin O to g ($p > 0$ if O lies to the left of g, $p < 0$ if O is to the right). We may therefore write

$$g = (\varphi, p).$$

In this way the space $\overline{\mathbb{G}}$ attains the product structure

$$\overline{\mathbb{G}} = \mathbb{S}_1 \times \mathbb{R}, \tag{2.1.1}$$

where $\mathbb{S}_1$ is the unit circle. Because of the correspondence

$$g \to \text{the right halfplane bounded by } g$$

the space (2.1.1) can also be considered as the space of closed halfplanes. In this sense the topology of $\overline{\mathbb{G}}$ complies with the space of closed sets on the plane; see §1.5.

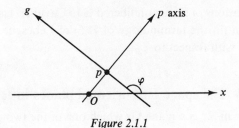

Figure 2.1.1

The group $\mathbb{T}_2$ of parallel shifts of $\mathbb{R}^2$ induces a group of transformations of $\overline{\mathbb{G}}$ (which we also call $\mathbb{T}_2$). A $t \in \mathbb{T}_2$ acts on a $y \in \overline{\mathbb{G}}$ in such a way that φ remains unchanged (sends a generator into itself). The generators of the cylinder represent the sets of parallel lines. The distances between the parallel lines under any $t \in \mathbb{T}_2$ remain unchanged. This means that $\mathbb{T}_2$ is a *group of Cavalieri transformations of* $\mathbb{S}_1 \times \mathbb{R}$. Therefore, product measures with Lebesgue factor on $\mathbb{R}$ are $\mathbb{T}_2$-invariant.

2.2 The space $\mathbb{G}$ of (non-directed) lines in $\mathbb{R}^2$

This space may be obtained from $\overline{\mathbb{G}}$ by erasing the arrow on each $g \in \overline{\mathbb{G}}$ (under which operation every two distinct directed lines, coincident but with opposite arrows, map into a non-directed line). We obtain a topology which complies with the space of closed sets in $\mathbb{R}^2$.

We can also consider $\mathbb{G}$ as a fibered space. Here the fibers are sets of parallel lines, and thus a fixed fiber model is $\mathbb{R}$. A fiber is represented by the line which passes through the origin $O \in \mathbb{R}^2$ and is orthogonal to the lines of the fiber. We label the lines through O by the points $\varphi \in \mathbf{E}_1$ (see §1.5), and thus call them φ-lines.

We represent the elements of $\mathbb{G}$ as

$$g = (\varphi, \mathscr{P}), \quad \text{where } \mathscr{P} = g \cap (\varphi\text{-line})$$

A sequence $g_n = (\varphi_n, \mathscr{P}_n)$ converges to $g = (\varphi, \mathscr{P})$ if and only if

(1) $\varphi = \lim \varphi_n$ in the topology of $\mathbf{E}_1$;
(2) $\mathscr{P} = \lim \mathscr{P}_n$ in the topology of $\mathbb{R}^2$.

Note that if $\mathscr{P} \neq O$ then (2) implies (1); yet $\lim \mathscr{P}_n = O$ does not guarantee convergence of $(\varphi_n, \mathscr{P}_n)$ in $\mathbb{G}$ (because φ_n can violate (1)). This hints that the topology of $\mathbb{G}$ is different from that of $\mathbb{R}^2$: one can show that $\mathbb{G}$ is homeomorphic to a Möbius band (see also §5.3).

Our fibering has the form

$$\pi : \mathbb{G} \to \mathbf{E}_1, \quad \pi(g) = \varphi.$$

In this way translations of $\mathbb{R}^2$ act on fibered $\mathbb{G}$ in Cavalieri fashion. Therefore every composition (in the terminology of §1.5) of Lebesgue measure L_1 on fibers is invariant with respect to $\mathbb{T}_2$.

2.3 The space $\overline{\mathbb{E}}$ of oriented planes in $\mathbb{R}^3$

An oriented plane in $\mathbb{R}^3$ is a plane for which one of the two possible normal directions is called positive. Let e be an oriented plane. The closed halfspace bounded by e, in which the positive normal of e lies, will be denoted by e^+ (see fig. 2.3.1). Owing to the natural correspondence

$$e \to e^+$$

we can consider $\overline{\mathbb{E}}$ also as the space of halfspaces.

We introduce coordinates on $\overline{\mathbb{E}}$ of the form (ω, p), where $-\infty < p < \infty$ and ω is a spatial direction in $\mathbb{R}^3$, i.e. ω belongs to the unit sphere $\mathbb{S}_2$. For a given plane $e \in \overline{\mathbb{E}}$, p is the signed distance from the origin O in $\mathbb{R}^3$ to e ($p > 0$ if $O \,\overline{\in}\, e^+$; $p < 0$ if $O \in e^+$; and $p = 0$ if $O \in e$) and ω is the positive normal direction. The space of pairs (ω, p) is the product $\mathbb{S}_2 \times \mathbb{R}$.

The topological structure on $\overline{\mathbb{E}}$ determined by the product

$$\overline{\mathbb{E}} = \mathbb{S}_2 \times \mathbb{R} \qquad\qquad (2.3.1)$$

complies with the topology induced on the set of halfspaces by closed sets in $\mathbb{R}^3$ (see §1.4).

Since the parallel shifts of $\mathbb{R}^3$ preserve the orientations of the planes, as well as the distances between pairs of parallel planes, we conclude that the group $\mathbb{T}_3$ of parallel shifts of $\mathbb{R}^3$ is a *group of Cavalieri transformations on the product* (2.3.1). Therefore product measures with Lebesgue factors on $\mathbb{R}$ are $\mathbb{T}_3$-invariant.

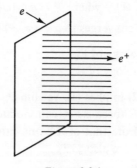

Figure 2.3.1

2.4 The space $\mathbb{E}$ of planes in $\mathbb{R}^3$

This space may be obtained from $\bar{\mathbb{E}}$ by mapping every two distinct half-spaces with a common boundary into their common boundary. This yields a topology which complies with the space of closed sets in $\mathbb{R}^3$.

We can also consider $\mathbb{E}$ as a fibered space. Now the fibers are sets of parallel planes and thus a fixed fiber model is $\mathbb{R}$. A fiber is represented by the *line* which passes through the origin O in $\mathbb{R}^3$ and is orthogonal to the planes of the fiber. We label the lines through O by points $\omega \in \mathbf{E}_2$ (the elliptical plane, see §1.5), and thus we call them ω-lines.

The elements of $\mathbb{E}$ are now represented as

$$e = (\omega, \mathscr{P}) \quad \text{where } \mathscr{P} = e \cap (\omega\text{-line}).$$

The topology on $\mathbb{E}$ is determined by the following convergence rule. A sequence $e_n = (\omega_n, \mathscr{P}_n)$ converges to $e = (\omega, \mathscr{P})$ if and only if

(1) $\omega = \lim \omega_n$ in the topology of $\mathbf{E}_2$;
(2) $\mathscr{P} = \lim \mathscr{P}_n$ in the topology of $\mathbb{R}^3$.

Note that if $\mathscr{P} \neq O$, then (2) implies (1); yet $\lim \mathscr{P}_n = O$ in $\mathbb{R}^3$ does not guarantee convergence of $(\omega_n, \mathscr{P}_n)$ in $\mathbb{E}$ (ω_n can violate (1)). We note that with this topology $\mathbb{E}$ is homeomorphic to elliptical (projective) 3-space with one point deleted.

Our fibering has the form

$$\pi : \mathbb{E} \to \mathbf{E}_2, \quad \pi(e) = \omega.$$

Translations of $\mathbb{R}^3$ act on this fibered $\mathbb{E}$ in Cavalieri fashion. Therefore compositions of L_1 planted on fibers are all invariant with respect to $\mathbb{T}_3$.

2.5 The space $\bar{\Gamma}$ of directed lines in $\mathbb{R}^3$

A directed line in $\mathbb{R}^3$ is a line with an arrow on it. In order to describe a $\gamma \in \bar{\Gamma}$ we first determine the direction Ω of γ, $\Omega \in \mathbb{S}_2$, and then the point $\mathscr{P}$ where γ hits the plane tangent to $\mathbb{S}_2$ at Ω (we assume $\mathbb{S}_2$ to be centered at O). This tangent plane is orthogonal to Ω.

The representation

$$\gamma = (\Omega, \mathscr{P})$$

maps $\bar{\Gamma}$ in a one-to-one way onto the tangent bundle of $\mathbb{S}_2$ described in §1.5 and induces the topology of the latter space on $\bar{\Gamma}$.

Thus we consider $\bar{\Gamma}$ as a fibered space:

$$\pi : \bar{\Gamma} \to \mathbb{S}_2, \quad \pi(\gamma) = \Omega.$$

We stress that the fibers in $\bar{\Gamma}$ are the sets of parallel and equidirected lines; the fixed fiber model is $\mathbb{R}^2$.

Parallel shifts of $\mathbb{R}^3$ leave $\mathbb{S}_2$ intact and act as parallel shifts of $\mathbb{R}^2$ on each fiber. Therefore $\mathbb{T}_3$ is a group of Cavalieri transformations of the space $\bar{\Gamma}$.

We conclude that composition measures on $\bar{\Gamma}$ which have the two-dimensional Lebesgue measure L_2 on fibers are all $\mathbb{T}_3$-invariant.

2.6 The space Γ of (non-directed) lines in $\mathbb{R}^3$

This space can be obtained from $\bar{\Gamma}$ by identifying two distinct directed lines, coincident but with opposite arrows. The resulting topology complies with the space of closed sets in $\mathbb{R}^3$.

We can also describe Γ as a fibered space. The fibers now are sets of parallel lines; a fixed fiber model is $\mathbb{R}^2$. A fiber can be represented by a plane through the origin $O \in \mathbb{R}^3$ which is orthogonal to the lines of the fiber. We label the planes through O by points of the elliptical plane $\Omega \in \mathbf{E}_2$ and thus call them Ω-planes.

We represent the elements of Γ as

$$\gamma = (\Omega, \mathscr{P}), \quad \text{where } \mathscr{P} = \gamma \cap (\text{the } \Omega\text{-plane}).$$

Using the topologies on $\mathbf{E}_2$ and $\mathbb{R}^3$ we define a topology on Γ by means of component-wise convergence. The result complies with the topology of closed sets in $\mathbb{R}^3$.

Our fibering has the form

$$\pi : \Gamma \to \mathbf{E}_2, \quad \pi(\gamma) = \Omega.$$

Translations of $\mathbb{R}^3$ act on Γ in Cavalieri fashion. Therefore composition measures with L_2 on fibers are $\mathbb{T}_3$-invariant.

2.7 Measure-representing product models

I Let $\mathbb{G}_x$ be the space of lines on $\mathbb{R}^2$ which hit the Ox axis (lines parallel to the Ox axis are excluded).

Each line $g \in \mathbb{G}_x$ can be described by two parameters, x, ψ:

x is the abscissa of the point $g \cap (Ox \text{ axis})$;
ψ is the angle of intersection of g with Ox.

Thus $x \in \mathbb{R}$, $\psi \in (0, \pi)$ (see fig. 2.7.1).

The map

$$g \to (\psi, x)$$

induces a product structure on $\mathbb{G}_x$:

$$\mathbb{G}_x = (0, \pi) \times Ox \text{ axis}.$$

On the other hand, in $\mathbb{G}_x$ we can use the (φ, p) coordinates described in §2.1. For this purpose it is enough to convert each line from $\mathbb{G}_x$ into a directed line

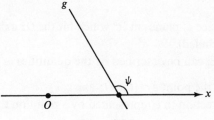

Figure 2.7.1

by putting an arrow on it in some way. We place the arrow so as to have

$$\varphi = \psi, \qquad (2.7.1)$$

i.e. each time the arrow points toward the *upper* halfplane. This, together with

$$p = x \sin \psi, \qquad (2.7.2)$$

describes the transition from (ψ, x) to (φ, p).

Clearly the above amounts to

$$\mathbb{G} \approx \mathbf{E}_1 \times \mathbb{R}. \qquad (2.7.3)$$

We can set a one-to-one mapping between $\mathbb{G}$ and $\mathbf{E}_1 \times \mathbb{R}$. In addition to φ, p coordinates on $\mathbb{G}_x$ we introduce, for this purpose, a p coordinate on the set of horizontal lines, i.e. on the fiber corresponding to the endpoints of $(0, \pi)$.

For the horizontal line with the equation

$$y = c$$

we put

$$p = c,$$

and this completes the construction of our map. It is discontinuous, yet it has the property that

the image of every compact set in $\mathbb{G}$ can be covered by a compact set in $\mathbf{E}_1 \times \mathbb{R}$ and vice versa.

The proof follows easily from the following observations. Since the distance of a line g from O always equals $|p|$, the image of the set (compact in $\mathbb{G}$)

$$C_x = \{g \in \mathbb{G} : \text{the distance of } g \text{ from } O \leqslant x\}$$

is a product set (compact in $\mathbf{E}_1 \times \mathbb{R}$)

$$C'_x = \mathbf{E}_1 \times \{p \in (-\infty, \infty) : |p| \leqslant x\}.$$

The image of any $A \subset C_x$ belongs to C'_x and that of any $B \subset C'_x$ belongs to C_x. Lastly, any compact $A \subset \mathbb{G}$ can be covered by some C_x, and any compact $B \subset \mathbf{E}_1 \times \mathbb{R}$ can be covered by some C'_x.

Corollary Under our map $\mathbf{E}_1 \times \mathbb{R}$ is a measure-representing model of $\mathbb{G}$ (see §1.6)

II Let $\mathbb{E}_z$ be the space of planes in $\mathbb{R}^3$ which hit the Oz axis (planes parallel to the Oz axis are excluded).

Each plane $e \in \mathbb{E}_z$ can be described by the quantities ω and z:

z is the abscissa of the point $e \cap$ (the Oz axis);

ω is the normal direction to e represented by a point on the open hemisphere $\mathbb{S}_2/2$.

Here $\mathbb{S}_2/2$ denotes the part of the unit sphere centered at O which lies in the halfspace $z > 0$.

The map

$$e \to (\omega, z)$$

induces a product structure on $\mathbb{E}_z$:

$$\mathbb{E}_z = \mathbb{S}_2/2 \times Oz \text{ axis}.$$

Since we have now defined ω to be a point in $\mathbb{S}_2$, we have thus mapped $\mathbb{E}_z$ on a part of $\bar{\mathbb{E}}$. Therefore, instead of (ω, z) we can use in $\mathbb{E}_z$ the coordinates (ω, p) defined in §2.3.

The relation between p and z is

$$p = z|\cos v|, \tag{2.7.4}$$

where v is the angle between ω and the Oz axis.

Clearly the above amounts to

$$\mathbb{E} \approx \mathbf{E}_2 \times \mathbb{R}. \tag{2.7.5}$$

We can extend the (ω, p) coordinates to the set of planes parallel to the Oz axis. Each plane of the latter set is determined by the line trace it leaves on the xOy plane. Therefore the problem is essentially that of introducing (φ, p) coordinates in the set of these lines assuming that $\varphi \in \mathbf{E}_1$ describes a pair of centrally-symmetrical points on the boundary of $\mathbb{S}_2/2$. We have just solved this problem in subsection I above. After applying this solution here we obtain a one-to-one (discontinuous) map of $\mathbb{E}$ onto $\mathbf{E}_2 \times \mathbb{R}$. By an argument similar to that we used in I above we conclude that *under our map* $\mathbf{E}_2 \times \mathbb{R}$ *is a measure-representing model of* $\mathbb{E}$ (see §1.6).

III Let $\mathbf{\Gamma}_{xy}$ be the space of lines in $\mathbb{R}^3$ which hit the xOy plane (lines parallel to the xOy plane are excluded). Each line $\gamma \in \mathbf{\Gamma}_{xy}$ can be described by

Q – the point of intersection of γ with the xOy plane;

Ω – the point where γ hits $\mathbb{S}_2/2$.

The map

$$\gamma \to (\Omega, Q)$$

induces a product structure on $\mathbf{\Gamma}_{xy}$:

$$\Gamma_{xy} = \mathbb{S}_2/2 \times xOy \text{ plane.}$$

Since we have now defined Ω to be a point in $\mathbb{S}_2$, we have thus mapped Γ_{xy} on a part of $\bar{\Gamma}$. Therefore in Γ_{xy} we can also use the coordinates $(\Omega, \mathscr{P})$ defined in §2.5.

For fixed Ω, $\mathscr{P}$ is the projection of Q on the plane perpendicular to γ. The image of the Lebesgue measure dQ will be

$$d\mathscr{P} = |\cos v| \, dQ, \tag{2.7.6}$$

where v is the angle between γ and the z axis.

We have found that

$$\Gamma \approx \mathbf{E}_2 \times \mathbb{R}^2. \tag{2.7.7}$$

Our map can be extended in a way similar to that used the previous cases. The result will be that $\mathbf{E}_2 \times \mathbb{R}^2$ is a measure-representing model of Γ.

Remark The space $\bar{\Gamma}$ also possesses a measure-representing product model, namely the space $\mathbb{S}_2 \times \mathbb{R}^2$.

IV Translations act on the space $\mathbb{G}_x$, $\mathbb{E}_z$ and Γ_{xy} in Cavalieri fashion. For instance, for $t \in \mathbb{T}_2$ and (ψ, x) representing a line from $\mathbb{G}_x$ we have

$$t(\psi, x) = (\psi, t_1 x),$$

where t_1 is the projection of t on the Ox axis.

Therefore, product measures on these spaces with Lebesgue factors on the Ox axis, Oz axis or xOy plane, respectively, are translation-invariant.

Because of the one-to-one nature of the maps, the action of translations on our product models of the spaces $\mathbb{G}$, $\mathbb{E}$ and Γ is well defined. Any translation produces rigid shifts of the Euclidean generators of our models, i.e. is again Cavalieri. Therefore the product measures with Lebesgue factors on $\mathbb{R}^n$, $n = 1$, 2, are translation-invariant. These measures *correspond* to similar composition measures on $\mathbb{G}$, $\mathbb{E}$ or Γ viewed as fibered spaces. We will see below that here no other $\mathbb{T}$-invariant measures exist.

2.8 Factorization of measures on spaces with slits

In the previous section we constructed product representations for the slitted versions $\mathbb{G}_x$, $\mathbb{E}_z$ and Γ_{xy} of the spaces $\mathbb{G}$, $\mathbb{E}$ and Γ. We can directly apply Lebesgue factorization to these products. The results are in table 2.8.1. In the second column the table lists the groups under which the measures are assumed invariant. Note that these groups are subgroups of the groups of translations of the corresponding spaces $\mathbb{R}^n$, $n = 2, 3$.

The striking feature of these factorizations is that they imply invariance with respect to all shifts of corresponding $\mathbb{R}^n$ (see the remarks at the end of the previous section).

Table 2.8.1

Space	Invariance assumed	Factor measures	Invariance that follows
$\mathbb{G}_x = (0, \pi) \times$ (Ox axis)	Shifts of $\mathbb{R}^2$ parallel to Ox	Lebesgue L_1 on the Ox axis, some measure on $(0, \pi)$	All shifts of $\mathbb{R}^2$ (the group $\mathbb{T}_2$)
$\mathbb{E}_z = \mathbb{S}_2/2 \times$ (Oz axis)	Shifts of $\mathbb{R}^3$ parallel to Oz	Lebesgue L_1 on the Oz axis, some measure on $\mathbb{S}_2/2$	All shifts of $\mathbb{R}^3$ (the group $\mathbb{T}_3$)
$\Gamma_{xy} = \mathbb{S}_2/2 \times$ (xOy plane)	Shifts of $\mathbb{R}^3$ parallel to xOy	Lebesgue L_2 on the xOy plane, some measure on $\mathbb{S}_2/2$	All shifts of $\mathbb{R}^3$ (the group $\mathbb{T}_3$)

A similar table can be compiled for the spaces $\overline{\mathbb{G}}_x$, $\overline{\mathbb{E}}_z$ and $\overline{\Gamma}_{xy}$ which are obtained from $\overline{\mathbb{G}}$, $\overline{\mathbb{E}}$ and $\overline{\Gamma}$ by deleting the elements which do not hit the corresponding line or plane. The spaces $(0, \pi)$ or $\mathbb{S}_2/2$ will be replaced by slitted versions of $\mathbb{S}_1$ or $\mathbb{S}_2$.

2.9 Dispensing with slits

Let μ be a $\overline{\mathbb{T}}_2$-invariant measure on $\mathbb{G}$ or equivalently on $\mathbb{E}_1 \times \mathbb{R}$ (see §2.7). Because for any $a > 0$

$$\mu(\{g = (\varphi, p): |p| < a\}) < \infty$$

the measure μ cannot charge more than a countable number of (linear) generators of the cylinder $\mathbb{E}_1 \times \mathbb{R}$. Therefore we can locate an axis g_0 through O with the property

$$\mu(\{\text{lines parallel to } g_0\}) = 0. \tag{2.9.1}$$

Without loss of generality we can (and do) take g_0 to be the Ox axis. Because of (2.9.1), μ now can be considered as a measure defined on $\mathbb{G}_x$. It is invariant with respect to shifts parallel to *any* axis, and also therefore to $g_0 = Ox$ axis. We can apply the first line of table 2.8.1: in ψ, x coordinates on $\mathbb{G}_x$, μ necessarily has the factorized form

$$m(\mathrm{d}\psi)\,\mathrm{d}x.$$

Using (2.7.1) and (2.7.2) we change to φ, p coordinates on $\mathbb{G}_x$. In terms of these coordinates μ will again have a factorized form, namely

$$(\sin \varphi)^{-1} m(\mathrm{d}\varphi)\,\mathrm{d}p. \tag{2.9.2}$$

Because μ was originally defined as a (locally-finite) measure on $\mathbb{E}_1 \times \mathbb{R}$ we necessarily have

$$\int_0^\pi (\sin \varphi)^{-1} m(\mathrm{d}\varphi) < \infty. \tag{2.9.3}$$

Table 2.9.1

Space	Measure representing product model	Group	Necessary product representation of invariant measures
G	$\mathbf{E}_1 \times \mathbb{R}$	$\mathbb{T}_2$	$m(d\varphi)\, dp$, where dp is Lebesgue L_1 on $\mathbb{R}$; m is a totally-finite measure on $\mathbf{E}_1$
E	$\mathbf{E}_2 \times \mathbb{R}$	$\mathbb{T}_3$	$m(d\omega)\, dp$, where dp is Lebesgue L_1 on $\mathbb{R}$; m is a totally-finite measure on $\mathbf{E}_2$
Γ	$\mathbf{E}_2 \times \mathbb{R}^2$	$\mathbb{T}_3$	$m(d\Omega)\, d\mathscr{P}$, where $d\mathscr{P}$ is Lebesgue L_2 on $\mathbb{R}^2$; m is a totally-finite measure on $\mathbf{E}_2$

In view of (2.9.1), the expression (2.9.2) provides a complete description of μ as a measure on $\mathbf{E}_1 \times \mathbb{R}$. We conclude that μ *is necessarily a product measure.*

A similar argument holds for the spaces $\mathbb{E}$ and Γ. We can compile the following factorization table 2.9.1. A similar table can be compiled for the space $\overline{\mathbb{G}}$, $\overline{\mathbb{E}}$ and $\overline{\Gamma}$.

2.10 Roses of directions and roses of hits

The measures in the spaces $\mathbf{E}_1$ and $\mathbf{E}_2$, which are listed in the last column of table 2.9.1 are called the *roses of directions* of the corresponding measures in the spaces $\mathbb{G}$, $\mathbb{E}$ or Γ. Accordingly we call the points of $\mathbf{E}_1$ and $\mathbf{E}_2$ *directions*. We denote translation-invariant measures in $\mathbb{G}$, $\mathbb{E}$ or Γ by μ in the following.

The above factorizations imply that the values of the measures μ on the sets of the type

$$[\delta] = \{g \in \mathbb{G} : g \text{ hits a line segment } \delta \subset \mathbb{R}^2\}$$

$$[f] = \{\gamma \in \Gamma : \gamma \text{ hits a flat } f \subset \mathbb{R}^3\}$$

(a 'flat' is a bounded part of a plane in $\mathbb{R}^3$ (see §2.13)), and

$$[s] = \{e \in \mathbb{E} : e \text{ hits a line segment } s \subset \mathbb{R}^3\}$$

possess a rather special structure.

We have

$$\mu([\delta]) = \int_{[\delta]} dp\, m(d\varphi) = \int_{\mathbf{E}_1} \text{proj}(\delta) m(d\varphi),$$

where proj(δ) denotes the length of the projection of δ on the p axis (the axis perpendicular to the direction φ). If the direction of δ is α then

$$\text{proj}(\delta) = \left| \cos\left(\frac{\pi}{2} + \varphi - \alpha\right) \right| \cdot |\delta| = |\sin(\varphi - \alpha)| \cdot |\delta|;$$

$|\delta|$ stands for the length of δ. Thus

$$\mu([\delta]) = \lambda(\alpha) \cdot |\delta|.$$

The function

$$\lambda(\alpha) = \int_{\mathbb{E}_1} |\sin(\varphi - \alpha)| m(\mathrm{d}\varphi), \quad \alpha \in \mathbb{E}_1 \tag{2.10.1}$$

is called the *rose of hits* of the measure μ on $\mathbb{G}$ (this terminology was introduced in [17]). We could obtain a more usual expression with cos instead of sin if in §2.1 we defined φ to be the direction *perpendicular* to the line g.

A similar result holds for $\mu([f])$. We have

$$\mu([f]) = \int_{[f]} m(\mathrm{d}\omega) \, \mathrm{d}\mathscr{P} = \int_{\mathbb{E}_2} \text{proj}(f) m(\mathrm{d}\omega),$$

where proj(f) is now the area of the projection of f on the plane carrying $\mathrm{d}\mathscr{P}$ (whose normal direction is ω). Clearly

$$\text{proj}(f) = \|f\| \cdot |\cos(\omega, \xi)|,$$

where ξ is the spatial direction normal to f, (ω, ξ) is the angle between the two directions, and $\|f\|$ is the area of f. Thus

$$\mu([f]) = \lambda(\xi) \cdot \|f\|,$$

where

$$\lambda(\xi) = \int_{\mathbb{E}_2} |\cos(\omega, \xi)| m(\mathrm{d}\omega), \quad \xi \in \mathbb{E}_2. \tag{2.10.2}$$

This function $\lambda(\xi)$ is called the *rose of* hits of the measure μ on Γ.

For $\mathbb{T}_3$-invariant measures on $\mathbb{E}$ we have

$$\mu([s]) = \lambda(\xi)|s|, \tag{2.10.3}$$

where ξ is the spatial direction of the segment s and $|s|$ is its length. It is not difficult to prove that $\lambda(\xi)$ in (2.10.3) necessarily possesses the representation (2.10.2). The function $\lambda(\xi)$ as defined by (2.10.3) is called the rose of hits of a $\mathbb{T}_3$-invariant measure on $\mathbb{E}$.

2.11 Density and curvature

Let μ be a $\mathbb{T}_2$-invariant measure on $\mathbb{G}$ and let $\lambda(\varphi)$ be its rose of hits. We assume now that the rose of directions has a continuous density $f(\varphi)$, i.e.

$$\mathrm{d}\mu = f(\varphi) \, \mathrm{d}\varphi \, \mathrm{d}p.$$

By (1.7.2) and the second part of Minkowski's proposition in §1.7 we conclude that $\lambda(\varphi)$ coincides with the breadth function $b(\varphi)$ of some bounded symmetrical convex contour:

$$\lambda(\varphi) = b(\varphi).$$

What is the geometrical meaning of $f(\varphi)$ in terms of this contour?

Using the formulae (3.6.2) and (3.7.4), we find

$$f(\varphi) = \lim_{l \to 0} l^{-2} \mu([\delta_1] \cap [\delta_2]), \tag{2.11.1}$$

where the infinitesimal segments δ_1 and δ_2 are shown in fig. 1.8.2,

$$[\delta_i] = \{\text{the lines that hit the segment } \delta_i\}.$$

We assume that δ_1 and δ_2 are perpendicular to s_1 and have a common length l, s_1 has direction φ, and the length of s_1 is 1.

On the other hand, by a simple application of the combinatorial formula (1.8.3) we find (using the notation of fig. 1.8.2)

$$2\mu([\delta_1] \cap [\delta_2]) = \mu([d_1]) + \mu([d_2]) - \mu([s_1]) - \mu([s_2])$$
$$\approx \lambda(\varphi + l)\sqrt{(1 + l^2)} + \lambda(\varphi - l)\sqrt{(1 + l^2)} - 2\lambda(\varphi)$$

In conjunction with (2.11.1) this yields

$$2f(\varphi) = \frac{d^2}{dl^2}(\lambda(\varphi + l)\sqrt{(1 + l^2)})|_{l=0} = \lambda(\varphi) + \lambda''(\varphi).$$

We compare the previous result with the following well known formula of differential geometry (see [2], p. 3)

$$2R(\varphi) = b(\varphi) + b''(\varphi), \tag{2.11.2}$$

which expresses the curvature radius R of a symmetrical convex contour at the point with normal direction φ in terms of the breadth function. We conclude that

$$R(\varphi) = f(\varphi). \tag{2.11.3}$$

Conversely, (2.11.2) can be resolved (see [6] for details) to yield

$$b(\varphi) = \int_0^\pi |\sin(\varphi - \alpha)| R(\alpha)\, d\alpha.$$

This integral is a version of (2.10.1). Therefore $b(\varphi)$ happens to be the rose of hits for the rose of directions $R(\varphi)\, d\varphi$. In this way we get an independent partial proof of the assertion in §1.7 about the generation of metrics by measures in $\mathbb{G}$.

2.12 The roses of $\mathbb{T}_3$-invariant measures on $\mathbb{E}$

Let μ be a $\mathbb{T}_3$-invariant measure on $\mathbb{E}$. It generates a pseudometric ρ in $\mathbb{R}^3$:

$$\rho(\mathscr{P}_1, \mathscr{P}_2) = \mu(\mathscr{P}_1|\mathscr{P}_2) = \lambda(\xi)|\mathscr{P}_1, \mathscr{P}_2|, \tag{2.12.1}$$

where ξ is the direction of the segment $\mathscr{P}_1\mathscr{P}_2$, $|\mathscr{P}_1\mathscr{P}_2|$ is its length, and

$$\mathcal{P}_1 | \mathcal{P}_2 = \{e \in \mathbb{E} : e \text{ separates } \mathcal{P}_1 \text{ from } \mathcal{P}_2\}.$$

If μ is not concentrated on a bundle of planes parallel to some fixed direction, then ρ is a metric.

Clearly $\lambda(\xi)$ is the rose of hits corresponding to μ. A proof that ρ as given by (2.12.1) is a pseudometric in $\mathbb{R}^3$ follows from the planar result of §1.7, since the image of μ under the map

$$e \to \text{the line } e \cap e_0 \tag{2.12.2}$$

where e_0 is some plane in $\mathbb{R}^3$, is a measure in the space of lines on e_0.

The converse problem: does every $\mathbb{T}_3$-invariant continuous, linearly-additive metric ρ in $\mathbb{R}^3$ permit a representation (2.12.1) with some measure μ on $\mathbb{E}$? has a *negative* solution which we outline below. Let us fix a plane $e_0 \subset \mathbb{R}^3$ and identify the space of lines on e_0 with $\mathbb{G}$. Let μ_0 be the measure on $\mathbb{G}$ which is the image of μ under (2.12.2).

If we assume the existence of the density:

$$d\mu = f(\omega)\, d\omega\, dp$$

(we use the notation of §2.7), then by the results of §3.6, §3.11 and (3.12.4) we find

$$d\mu_0 = F(\varphi)\, d\varphi\, dp', \tag{2.12.3}$$

where φ and p' are the coordinates in $\mathbb{G}$ as explained in §2.7, and

$$F(\varphi) = F(\Omega) = \int_{\langle \Omega \rangle} f \sin^2 \psi\, d\psi. \tag{2.12.4}$$

We stress that the direction φ on e_0 determines a direction Ω in $\mathbb{R}^3$, and in (2.12.4) integration is over the great semicircle $\langle \Omega \rangle$ orthogonal to the direction Ω. The angle ψ is between $\omega \in \langle \Omega \rangle$ and the direction normal to e_0 (the situation is in a sense dual to that shown in fig. 1.7.2). The condition

$$F(\Omega) = \int_{\langle \Omega \rangle} f \sin^2 \psi\, d\psi > 0 \quad \text{for every } e_0, \psi \tag{2.12.5}$$

suffices to have a continuous, linearly-additive metric in $\mathbb{R}^3$ which on *every* plane e_0 corresponds to the measure on $\mathbb{G}$ given by (2.12.3). However, the condition (2.12.5) can be met by smooth functions f which are not everywhere non-negative. For instance, let us take $f_0 = 1$ on the whole hemisphere with the exception of a circle of radius ε; in the center of this circle, say, let $f_0 = -1$ and let $|f_0| \leqslant 1$ everywhere. We leave it to the reader to prove that by appropriate small choice of ε any such (smooth) f_0 will satisfy (2.12.5). Suppose in (2.12.4) we insert $f = f_0$. Then the corresponding measure μ_0 will define a metric $\rho_{e_0}(\mathcal{P}_1, \mathcal{P}_2) > 0$ in the plane e_0. Clearly

$$\rho_{e_0}(\mathcal{P}_1, \mathcal{P}_2) = \int_{\mathcal{P}_1 | \mathcal{P}_2} f_0\, d\omega\, dp,$$

and this value does not depend on the choice of e_0 through $\mathcal{P}_1, \mathcal{P}_2$. Thus,

$$\rho_{e_0}(\mathscr{P}_1, \mathscr{P}_2) = \rho_0(\mathscr{P}_1, \mathscr{P}_2)$$

will be a metric in $\mathbb{R}^3$. It follows from a uniqueness result discussed in §6.2 that ρ_0 *cannot* be generated by some measure μ on $\mathbb{E}$ by means of (2.12.1).

By Minkowski's proposition (which is valid in $\mathbb{R}^3$, see §1.7) the ratio $|\mathscr{P}_1, \mathscr{P}_2|^{-1} \rho_0(\mathscr{P}_1, \mathscr{P}_2)$ happens to be the breadth function of a convex body in $\mathbb{R}^3$. Since this ratio cannot be a rose of hits, we conclude that the body fails to be a zonoid. Examples of non-zonoidal polyhedrons can be easily constructed using the criterion of §5.10. Here we have constructed a non-zonoidal convex body, which is not a polyhedron.

Now let us return to (2.12.1) assuming the existence of density $f(\omega)$. By Minkowski's proposition, $\lambda(\xi)$ coincides with the breadth function $b(\xi)$ of some symmetrical convex body $D \subset \mathbb{R}^3$, i.e.

$$\lambda(\xi) = b(\xi). \tag{2.12.6}$$

Now the purpose is to relate the curvatures of ∂D with f.

The restriction of $\lambda(\xi)$ to directions in e_0 is the breadth function of the projection of D on the plane e_0 (see the end of §1.7). By (2.11.2) and (2.11.3) the expression

$$2R = \lambda''(\Omega) + \lambda(\Omega) = 2\int_{\langle\Omega\rangle} f \sin^2 \psi \, d\psi \tag{2.12.7}$$

gives the doubled curvature radius R of the boundary of the aforementioned projection at the point having planar normal direction φ (which corresponds to Ω). We stress that the double differentiation above is within the set of directions which belong to the plane e_0.

Now we fix the direction Ω and rotate e_0 around Ω. The quantities in (2.12.7) depend on Φ, the angle of rotation. We average (2.12.7) with respect to Φ. We have

$$\pi^{-1} \int_0^\pi \lambda''(\Omega) \, d\Phi = \frac{1}{2}\Delta_2 \lambda(\Omega) \tag{2.12.8}$$

where Δ_2 is the Laplace operator.

By an interchange of the order of integration we readily find

$$\pi^{-1} \int_0^\pi d\Phi \int_{\langle\Omega\rangle} f \sin^2 \psi \, d\psi = \frac{1}{2} \int_{\langle\Omega\rangle} f \, d\Phi.$$

Thus the result of averaging (2.12.7) over $(0, \pi)$ will be

$$\frac{1}{2}\Delta_2 \lambda(\Omega) + \lambda(\Omega) = \int_{\langle\Omega\rangle} f \, d\Phi \tag{2.12.9}$$

The expression $\frac{1}{2}\Delta_2 \lambda(\Omega) + \lambda(\Omega)$ is well known in differential geometry [6]. It equals the sum $R_1 + R_2$ of the main curvature radii. Thus we have

$$R_1 + R_2 = \int_{\langle\Omega\rangle} f \, d\Phi. \tag{2.12.10}$$

As a byproduct the following result concerning the mean values of curvature radii is obtained

$$\bar{R} = \frac{1}{\pi} \int R(\Phi)\, d\Phi = \frac{R_1 + R_2}{2}$$

2.13 Spaces of segments and flats

I Let Δ_2^* denote the space of unit length directed segments on $\mathbb{R}^2$ (a directed segment is a linear segment, one of whose endpoints is called a source). A segment $\delta \in \Delta_2^*$ can be described as

$$\delta = (\varphi, Q),$$

where $Q \in \mathbb{R}^2$ is the source, and $\varphi \in \mathbb{S}_1$ is the direction of δ (see fig. 2.13.1). Therefore we have

$$\Delta_2^* = \mathbb{S}_1 \times \mathbb{R}^2. \tag{2.13.1}$$

The space Δ_2^* also has the following important interpretation:

$$\Delta_2^* = \mathbb{M}_2,$$

where $\mathbb{M}_2$ is the group of all Euclidean motions of the plane. Indeed each motion $M \in \mathbb{M}_2$ can be identified with $\delta = M\delta_0$, where δ_0 is the segment with the source at O (the origin in $\mathbb{R}^2$) and $\varphi = 0$. Lastly

$$\Delta_2^* = \overline{\mathbb{G}} \times \mathbb{R} \tag{2.13.2}$$

since each δ can be determined by its carrying line $g \in \overline{\mathbb{G}}$ and one-dimentional coordinate x of its source on g. It is also possible to define Δ_2^* to be the space of directed segments of arbitrary fixed length $|\delta| = \text{const}$.

II The space of all directed segments on $\mathbb{R}^2$ (with varying lengths) can be described as

$$\Delta_2 = \Delta_2^* \times (0, \infty) \tag{2.13.3}$$

(as compared with (2.13.1) one dimension is added to indicate the length l of the segment). We again denote elements of Δ_2 by δ.

III The space Δ_3^* of unit length directed segments in $\mathbb{R}^3$ has the following representations:

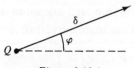

Figure 2.13.1

$$\Delta_3^* = \mathbb{S}_2 \times \mathbb{R}^3 \qquad (2.13.4)$$

and

$$\Delta_3^* = \overline{\Gamma} \times \mathbb{R}, \qquad (2.13.5)$$

which are similar to the representations for Δ_2^*.

Warning: we can describe a $\delta \in \Delta_3^*$ as

$$\delta = (e, \mathscr{P})$$

where $e \in \overline{\mathbb{E}}$ is a plane through the source of δ with positive normal vector parallel to δ, and $\mathscr{P} \in \mathbb{R}^2$ determines the position of the source of δ on e. However, Δ_3^* is not homeomorphic to $\overline{\mathbb{E}} \times \mathbb{R}^2$. The reader may try to prove that $\overline{\mathbb{E}} \times \mathbb{R}^2$ is a measure-representing model of Δ_3^*.

Neither can Δ_3^* be identified with M_3, the group of Euclidean motions of $\mathbb{R}^3$. In fact, Δ_3^* is less than M_3 *in dimension* (see §3.13).

As for Δ_3, the space of all directed segments in $\mathbb{R}^3$ (varying lengths), we have

$$\Delta_3 = \Delta_3^* \times (0, \infty). \qquad (2.13.6)$$

IV By definition a *flat* in $\mathbb{R}^3$ is a bounded part of a plane. For simplicity we take flats to be discs of unit radius. Such a disc can be described by its center $\mathscr{P} \in \mathbb{R}^3$ and the direction $\omega \in \mathbf{E}_2$ normal to its plane. The space $\mathbb{F}$ of such flats can be represented as

$$\mathbb{F} = \mathbf{E}_2 \times \mathbb{R}^3.$$

The result of §3.16, IV, can be considered as referring to triangular flats in $\mathbb{R}^3$.

A direct application of Lebesgue factorization yields table 2.13.1 for $\mathbb{T}$-invariant measures on the above spaces

Table 2.13.1

Space	Group	Factor measures
$\Delta_2^* = \mathbb{S}_1 \times \mathbb{R}^2$	$\mathbb{T}_2$	Lebesgue measure on $\mathbb{R}$; a finite measure on $\mathbb{S}_1$
$\Delta_2 = \mathbb{S}_1 \times (0, \infty) \times \mathbb{R}^2$	$\mathbb{T}_2$	Lebesgue measure on $\mathbb{R}^2$; a measure on $\mathbb{S}_1 \times (0, \infty)$
$\Delta_3^* = \mathbb{S}_2 \times \mathbb{R}^3$	$\mathbb{T}_3$	Lebesgue measure on $\mathbb{R}^3$; a finite measure on $\mathbb{S}_2$
$\Delta_3 = \mathbb{S}_2 \times (0, \infty) \times \mathbb{R}^3$	$\mathbb{T}_3$	Lebesgue measure on $\mathbb{R}^3$; a measure on $\mathbb{S}_2 \times (0, \infty)$
$\mathbb{F} = \mathbf{E}_2 \times \mathbb{R}^3$	$\mathbb{T}_3$	Lebesgue measure on $\mathbb{R}^3$; a finite measure on $\mathbf{E}_2$

2.14 Product spaces with slits

The translations of $\mathbb{R}^2$ or $\mathbb{R}^3$ induce transformations of the product spaces $\overline{G} \times \overline{G}$, $\overline{E} \times \overline{E}$, $\overline{\Gamma} \times \overline{\Gamma}$. What can we state about translation-invariant measures on these product spaces?

For simplicity we consider measures which are concentrated on the slitted versions $\mathbb{X}_1$, $\mathbb{X}_2$ and $\mathbb{X}_3$ of these spaces.

I

$$\mathbb{X}_1 = (\overline{G} \times \overline{G}) \backslash Z_1,$$

where

$$Z_1 = \{\text{pairs of parallel or antiparallel lines}\}.$$

A pair $(g_1, g_2) \in \mathbb{X}_1$ can be represented as follows:

$$(g_1, g_2) = (\varphi_1, \varphi_2, \mathscr{P}),$$

where $\mathscr{P} = g_1 \cap g_2$ (the intersection point), and φ_i is the direction of g_i. Thus we have

$$\mathbb{X}_1 = \{(\varphi_1, \varphi_2): \varphi_1 \neq \varphi_2 \text{ and } \varphi_1 \neq \varphi_2 + \pi\} \times \mathbb{R}^2 = \mathbb{Y}_1 \times \mathbb{R}^2.$$

II

$$\mathbb{X}_2 = (\overline{E} \times \overline{E}) \backslash Z_2$$

where

$$Z_2 = \{(e_1, e_2): e_1, e_2 \text{ are parallel or antiparallel or intersect by a line} \\ \text{parallel to the } XOY \text{ plane}\}$$

The line $\gamma = e_1 \cap e_2$ together with planar directions Φ_1 and Φ_2 of the vectors normal to e_1 and e_2 (these vectors lie in the plane orthogonal to γ), determine the pair e_1 and e_2 completely. Thus

$$\mathbb{X}_2 = \{(\Phi_1, \Phi_2): \Phi_1 \neq \Phi_2 \text{ and } \Phi_1 \neq \Phi_2 + \pi\} \times \Gamma_{xy}.$$

Using the representation of §2.7, III we find

$$\mathbb{X}_2 = \{(\Phi_1, \Phi_2): \Phi_1 \neq \Phi_2 \text{ and } \Phi_1 \neq \Phi_2 + \pi\} \times \mathbb{S}_2/2 \times \mathbb{R}^2 = \mathbb{Y}_2 \times \mathbb{R}^2.$$
$$(2.14.1)$$

Here we identify $\mathbb{R}^2$ with the xOy plane.

III

$$\mathbb{X}_3 = (\overline{\Gamma} \times \overline{\Gamma}) \backslash Z_3$$

where

$$Z_3 = \{\text{pairs of parallel or antiparallel lines}\}.$$

For every $(\gamma_1, \gamma_2) \in \mathbb{X}_3$ there exists a unique line segment s with endpoints on γ_1 and γ_2 which has the *shortest length* among similar segments. Let

Table 2.14.1

Space	Assume invariance of the measure μ	Additional condition	Factor measures
$\overline{\mathbb{G}} \times \overline{\mathbb{G}}$	$\mathbb{T}_2$ (shifts of the plane)	$\mu(Z_1) = 0$	Lebesgue L_2 on $\mathbb{R}^2$; a finite measure on $\mathbb{Y}_1$
$\overline{\mathbb{E}} \times \overline{\mathbb{E}}$	Shifts of $\mathbb{R}^3$ parallel to xOy plane	$\mu(Z_2) = 0$	Lebesgue L_2 on xOy; a finite measure on $\mathbb{Y}_2$
$\overline{\Gamma} \times \overline{\Gamma}$	$\mathbb{T}_3$ (shifts of $\mathbb{R}^3$)	$\mu(Z_3) = 0$	Lebesgue L_3 on $\mathbb{R}^3$; a measure on $\mathbb{Y}_3$ which is finite on every $\mathbb{Y}_3(a)$

Q be the midpoint of s,

h be the length of s,

ω_i be the spatial direction of γ_i, $i = 1, 2$.

The parameters $(Q, \omega_1, \omega_2, h)$ uniquely determine any pair $(\gamma_1, \gamma_2) \in \mathbb{X}_3$. We have the representation

$$\mathbb{X}_3 = \{(\omega_1, \omega_2): \omega_1 \neq \omega_2, \omega_1 \neq -\omega_2\} \times \{0, \infty) \times \mathbb{R}^3 = \mathbb{Y}_3 \times \mathbb{R}^3.$$

We will also use the notation

$$\mathbb{Y}_3(a) = \{(\omega_1, \omega_2): \omega_1 \neq \omega_2, \omega_1 \neq -\omega_2\} \times (0, a).$$

We conclude by Lebesgue factorization that any $\mathbb{T}_2$-invariant measure on $\mathbb{X}_1$ is a product of the Lebesgue measure L_2 and a measure on $\mathbb{Y}_1$. But, if a measure on $\mathbb{X}_1$ has to correspond to a measure μ on $\overline{\mathbb{G}} \times \overline{\mathbb{G}}$ for which $\mu(Z_1) = 0$, then we must *exclude* the totally-infinite measures on $\mathbb{Y}_1$. Similar remarks are valid for other cases as well, and we obtain the factorization table 2.14.1. We apply these factorizations in §2.16.

2.15 Almost sure *T*-invariance of random measures

In 1968 a Cambridge mathematician Rollo Davidson (who died in an accident at the age of 26) wrote a remarkable thesis which dealt with line and flat processes. A phenomenon discovered by Davidson was the fact that, under rather broad conditions, the invariance of the first (μ_1) and the second (μ_2) moment measures of a random measure η on the space $\overline{\mathbb{G}}$ of lines implies that η is invariant with probability 1 (i.e. almost sure). Davidson required invariance of μ_1 and μ_2 with respect to the (Euclidean) group of motions of the plane, but later this condition was replaced by $\mathbb{T}_2$-invariance. Davidson's original result ([43]) was extended in various directions by Krickeberg [8], Kallenberg [9], Papangelou [10], and Ambartzumian [34]. Below we treat this topic using the factorizations of §2.14.

By definition, a random measure η in some space $\mathbb{X}$ is a map

$$\eta : \Omega \to \mathscr{N}_{\mathbb{X}},$$

where Ω is a probability space and $\mathscr{N}_{\mathbb{X}}$ is the set of measures on $\mathbb{X}$. We introduce a natural σ-algebra on $\mathscr{N}_{\mathbb{X}}$ as the minimal one containing the sets

$$\{\eta \in \mathscr{N}_{\mathbb{X}} : \eta(A) < x\}$$

for the Borel $A \subset \mathbb{X}$, and all $x \in \mathbb{R}$. As usual we require that the map η be measurable (i.e. for every Borel, A $\eta(A)$ should be a random variable).

The first and second moment measures of η are said to exist if

$$\mu_1 = \mathrm{E}\eta, \quad \mu_2 = \mathrm{E}(\eta \times \eta)$$

(where E stands for expectation) happen to be locally-finite measures (or simply measures in our terminology) in the spaces $\mathbb{X}$ and $\mathbb{X} \times \mathbb{X}$, respectively.

2.16 Random measures on $\overline{\mathbb{G}}$

Let η be a random measure on $\overline{\mathbb{G}}$ with $\mathbb{T}_2$-invariant moment measures μ_1 and μ_2.

We will require additionally that

(a) $\mu_2(Z_1) = 0$, where Z_1 is the set of parallel or antiparallel pairs of lines in the plane.

By virtue of (a) we can consider μ_2 as essentially a measure on $\overline{\mathbb{G}} \times \overline{\mathbb{G}} \backslash Z_1$, and according to the first line of table 2.14.1 we have

$$\mu_2(\mathrm{d}g_1 \, \mathrm{d}g_2) = \mathrm{d}\mathscr{P}\, m(\mathrm{d}\varphi_1 \, \mathrm{d}\varphi_2),$$

where $\mathrm{d}\mathscr{P}$ is the Lebesgue measure and m is some measure on $\mathbb{Y}_1$. We now apply the (φ, p) representation of lines described in §2.1. Let u be a shift of the cylinder $\overline{\mathbb{G}} = \mathbb{S}_1 \times \mathbb{R}$, i.e. a transformation

$$\varphi' = \varphi$$
$$p' = p + u,$$

and let $\mathbb{U}^2$ be the group of 'independent shifts' of the factors in the product $\overline{\mathbb{G}} \times \overline{\mathbb{G}}$ (an element of $\mathbb{U}^2$ is (u_1, u_2)).

The measure μ_2 is invariant under transformations from $\mathbb{U}^2$.

Proof The following formula is clear from fig. 2.16.1:

$$\mathrm{d}\mathscr{P} = |\sin(\varphi_1 - \varphi_2)|^{-1} \, \mathrm{d}p_1 \, \mathrm{d}p_2.$$

Therefore

$$\mu_2(\mathrm{d}g_1 \, \mathrm{d}g_2) = |\sin(\varphi_1 - \varphi_2)|^{-1} m(\mathrm{d}\varphi_1 \, \mathrm{d}\varphi_2) \, \mathrm{d}p_1 \, \mathrm{d}p_2.$$

Clearly, this measure remains invariant under the group $\mathbb{U}^2$.

In particular for every Borel $A \subset \overline{\mathbb{G}}$ we have

$$\mu_2(A \times A) = \mu_2(uA \times uA)$$
$$\mu_2(A \times A) = \mu_2(A \times uA).$$

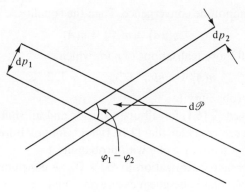

Figure 2.16.1

These equations imply probability 1 invariance with respect to shifts of the cylinder. Indeed by the Schwartz inequality

$$\mu_2(A \times uA) = E\eta(A)\eta(uA) \leqslant \sqrt{(E\eta^2(A)E\eta^2(uA))}$$
$$= E\eta^2(A) = \mu_2(A \times A),$$

and, since here we actually have an equality, we conclude that, for fixed A and u,

$$\eta(A) = c \cdot \eta(uA) \quad \text{with probability 1,}$$

$c = c(A, u)$ is non-random. Progressing to expectations, we find

$$\mu_1(A) = c\mu_1(uA).$$

$\mathbb{T}_2$-invariance implies u-invariance of μ_1 (by §2.9 and the Cavalieri principle). Thus $c = 1$.

We have proved that for any fixed $A \subset \overline{\mathsf{G}}$ and u

$$\eta(A) = \eta(uA) \quad \text{with probability 1.} \tag{2.16.1}$$

However, the subset of Ω, where (2.16.1) holds, may depend on A and u. Since they are more than countable in number, we cannot directly conclude that (2.16.1) holds for every A and u simultaneously (which is our aim for the moment).

The remedy lies in considering appropriate 'dense' countable sets. We will use 'shields': a shield is a product of an arc from $\mathbb{S}_1$, and an interval from $\mathbb{R}$. Shields with rational vertices we will denote as $A^{(r)}$; similarly $u^{(r)}$ will denote rational shifts. It follows from (2.16.1) that, for any sequences $\{A_i^{(r)}\}$ and $\{u_i^{(r)}\}$,

$$P\{\eta(A_i^{(r)}) = \eta(u_i^{(r)}A_i^{(r)}), i = 1, 2, \ldots\} = 1.$$

Given any two open shields $A, A' \subset \overline{\mathsf{G}}$ such that

$$A' = uA$$

we can find sequences $\{A_i^{(r)}\}$ and $\{u_i^{(r)}\}$ such that *simultaneously*

$$A = \lim_{i \to \infty} A_i^{(r)}, \quad A' = \lim_{i \to \infty} u_i^{(r)} A_i^{(r)},$$

in the sense of monotone convergence. Then the equality

$$\eta(uA) \equiv \eta(A') = \eta(A)$$

will follow for all those realisations of η for which

$$\eta(A_i^{(r)}) = \eta(u_i^{(r)} A_i^{(r)}), \qquad i = 1, 2, \ldots$$

holds, i.e. with probability 1.

We have proved (2.16.1) for all open shields and all shifts u. By measure continuation we can conclude that (2.16.1) holds for any Borel A and all shifts u. This implies $\mathbb{T}_2$-invariance of η with probability 1.

Applying Lebesgue factorization to $\mathbb{S}_1 \times \mathbb{R}$, we conclude that η is probability 1, a product of a Lebesgue measure on $\mathbb{R}$ and a measure on $\mathbb{S}_1$. By the Cavalieri principle, η is then $\mathbb{T}_2$-invariant (with probability 1).

We summarize:

Let η be a random measure on $\overline{\mathbb{G}} = \mathbb{S}_1 \times \mathbb{R}$ which has $\mathbb{T}_2$-invariant moment measures μ_1, μ_2. If condition (a) is satisfied then η is $\mathbb{T}_2$-invariant with probability 1. It follows from the results of §2.8 that η with probability 1 factorizes into a product of a Lebesgue measure on $\mathbb{R}$ and a finite random measure on $\mathbb{S}_1$.

A similar probability 1 invariance-factorization result holds also for random measures on $\mathbb{G}$.

2.17 Random measures on $\overline{\mathbb{E}}$

Let η be a random measure in the space $\overline{\mathbb{E}}$ with $\mathbb{T}_3$-invariant first and second moment measures μ_1 and μ_2. We assume additionally that $\mu_2(Z_2) = 0$ see §2.14.

Then η is $\mathbb{T}_3$-invariant with probability 1 and therefore η factorizes into a product of the Lebesgue measure on $\mathbb{R}$ and a random finite measure on $\mathbb{S}_2$.

The proof follows the ideas of §2.16. We outline the main steps. Using the (ω, p) representation of the planes (see §2.3) we define a shift u of $\overline{\mathbb{E}}$ as follows:

$$\omega' = \omega$$

$$p' = p + u.$$

Let $d\mathscr{P}$ be the Lebesgue measure on $\mathbb{R}^2$ in the factor representation (2.14.1). We have

$$d\mathscr{P} = f(\omega_1, \omega_2) \, dp_1 \, dp_2,$$

where f depends only on the orientations of the planes

$$e_1 = (\omega_1, p_1), \quad e_2 = (\omega_2, p_2).$$

The determination of the exact form of the function f is left to the reader. This, together with the factorization of μ (see table 2.14.1), implies that μ_2 is invariant with respect to *independent* shifts of the factor spaces in the product $\bar{\mathbb{E}} \times \bar{\mathbb{E}}$.

We then follow essentially the same steps as in the derivation of the previous section. A similar result also holds for random measures on $\mathbb{E}$.

2.18 Random measures on $\bar{\Gamma}$

In this section we use the notation of §2.5 and §2.14, III.

Let η be a random measure on $\bar{\Gamma}$ possessing $\mathbb{T}_3$-invariant moment measures μ_1 and μ_2 (in particular, μ_1 factorizes according to §2.9). We assume additionally that

(a) $\mu_2 (Z_3) = 0$.

According to §2.14, (a) implies that

$$\mu_2(d\gamma_1 \, d\gamma_2) = dQ \, m_1(dh \, d\omega_1 \, d\omega_2), \tag{2.18.1}$$

where dQ is Lebesgue on $\mathbb{R}^3$ and m_1 is some measure on $\mathbb{Y}_3$.

If m_1 factorizes into a product of Lebesgue measure on $(0, \infty)$ and some measure m_2 on $\mathbb{S}_2 \times \mathbb{S}_2$, i.e. if

$$m_1(dh \, d\omega_1 \, d\omega_2) = dh \, m_2(d\omega_1 \, d\omega_2) \tag{2.18.2}$$

then η is $\mathbb{T}_3$-invariant with probability 1. The converse is also true.

Sketch proof From (2.18.1) and (2.18.2) we have

$$\mu_2(dQ \, dh \, d\omega_1 \, d\omega_2) = dQ \, dh \, m_2(d\omega_1 \, d\omega_2). \tag{2.18.3}$$

Let us now *fix the orientations* ω_1 and ω_2 of the lines. Let $d\mathscr{P}_1$ and $d\mathscr{P}_2$ be planar Lebesgue measures on planes through O orthogonal to the directions ω_1 and ω_2 (see §2.5). We have

$$dQ \, dh = \sin \theta \, d\mathscr{P}_1 \, d\mathscr{P}_2, \tag{2.18.4}$$

where θ is the flat angle between the planes carrying $d\mathscr{P}_1$ and $d\mathscr{P}_2$. This result strongly depends on the fact that the segment h always remains parallel to the line of intersection of the planes containing $d\mathscr{P}_1$ and $d\mathscr{P}_2$. Let us denote by u transformations of $\bar{\Gamma}$ which correspond to parallel shifts of $\mathbb{R}^2$ in the representation

$$\mathbb{S}_2 \times \mathbb{R}^2$$

of $\bar{\Gamma}$ discussed in §2.7, III. The directions of lines under u remain unchanged. The measure

$$m_2(d\omega_1 \, d\omega_2) \sin \theta \, d\mathscr{P}_1 \, d\mathscr{P}_2$$

does not change under application of independent transformations u_1 and u_2 to the factor spaces in $\bar{\Gamma} \times \bar{\Gamma}$. In particular for every $A \subset \bar{\Gamma}$ we have

$$\mu_2(uA \times uA) = \mu_2(A \times A),$$

$$\mu_2(A \times uA) = \mu_2(A \times A).$$

Hence our assertion follows by essentially repeating the reasoning outlined in §2.16.

The necessity of the condition (2.18.2) can be shown as follows. The probability 1 $\mathbb{T}_3$-invariance implies, according to §2.9, that η factorizes with probability 1, i.e.

$$\eta = L_2 \times v,$$

where L_2 is Lebesgue on $\mathbb{R}^2$ (and therefore non-random) and v is some random measure on $\mathbb{S}_2$. By the assumption (a), $\eta \times \eta$ will not charge the set Z_3 (probability 1). Hence we can use (2.18.4) and therefore the product measure $\eta \times \eta$ will possess a non-random measure factor $dQ\,dh$. As a result, μ_2 will also have this factor. This completes the proof.

3

Measures invariant with respect to Euclidean motions

The requirement of invariance with respect to both rotations and translations of the basic space (i.e. invariance with respect to Euclidean motions) in fact enables us to determine measures in some of the spaces considered in chapter 2 in a unique way. Below we consider these questions, starting with a description of Haar measures on the groups of rotations of $\mathbb{R}^2$ and $\mathbb{R}^3$. To derive Haar measures on Euclidean groups (or kinematic measures) we apply the method of Haar factorization.

A natural application of the factorization ideas leads to position–size–shape factorizations which we consider in the concluding sections. Some of these factorizations reappear in chapter 4 in a different context.

There is a conceptual difference between the 'shapes' we consider in this chapter and the 'affine shapes' of chapter 4. Following the thinking which led to the term 'affine shape', the shapes of the present chapter should be termed 'Euclidean shapes'. However we cling to the shorter term 'shape' which is now widely used, [34], [41], [63].

3.1 The group $\mathbb{W}_2$ of rotations of $\mathbb{R}^2$

Rotations of $\mathbb{R}^2$ around the origin O can be represented by points on the unit circle $\mathbb{S}_1$: a point $\varphi \in \mathbb{S}_1$ corresponds to the (anticlockwise, say) rotation by the angle φ. The product of rotations corresponds to addition (mod 2π).

There is a natural measure on $\mathbb{S}_1$ which ascribes to each arc $a \subset \mathbb{S}_1$ its length $|a|$. We call this the *arc length measure* and denote it by $d\varphi$.

The arc length measure is invariant under rotations (because the lengths of the arcs remain invariant). We denote the group of rotations of $\mathbb{R}^2$ around O by $\mathbb{W}_2$, $w \in \mathbb{W}_2$. By the correspondence

$$\mathbb{S}_1 \to \mathbb{W}_2$$

the arc length measure can be considered as a measure on W_2. It is finite and both left- and right-invariant (the group W_2 is commutative). From the remarks in §1.4 it follows that

every Haar measure on the group W_2 of rotations of $\mathbb{R}^2$ is proportional to the arc length measure.

3.2 Rotations of $\mathbb{R}^3$

We denote the group of rotations of $\mathbb{R}^3$ by W_3. A rotation is a Euclidean motion of $\mathbb{R}^3$ which keeps the origin O intact. We describe two different representations of elements from W_3.

I Directed flags We call a figure which consists of a directed line γ through the origin O and a halfplane $h \subset \mathbb{R}^3$ bounded by γ a *directed flag* (non-directed flags will be considered in chapter 5). An example of a directed flag yields the pair (γ_0, h_0) which we describe in terms of the usual Cartesian coordinates in $\mathbb{R}^3$:

γ_0 – the axis Ox;
h_0 – the halfplane $y > 0$ of the $z = 0$ plane.

Every $w \in W_3$ can be completely described by the directed flag $w(\gamma_0, h_0)$, i.e. by the image of (γ_0, h_0) under w. In fact, given $w\gamma_0$ (the image of the Ox axis under w) we still have the freedom of rotation around the $w\gamma_0$ axis. We remove this freedom by fixing the position of wh_0.

Clearly the range of $w\gamma_0$ is the unit sphere S_2; with $w\gamma_0$ fixed, the range of wh_0 is S_1. However, an attempt to ascribe the topology of $S_2 \times S_1$ to W_3 *fails* (there will be pairs of flags near to each other both as sets and in directions of γ and yet far apart as points in $S_2 \times S_1$).

We can represent a rotation as

$$w = (\Omega, \mathscr{P}),$$

where

$\Omega \in S_2$ is the spatial direction of $w\gamma_0$;
$\mathscr{P} \in \mathbb{R}^3$ is a point on the plane $t(\Omega)$ tangent to S_2 at Ω;
$\mathscr{P}$ belongs to the unit circle $C(\Omega) \subset t(\Omega)$ centered at the point of tangency,
   $\mathscr{P} = wh_0 \cap C(\Omega)$.

The topology in W_3 is now defined by the convergence rule

$$\lim(\Omega_n, \mathscr{P}_n) = (\Omega, \mathscr{P})$$

if and only if

$$\lim \Omega_n = \Omega \quad \text{in } S_2$$

and
$$\lim \mathscr{P}_n = \mathscr{P} \quad \text{in } \mathbb{R}^3.$$

Actually W_3 is a fibered space:
$$\pi(\Omega, \mathscr{P}) = \Omega \in S_2;$$

S_1 is a fixed fiber model.

Remark If we delete the fibers corresponding to

Ω_1 – direction of the Ox axis, and
Ω_2 – the opposite direction,

then this slitted W_3 becomes identical to the product
$$(S_2 \backslash \{\Omega_1, \Omega_2\}) \times S_1,$$
i.e. we have
$$W_3 \approx S_2 \times S_1$$

in the sense of §1.6. It is easy to complement this map to become (discontinuous) one-to-one. Then $S_2 \times S_1$ becomes a measure-representing model of W_3. (This follows from the fact that both spaces are compact.) Having this map in mind we will label rotations by points of $S_2 \times S_1$, i.e. we will use

$$w = (\Omega, \Phi), \quad \Omega \in S_2, \quad \Phi \in S_1. \tag{3.2.1}$$

II Dual representation Parametrization of rotations by points from $S_2 \times S_1$ can be achieved in a different way from (3.2.1) owing to the possibility of the dual description (ω, φ) of a directed flag (γ, h). Here $\omega \in S_2$ is a spatial direction defined by two conditions:

(a) that ω is normal to the plane containing h;
(b) an observer whose feet are at the centre of S_2 and whose head is at ω looking in the direction of γ will find h on his left hand.

As soon as ω is determined, we define $\varphi \in S_1$ to be the planar direction of γ in the plane orthogonal to ω. In this way we get a parametrization

$$w = (\omega, \varphi) \tag{3.2.2}$$

which we call *dual* to (Ω, Φ) in (3.2.1).

3.3 The Haar measure on W_3

We consider on the product space $S_2 \times S_1$ the measure which is a product of the area measure on S_2 and the arc length measure on S_1. This measure has two images on W_3 generated by the maps (3.2.1) and (3.2.2). We naturally denote these images by

$d\Omega \, d\Phi$, corresponds to (3.2.1), and
$d\omega \, d\varphi$, corresponds to (3.2.2).

Let us consider a map

$$(\Omega, \Phi) \to w \circ (\Omega, \Phi) \qquad (3.3.1)$$

where $w \circ (\Omega, \Phi)$ is the result of an action of $w \in W_3$ on the flag (Ω, Φ). By a kind of Cavalieri principle the measure $d\Omega\, d\Phi$ is invariant under the transformation (3.3.1). (This follows from invariance of area and arc lengths under rotations.) But (3.3.1) corresponds to multiplication in W_3:

$$w \circ (\Omega, \Phi) = ww_1 \quad \text{with } w_1 = (\Omega, \Phi).$$

It follows that $d\Omega\, d\Phi$ is a left-invariant Haar measure on W_3 with total value $2\pi \cdot 4\pi$. By the remarks in §1.4 $d\Omega\, d\Phi$ is bi-invariant.

A similar argument applies to the measure $d\omega\, d\varphi$. We come to the conclusion that $d\omega\, d\varphi$ is also bi-invariant Haar. Since the total values of $d\Omega\, d\Phi$ and $d\omega\, d\varphi$ coincide, we have by unicity

$$dw = d\Omega\, d\Phi = d\omega\, d\varphi, \qquad (3.3.2)$$

a result often used in geometric probability [13].

It remains to add that the corresponding measure on W_3 when viewed as a fibered space is a composition (see §1.6) of arc length measures on the circular fibers via the area measure on S_2.

3.4 Geodesic lines on a sphere

By 'polar mapping', any point $\omega \in S_2$ determines a directed geodesic line g on S_2: by definition, the plane containing g passes through the center of S_2 and is orthogonal to ω; the direction of g is chosen in such a way that it is seen anticlockwise from the point ω. Therefore the space $\overline{G}_0$ of directed lines on S_2 can be identified with S_2 itself:

$$\overline{G}_0 = S_2.$$

Let us denote by μ the measure on $\overline{G}_0$ which is the image of the area measure $d\omega$ under the polar mapping. Clearly μ is invariant under rotations of S_2 (because $d\omega$ is), but is μ unique? This question can be given an affirmative answer by using the constructions of §3.2.

Let us consider the space $\overline{G}_0 \times S_1$ of pairs

$$(g, \varphi) = (\omega, \varphi),$$

where $g \in \overline{G}_0$ and φ is a point on g. Every (g, φ) can be identified with a rotation (see §3.2) and therefore there is only one finite measure on the space $\overline{G}_0 \times S_1$ which is invariant under W_3, namely

$$d\omega\, d\varphi.$$

Let m be any measure on $\overline{G}_0$ which is finite and invariant under W_3. On the space $\overline{G}_0 \times S_1$ we consider the product measure

$$m(dg) \cdot d\varphi.$$

By the Cavalieri principle, this measure is $\mathbb{W}_3$-invariant. Therefore,

$$m(dg)\, d\varphi = c\, d\omega\, d\varphi \quad \text{for some } c \geqslant 0.$$

Eliminating the measure factor $d\varphi$, we find that

$$m(dg) = d\omega \cdot c.$$

This means that m is necessarily proportional to μ. Note that we have also proved the uniqueness of the measure $d\omega$.

3.5 Bi-invariance of Haar measures on Euclidean groups

We denote by $\mathbb{M}_n$ the group of all Euclidean (rigid) motions of $\mathbb{R}^n$ (we actually consider the cases $n = 2, 3$), $M \in \mathbb{M}_n$. Let us see how the criterion of §1.3 applies to $\mathbb{M}_n$.

The subgroups $\mathbb{U}$ and $\mathbb{V}$ are

$$\mathbb{U} = \mathbb{T}_n, \quad \text{the translations of } \mathbb{R}^n$$

and

$$\mathbb{V} = \mathbb{W}_n, \quad \text{the rotations of } \mathbb{R}^n.$$

Both representations of (1.3.4) are known to exist; $h_\mathbb{U}$ is the Lebesgue measure (also denoted as dt); $h_\mathbb{V}$ is the Haar measure on $\mathbb{W}_n$ (also denoted as dw).

The equation

$$t_l w_r = w_l t_r$$

has the following solution:

$$t_l = w_l t_r w_l^{-1},$$
$$w_r = w_l. \tag{3.5.1}$$

The above is a Cavalieri-type transformation of $\mathbb{W}_n \times \mathbb{T}_n$. Indeed, for each w_l, the transformation given by (3.5.1) preserves the Lebesgue measure on $\mathbb{T}_n$ (this follows from rotation-invariance of Lebesgue measure in $\mathbb{R}^n$). Thus

$$dM = dt_l\, dw_r = dt_r\, dw_l \tag{3.5.2}$$

is a bi-invariant Haar measure on $\mathbb{M}_n$.

In the sections that follow we also denote by $\mathbb{M}_n$ the groups of transformations of the spaces of lines and planes in $\mathbb{R}^n$ ($n = 2, 3$) *induced* by the corresponding Euclidean groups.

3.6 The invariant measure on $\overline{\mathbb{G}}$ and $\mathbb{G}$

Any $\mathbb{M}_2$-invariant measure μ on $\overline{\mathbb{G}} = \mathbb{S}_1 \times \mathbb{R}$ is proportional to the product measure

$$l \times L_1, \tag{3.6.1}$$

where l is the arc length measure on $\mathbb{S}_1$ and L_1 is a Lebesgue measure on $\mathbb{R}$.

We will also denote the above measure by dg. Thus, in the coordinates introduced in §2.1,

$$dg = d\varphi\, dp. \tag{3.6.2}$$

To prove the assertion we consider a product set

$$A \times B \quad \text{with } A \subset \mathbb{S}_1, B \subset \mathbb{R}.$$

Since $\mathbb{T}_2$ is a subgroup of $\mathbb{M}_2$, μ necessarily must factorize as in §2.9, i.e.

$$\mu(A \times B) = m(A) \cdot L_1(B). \tag{3.6.3}$$

Let us now apply a rotation w to $A \times B$. We have

$$w(A \times B) = wA \times B.$$

Taking the measure μ from both sides and applying (3.6.3) we find that necessarily

$$m(wA) = m(A),$$

i.e. m should be invariant with respect to rotations, i.e. up to constant factor

$$m = l.$$

It remains to check that the measure dg is indeed invariant with respect to $\mathbb{M}_2$, but this follows from the possibility of presenting any $M \in \mathbb{M}_2$ as a product of a parallel shift and a rotation.

Using the factorization described in §2.9 it is also easy to prove that any $\mathbb{M}_2$-invariant measure on $\mathbb{G}$ corresponds to a product measure on the model $\mathbb{E}_1 \times \mathbb{R}$ of this space (see §2.7).

The factor measures are (up to constant factors)

on $\mathbb{E}_1$ – the arc length measure $d\varphi$,
on $\mathbb{R}$ – the Lebesgue measure dp.

3.7 The form of dg in two other parametrizations of lines

Since we have dg in (φ, p) coordinates, the form of dg in terms of other parameters can be found by Jacobian calculations. However, in the following we use the more geometrical 'symmetry principle'; the latter is especially useful when the dimensionality of the problem increases (see §3.10). We consider the following:

(a) The (ψ, x) coordinates as described in §2.7, 1. They are applicable because the invariant measure of $\mathbb{G} \backslash \mathbb{G}_x$ is zero.
(b) The (l_1, l_2) coordinates. Let us assume that two 'reference lines' g_1 and g_2 are fixed on $\mathbb{R}^2$. Let l_i, $i = 1, 2$, be the usual one-dimensional coordinates on g_i.

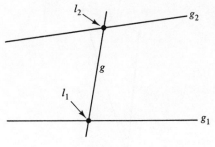

Figure 3.7.1

We can exclude lines parallel either to g_1 or to g_2. On the remaining set, the points of intersection

$$l_1 = g \cap g_1, \quad l_2 = g \cap g_2$$

completely determine the line g (fig. 3.7.1).

Let us write using (a)

$$dg = F \, dl \, d\psi. \tag{3.7.1}$$

Our problem is to find that particular F which yields the required invariance property.

Geometrically, the element $F \, dl \, d\psi$ has the meaning of the value of the measure of set of lines

$$\{g \in \mathbb{G} : g \text{ intersects an interval } (l, l + dl) \subset g_0, \atop \text{the intersection angle } \psi \text{ lies in } (\psi, \psi + d\psi)\}. \tag{3.7.2}$$

We choose two arbitrary linear elements, dl_1 and dl_2, on $\mathbb{R}^2$ (necessarily on different lines but of equal lengths).

The corresponding sets (3.7.2) (with the same ψ, $\psi + d\psi$) will be congruent. From this we conclude that

(1) the function F in (3.7.1) is *universal*, i.e. it does not depend on the choice of reference line g_0;

(2) F does not depend on l, i.e.

$$F = F(\psi).$$

To determine F let us choose a pair of lines g_1 and g_2 and elementary intervals $dl_1 \subset g_1$ and $dl_2 \subset g_2$, and let

$$A = \{g \in \mathbb{G} : g \text{ intersects both } dl_1 \text{ and } dl_2\}.$$

By first choosing g_1 to be the reference line in the representation (a) we write (μ is another notation for the measure dg)

$$\mu(A) = F(\psi_1) \, dl_1 \, d\psi_1.$$

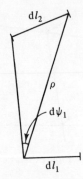

Figure 3.7.2 ρ is the distance between dl_1 and dl_2

In the notation of fig. 3.7.2

$$d\psi_1 = \frac{\sin \psi_2}{\rho}\, dl_2.$$

Thus

$$\mu(A) = F(\psi_1)\rho^{-1}\sin\psi_2\, dl_2\, dl_1.$$

This expression should be symmetrical in the indices 1 and 2 (since we could start the derivation from g_2). The variables ψ_1 and ψ_2 are independent. Therefore we have to conclude that

$$F(\psi) = c\cdot\sin\psi.$$

The calculation of the measure of lines which hit a unit disc (say) shows that, in order to get $d\varphi\, dp$, we have to choose $c = 1$. Thus we have found the following expressions:

$$dg = \sin\psi\, dl\, d\psi \qquad (3.7.3)$$

and

$$dg = \rho^{-1}\sin\psi_1\sin\psi_2\, dl_1\, dl_2. \qquad (3.7.4)$$

Remark The above can be considered as an independent proof of the uniqueness of invariant measure on $\mathbb{G}$, in the class of measures possessing densities of the form (3.7.1).

3.8 Other parametrizations of geodesic lines on a sphere

Here we consider the space $\mathbb{G}_0$ of non-directed great circles.

(a) Let us assume that a 'reference' great circle g_0 is fixed on $\mathbb{S}_2$ and let g_0^+ denote a semicircle in some way specified on g_0. If we ignore the geodesics which pass through the endpoints of g_0^+, then every $g \in \mathbb{G}_0$ is determined by the unique point l of its intersection with g_0^+ and by the angle ψ of intersection of g with g_0^+ at l. Thus

$$g = (l, \psi), \quad l \in (0, \pi), \quad \psi \in (0, \pi).$$

(b) Let two 'reference' great semicircles g_1^+ and g_2^+ which lie on different great circles be fixed on $\mathbb{S}_2$. Except for the geodesics which pass through the point $g_1^+ \cap g_2^+$ each $g \in \mathbb{G}$ is determined by the pair (l_1, l_2), where l_i is the intersection point $l_i = g \cap g_i^+$. Thus,

$$g = (l_1, l_2), \quad \text{each } l_i \in (0, \pi).$$

The form of the invariant measure dg in the coordinates (l, ψ) and (l_1, l_2) can be found by Jacobian calculation starting from

$$d\omega = \sin \theta \, d\theta \, d\Phi, \tag{3.8.1}$$

which is the usual expression of an area element in polar angular coordinates on $\mathbb{S}_2$. However, an approach in the style of §3.7 (the symmetry principle) is also possible. The result is

$$dg = \sin \psi \, dl \, d\psi \tag{3.8.2}$$

and

$$dg = (\sin \rho)^{-1} \sin \psi_1 \sin \psi_2 \, dl_1 \, dl_2. \tag{3.8.3}$$

Here dl_1 and dl_2 are length elements, ρ is the geodesic distance between dl_1 and dl_2, and ψ_i is the angle of intersection of g_i^+ by g.

3.9 The invariant measure on $\bar{\Gamma}$ and Γ

Any $\mathbb{M}_3$-invariant measure μ on $\bar{\Gamma}$ is proportional to the product measure

$$a \times L_2, \tag{3.9.1}$$

on the product model $\mathbb{S}_2 \times \mathbb{R}^2$ of the space $\bar{\Gamma}$ where a is the area measure on $\mathbb{S}_2$, and L_2 is the Lebesgue measure on $\mathbb{R}^2$. To obtain the proof we consider the set

$$A = \{\gamma \in \bar{\Gamma} : \Omega \in B, \mathscr{P} \text{ belongs to the unit disc } K_1 \text{ centered at } O\}$$

(see §2.5 for notation). For every rotation $w \in \mathbb{W}_3$, wA is a product set

$$wA = wB \times K_1.$$

Therefore, by the factorization result in §2.9 concerning $\mathbb{T}_3$-invariant measures on $\bar{\Gamma}$,

$$\mu(wA) = \mu(A)$$

amounts to $\mathbb{W}_3$-invariance of the factor measure on $\mathbb{S}_2$. Therefore the latter measure is necessarily proportional to the area measure a (the uniquess of a was essentially shown in §3.4).

Finally $\mathbb{M}_3$-invariance of (3.9.1) can be checked by applying the Cavalieri principle (see §1.5).

A similar result holds for the space Γ, in which case the area measure a is defined on $\mathbf{E}_2$ (see §2.7).

In terms of the parametrization described in §2.7 the measure (3.9.1) is written as

$$d\gamma = d\Omega \, d\mathscr{P}, \qquad (3.9.2)$$

where $d\gamma$ is an element of μ, $d\Omega$ is an element of a, and $d\mathscr{P}$ is an element of L_2. The representation (3.9.2) is also valid for the M_3-invariant measure on the space Γ.

3.10 Other parametrizations of lines in $\mathbb{R}^3$

Below we consider the form of $d\gamma$ in two other parametrizations of lines from Γ.

(a) Let e_0 be a 'reference plane' fixed in $\mathbb{R}^3$. If we ignore lines parallel to e_0 then any line γ can be described as

$$\gamma = (\Omega, Q),$$

where Q is the point in which γ hits e_0, and $\Omega \in \mathbb{S}_2/2$ is the direction of γ (compare with §2.7).

(b) Let us assume that two distinct reference planes e_1 and e_2 are fixed in $\mathbb{R}^3$. Ignoring those lines which are parallel either to e_1 or to e_2 or hit the line $e_1 \cap e_2$ we can write

$$\gamma = \{Q_1, Q_2\} \quad \text{(an unordered pair)},$$

where Q_i is the point where γ hits e_i.

Applying the symmetry principle (similar to that used in §3.7) requires the consideration of infinitesimal flats rather than segments. However, the main idea is the same. The result is as follows:

$$d\gamma = \cos v \, dQ \, d\Omega, \qquad (3.10.1)$$

where $d\Omega$ is an element of area measure on $\mathbb{S}_2$, dQ is an element of planar Lebesgue measure on e_0, v is the angle between the direction Ω and the direction normal to e_0, and

$$d\gamma = \frac{\cos v_1 \cos v_2}{\rho^2} \, dQ_1 \, dQ_2, \qquad (3.10.2)$$

where v_i and dQ_i have similar meanings with respect to e_i, $i = 1, 2$, and ρ is the distance between dQ_1 and dQ_2.

The basic property of $\mathbb{R}^3$ which is responsible for the final form of the above expressions is the formula giving the solid angle $d\Omega$ at which an infinitesimal flat of area dQ is seen from a point distance ρ apart, namely

$$d\Omega = \rho^{-2} \cos v \, dQ,$$

where v is the angle between the direction of ρ and the direction normal to dQ.

3.11 The invariant measure in the spaces $\overline{\mathbb{E}}$ and $\mathbb{E}$

Any $\mathbb{M}_3$-invariant measure μ on $\overline{\mathbb{E}} = \mathbb{S}_2 \times \mathbb{R}$ (see §2.3) is proportional to the product measure

$$a \times L_1 \tag{3.11.1}$$

where a is the area measure on $\mathbb{S}_2$, and L_1 is the Lebesgue measure on $\mathbb{R}$.

The proof is similar to that used in §3.6 and §3.9 (i.e. using factorization described in §2.9 and the uniqueness of a on $\mathbb{S}_2$ as a $\mathbb{W}_3$-invariant measure).

A similar result holds for $\mathbb{E}$ in terms of its product model $\mathbb{E}_2 \times \mathbb{R}$ (see §2.7; the measure a is well-defined on $\mathbb{E}_2$).

In both the spaces $\overline{\mathbb{E}}$ and $\mathbb{E}$ the element de of the invariant measure is written in the form

$$de = d\omega \, dp, \tag{3.11.2}$$

where $d\omega$ is an element of a, and dp is an element of L_1.

3.12 Other parametrizations of planes in $\mathbb{R}^3$

Below we consider the form of de in three other parametrizations of planes.

(a) Let us assume that a reference axis γ_0 is fixed in $\mathbb{R}^3$. If we ignore planes parallel to γ_0, then any plane e can be described as

$$e = (\omega, x),$$

where x is the point where e hits γ_0, and ω is the direction normal to e.

(b) Let us assume three distinct axes γ_1, γ_2 and γ_3 are fixed in $\mathbb{R}^3$. Ignoring planes parallel to the axes, as well as planes through points of their intersection (if any exist), we can write

$$e = (x_1, x_2, x_3),$$

where x_i is the point where e hits γ_i.

(c) Let us assume that a reference plane e_0 is fixed in $\mathbb{R}^3$. Ignoring planes parallel to e_0 we can write

$$e = (g, \psi),$$

where g is the line of intersection of e with e_0, and ψ is the flat angle of intersection.

We can use the symmetry principle (in the same way as in §3.7) working with the parametrizations (a) and (b). The results will be

$$de = \cos v \, d\omega \, dx, \tag{3.12.1}$$

where v is the angle between ω and the direction of γ_0, dx is a length element, and $d\omega$ is an element of area measure on $\mathbb{S}_2$;

$$de = \frac{\cos v_1 \cos v_2 \cos v_3}{2|\Delta|} dx_1 \, dx_2 \, dx_3, \tag{3.12.2}$$

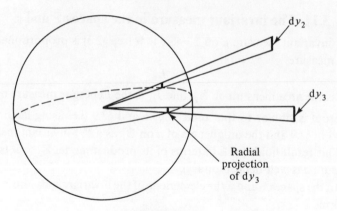

Figure 3.12.1

where v_i is as above but referred to γ_i, and $|\Delta|$ is the area of the triangle Δ which has its vertices on each of the elements dx_i.

The two above formulae are related by the value of $d\omega_1$, which is the solid angle subtended by the normals to the planes from the set {the planes through an endpoint of dx_1, which hit both dx_2 and dx_3}. We have

$$d\omega_1 = \frac{\cos v_2 \cos v_3}{2|\Delta|} \, dx_2 \, dx_3. \tag{3.12.3}$$

In proving (3.12.3) the formula (3.8.3) can be useful. Indeed, the value of the solid angle in question will not change if we replace dx_2 and dx_3 by their projections dy_2 and dy_3 on the lines which emerge from the corresponding vertices of Δ perpendicularly to the plane of Δ. The length of dy_i is $\cos v_i \, dx_i$, $i = 2, 3$. Next we note that $d\omega_1$ equals the invariant measure of the set of geodesic lines which hit both radial projections of dy_2 and dy_3 (see fig. 3.12.1) on the unit radius sphere centered at dx_1. The length of these projections is $dy_i \, h_i^{-1}$, where h_i is the distance between dx_1 and dx_i, $i = 2, 3$. Since our radial projections are perpendicular to the geodesic line joining them, (3.8.3) yields

$$d\omega_1 = \frac{\cos v_2 \cos v_3}{h_2 h_3 \sin \rho} \, dx_2 \, dx_3,$$

which is the same as (3.12.3).

From (3.12.2) the form of de in the parametrization (c) can be derived in the following way.

Let us assume that all three segments dx_1, dx_2 and dx_3 are parallel, and that both dx_1 and dx_2 lie in e_0 and are both orthogonal to a line g joining them. Let ψ be the flat angle between e_0 and a plane through dx_1, dx_2 and dx_3; let the distance between dx_1 and dx_2 be ρ; and let the distance from g to

$\mathrm{d}x_3$ be h. Then (3.12.2) yields

$$\mathrm{d}e = \frac{\sin^3 \psi \, \mathrm{d}x_1 \, \mathrm{d}x_2 \, \mathrm{d}x_3}{\rho h}.$$

Using

$$\frac{\mathrm{d}x_1 \, \mathrm{d}x_2}{\rho} = \mathrm{d}g$$

(see §3.7) and

$$\frac{\mathrm{d}x_3}{h} = \frac{\mathrm{d}\psi}{\sin \psi}$$

we find

$$\mathrm{d}e = \sin^2 \psi \, \mathrm{d}g \, \mathrm{d}\psi. \tag{3.12.4}$$

3.13 The kinematic measure

Let J be a 'figure' in $\mathbb{R}^n$. We denote by MJ the result of applying the motion $M \in \mathsf{M}_n$ to J. We will consider the space $\mathbb{J}$ of figures congruent to J, i.e.

$$\mathbb{J} = \{MJ : M \in \mathsf{M}_n\}.$$

Examples

(1) J is a ball in $\mathbb{R}^n$. Then $\mathbb{J} = \mathbb{R}^n$.
(2) J is a directed segment in $\mathbb{R}^3$. Then $\mathbb{J} = \Delta_3^*$ (see §2.13).
(3) J is a directed segment in $\mathbb{R}^2$. Then $\mathbb{J} = \mathsf{M}_2$.
(4) J is a pair of non-collinear segments of different lengths emerging from O. Then $\mathbb{J} = \mathsf{M}_3$.

The measure on $\mathbb{J}$ which is the image of the Haar measure on M_n under the mapping

$$M \to MJ$$

is called kinematic. Thus, the kinematic measure essentially differs from $\mathrm{d}M$ only in the cases where the above mapping is not one-to-one (as in the case in Examples (1) and (2)). In many cases the results of the previous sections imply the following proposition.

Any measure on $\mathbb{J}$ which is invariant with respect to the group M_n is proportional to the kinematic measure.

The proof is tautological whenever $\mathbb{J} = \mathsf{M}_n$. Let us outline the proof for the $\mathbb{J}$ of Example (2).

Let μ be an M_3-invariant measure on Δ_3^*. By the Cavalieri principle, the product measure $\mu(\mathrm{d}\delta) \, \mathrm{d}\Phi$ on the space

$$\Delta_3^* \times \mathbb{S}_1$$

is invariant with respect to $\mathbb{M}_3$. But the above product space is a measure-representing model of the group $\mathbb{M}_3$ (the corresponding construction is left to the reader). Any $\mathbb{M}_3$-invariant measure on this model is necessarily proportional to the Haar measure on $\mathbb{M}_3$ (by the uniqueness of the latter). Using the results of §3.5 and §3.2 we conclude that

$$\mu(\mathrm{d}\delta)\, \mathrm{d}\Omega = c\, \mathrm{d}t\, \mathrm{d}\Omega\, \mathrm{d}\Phi,$$

where c is a constant. Denoting the kinematic measure in the case $c = 1$ as $\mathrm{d}\delta$, we get by integration

$$\mathrm{d}\delta = 2\pi\, \mathrm{d}t\, \mathrm{d}\Omega. \tag{3.13.1}$$

Since every $\delta \in \Delta_3^*$ can be described by its continuation $\gamma \in \bar{\Gamma}$ and the one-dimensional coordinate x of its endpoint on γ, we have

$$\Delta_3^* = \bar{\Gamma} \times \mathbb{R}.$$

The kinematic measure on Δ_3^* is 2π times the product of the invariant measure on $\bar{\Gamma}$ and the Lebesgue measure on $\mathbb{R}$, i.e.

$$\mathrm{d}\delta = 2\pi\, \mathrm{d}\gamma\, \mathrm{d}x. \tag{3.13.2}$$

Proof By the Cavalieri principle, $\mathrm{d}\gamma\, \mathrm{d}x$ is $\mathbb{M}_3$-invariant and is therefore proportional to $\mathrm{d}\delta$. It remains to check that the proportionality constant is 2π.

There is a similar relation in the space Δ_2^*. We recall that

$$\Delta_2^* = \bar{\mathbb{G}} \times \mathbb{R} = \mathbb{M}_2$$

(see §2.13 and notation therein). We have

$$\mathrm{d}M = \mathrm{d}\delta = \mathrm{d}g\, \mathrm{d}x. \tag{3.13.3}$$

The same idea can be used to derive a useful representation for $\mathrm{d}M$, the element of the Haar measure on $\mathbb{M}_3$. Let s_1 and s_2 be the two segments emerging from O see (4) above. The image of the two segments under $M \in \mathbb{M}_3$ defines M completely. We can describe Ms_1 and Ms_2 by determining the plane e where they lie and by their position k on e. Let $\mathrm{d}k$ be the planar kinematic measure.

By the Cavalieri principle, $\mathrm{d}e\, \mathrm{d}k$ is $\mathbb{M}_3$-invariant and is therefore proportional to the Haar measure $\mathrm{d}M$ on $\mathbb{M}_3$. A simple check shows that for $\mathrm{d}M$ as in (3.5.2) the proportionality constant is 1. Thus we have

$$\mathrm{d}M = \mathrm{d}e\, \mathrm{d}k \tag{3.13.4}$$

which is an analog of (3.13.3).

Along similar lines we briefly derive one more useful result. Every $M \in \mathbb{M}_3$ can be described by the parameters

$$M = (e, g, x),$$

where e is the image of the xOy plane under M, g is the image of the x axis (therefore $g \subset e$), and x determines the position of MO on g. Because (g, x) can be identified with k (a position on e), from (3.13.3) and (3.13.4) we find

$$dM = de \, dg \, dx. \tag{3.13.5}$$

On the other hand, we can also describe M as

$$M = (\gamma, \Phi, x),$$

where $\gamma \in \bar{\Gamma}$ is the image of the Ox axis, $\Phi \in \mathbb{S}_1$ is the angle of rotation of e around γ, and x determines the position of MO on γ. By a kind of Cavalieri principle the product measure

$$d\gamma \, d\Phi \, dx$$

is a left-invariant Haar measure on $\mathbb{M}_3$. From the uniqueness of the latter we conclude that

$$d\gamma \, d\Phi \, dx = c \cdot dM,$$

or, using (3.13.5),

$$c \cdot de \, dg \, dx = d\gamma \, d\Phi \, dx.$$

Calculation of the measure of the set

$$\{M : MO \text{ belongs to the unit ball}\}$$

shows that $c = 1$. Elimination of the dx factor yields

$$de \, dg = d\gamma \, d\Phi. \tag{3.13.6}$$

We have obtained two expressions of the (unique) $\mathbb{M}_3$-invariant measure in the space of figures

$$J = (\text{a plane } e; \text{a line } g \text{ on } e).$$

Strictly speaking, this measure cannot be termed kinematic; indeed, the image of dM in the space of the above figures is not locally finite. We use (3.13.6) in §6.3.

3.14 Position–size factorizations

In this section (and in the next two) we consider products of the $\mathbb{M}$-invariant measures that we met in previous sections. It follows from our factorization proposition of §1.3 that these products always allow separation of the kinematic measure (i.e. the unique $\mathbb{M}$-invariant measure in the space of 'positions' of the geometrical figure in question). We are interested in the measure factor that remains; in this section we consider cases where this factor is a measure in the 'size' space (which is always $(0, \infty)$).

Below we write the product measures in differential form and derive the desired factorizations using simple 'chain' procedures. Justification of each step in these chains can be based upon the following two rules:

(1) The Jacobians of transformations of the form

$$y_1 = f_1(x_1, \ldots, x_m) \quad \text{and} \quad y_1 = f_1(x_1, \ldots, x_m)$$

$$\ldots\ldots\ldots\ldots\ldots \qquad\qquad \ldots\ldots\ldots\ldots\ldots$$

$$y_m = f_m(x_1, \ldots, x_m) \qquad\quad y_m = f_m(x_1, \ldots, x_m)$$

$$y_{m+1} = x_{m+1}$$

$$\ldots\ldots\ldots\ldots$$

$$y_n = x_n$$

coincide. This means that, where only a part of the variables are transformed and others are left intact, we have a gain in dimensionality. It is easier to calculate several Jacobians of reduced dimensionality than one of full dimension.

(2) The invariance properties of the measures yield 'universality' (see §3.7 for an example) of their differential representations. Therefore the applicability of the latter is not influenced by the dependence of 'reference' elements on parameters. The benefit is twofold: we apply ready differential formulae from the preceding sections: we also achieve in this way 'explanations' for the resulting densities.

I Pairs of points in $\mathbb{R}^2$ An ordered pair (Q_1, Q_2) with $Q_1, Q_2 \in \mathbb{R}^2$, $Q_1 \neq Q_2$, can be represented by a segment δ from Δ_2. Using (2.13.3) we get

$$(Q_1, Q_2) = (M, l), \quad \text{where } M \in \mathbb{M}_2, l \in (0, \infty)$$

(l is the distance between Q_1 and Q_2). Thus (see §1.6)

$$\mathbb{R}^2 \times \mathbb{R}^2 \approx \mathbb{M}_2 \times (0, \infty).$$

We denote by dQ_i planar Lebesgue measure elements. The measure $dQ_1\, dQ_2$ in $\mathbb{R}^2 \times \mathbb{R}^2$ is $\mathbb{M}_2$-invariant and therefore necessarily admits Haar factorization (or, equivalently, separation of kinematic measure).

If we take $dQ_1\, dQ_2$ to be as shown in fig. 3.14.1 we will have

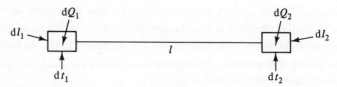

Figure 3.14.1 dQ_i, $i = 1, 2$, are right angles, the segment l joining them is perpendicular both to dl_1 and to dl_2

$$dQ_1 \, dQ_2 = dl_1 \, dl_2 \, dt_1 \, dt_2$$
$$= l \, dg \, dt_1 \, dt_2$$
$$= l \, dg \, dt \, dl$$
$$= dM \, l \, dl. \tag{3.14.1}$$

Also we used (3.7.4) (with $\psi_1 = \psi_2 = \pi/2$) and (3.13.3). Also

$$t = \tfrac{1}{2}(t_1 + t_2),$$
$$l = |t_2 - t_1|,$$

where

t_1 and t_2 are one-dimensional coordinates of Q_1 and Q_2 on the line joining these two points, dM is the Haar measure on $\mathbb{M}_2$ (or the kinematic on Δ_2^*). (3.14.1) is the desired factorization.

II Pairs of points on the unit sphere Similar arguments remain valid when we consider the pairs (ω_1, ω_2) of points on $\mathbb{S}_2$, the unit sphere. Following (3.8.3)

$$d\omega_1 \, d\omega_2 = d\Omega \, d\varphi_1 \, d\varphi_2 \sin \rho = \sin \rho \, dw \, d\rho, \tag{3.14.2}$$

where $d\omega_i$ are elementary solid angles, Ω is the direction perpendicular to both ω_1 and ω_2 ($d\Omega$ is the corresponding elementary solid angle), ρ is the angle between ω_1 and ω_2, φ_1 and φ_2 are the angles determining the position of ω_1 and ω_2 on the geodesic through ω_1, ω_2, and dw is an element of Haar measure on $\mathbb{W}_3$ (see §3.3). Strictly speaking, ρ cannot be called a 'size' parameter.

III Pairs of points in $\mathbb{R}^3$ Similarly, an ordered pair $(\mathscr{P}_1, \mathscr{P}_2)$, $\mathscr{P}_1 \neq \mathscr{P}_2$, of points from $\mathbb{R}^3$ can be represented as (δ, l) with $\delta \in \Delta_3^*$, $l \in (0, \infty)$ (l is the distance between $\mathscr{P}_1$ and $\mathscr{P}_2$). Thus

$$\mathbb{R}^3 \times \mathbb{R}^3 \approx \Delta_3^* \times (0, \infty).$$

We now denote by $d\mathscr{P}_i$ the elements of Lebesgue measure in $\mathbb{R}^3$. The product measure $d\mathscr{P}_1 \, d\mathscr{P}_2$ is $\mathbb{M}_3$-invariant, therefore it admits separation of the kinematic measure in Δ_3^* (see §3.13). Using (3.10.2) and (3.13.2), by transformations similar to those in I, we get the desired position–size factorization

$$d\mathscr{P}_1 \, d\mathscr{P}_2 = \frac{1}{2\pi} \, d\delta \, l^2 \, dl. \tag{3.14.3}$$

3.15 Position–shape factorizations

In this section we continue with separation of kinematic measure from products of $\mathbb{M}$-invariant measures (see the introductory remarks to §3.14). In the cases we consider here the non-kinematic factors are measures in different 'shape' spaces.

I Pairs of lines on $\mathbb{R}^2$ On $\overline{G} \times \overline{G}$ we consider the product measure $dg_1\, dg_2$, where the dg_i are elements of the M_2-invariant measure on $\overline{G}$ (see §3.6). It is not difficult to justify the use of (3.7.3) for dg_2, where we take g_1 for the reference line, i.e. (see fig. 3.15.1)

$$dg_1\, dg_2 = dg_1\, |\sin \psi|\, d\psi\, dx.$$

Writing, according to (3.6.2), $d\varphi_1\, dp_1$ for dg_1, we get

$$dg_1\, dg_2 = d\varphi_1\, dp_1\, |\sin \psi|\, d\psi\, dx.$$

Since dp_1 and dx are perpendicular, we have by (3.5.2) (the planar case)

$$d\varphi_1\, dp_1\, dx = dM;$$

hence the desired (position–shape) factorization

$$dg_1\, dg_2 = dM\, |\sin \psi|\, d\psi. \tag{3.15.1}$$

II Pairs of planes in $\mathbb{R}^3$ On $\overline{E} \times \overline{E}$ we consider the product measure $de_1\, de_2$, where de_i are elements of the M_3-invariant measure on $\overline{E}$ (see §3.11). Using e_1 for the reference plane, we put de_2 in the form (3.12.4):

$$de_1\, de_2 = de_1\, \sin^2 \psi\, dg\, d\psi,$$

where $\psi \in (0, \pi)$ is now the flat angle between e_1 and e_2, and dg is the invariant measure of directed lines on e_1.

Each pair (e, g) can be represented as (γ, Φ), where $\gamma \in \overline{\Gamma}$ coincides with g (viewed now as a line in $\mathbb{R}^3$), and the angle $\Phi \in \mathbb{S}_1$ determines the rotation of e_1 around γ. From (3.13.6)

$$de_1\, de_2 = d\gamma\, d\Phi\, \sin^2 \psi\, d\psi. \tag{3.15.2}$$

This is the desired (position–shape) factorization. Indeed, γ and Φ together determine the position of (e_1, e_2) in $\mathbb{R}^3$, and $d\gamma\, d\Phi$ is essentially the only M_3-invariant measure in the space of these positions. The shape parameter is the angle ψ.

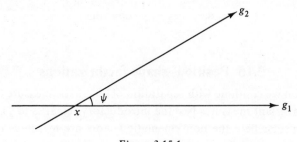

Figure 3.15.1

III Triads of planes in $\mathbb{R}^3$ We use the same notations as in II above. On $\bar{\mathbb{E}} \times \bar{\mathbb{E}} \times \bar{\mathbb{E}}$ we consider the product measure $de_1 \, de_2 \, de_3$. Using e_1 for the reference plane, we write, using (3.12.4),

$$de_1 \, de_2 \, de_3 = de_1 \sin^2 \psi_2 \, dg_2 \sin^2 \psi_3 \, dg_3 \, d\psi_2 \, d\psi_3$$

where g_i is the directed line of intersection of e_i with e_1, $i = 2, 3$, and ψ_i is the corresponding flat angle. Using (3.15.1) we find

$$de_1 \, de_2 \, de_3 = de_1 \, dM \, |\sin \alpha| \, d\alpha \sin^2 \psi_2 \, d\psi_2 \sin^2 \psi_3 \, d\psi_3, \quad (3.15.3)$$

where α denotes the angle between the lines g_1 and g_2. This is the desired position–shape factorization, since, according to (3.13.4), $de \, dM$ (where dM stands for the Haar measure on $\mathbb{M}_2$) gives the Haar measure on $\mathbb{M}_3$.

3.16 Position–size–shape factorizations

This section continues the work of the previous two (see the opening paragraph of §3.14). Here the non-kinematic factors are measures in 'size–shape' spaces. This factor permits further factorization, which results in the separation of measures in 'size' and 'shape' spaces. Although in each case the measure in the size–shape space is unique, the factor measure in the shape space may depend on the choice of the 'size' parameter.

I Pairs of lines in $\mathbb{R}^3$ We consider the product measure $d\gamma_1 \, d\gamma_2$ on the space $\bar{\Gamma} \times \bar{\Gamma}$, where $d\gamma$ is the $\mathbb{M}_3$-invariant measure on $\bar{\Gamma}$. Our aim is to obtain the position–size–shape factorization

$$d\gamma_1 \, d\gamma_2 = dM \cdot m(dh)\mu(d\sigma),$$

where dM is the Haar measure on $\mathbb{M}_3$, and h and σ are parameters describing the size and shape of a figure γ_1, γ_2. Here h is the (minimal) distance between the lines, and σ is the angle between the spatial directions ω_1 and ω_2 of the lines.

We start with the representation (see §3.9)

$$d\gamma_i = d\omega_i \, d\mathscr{P}_i,$$

where $d\omega_i$ is the solid angle element and $d\mathscr{P}_i$ is an area element on a plane perpendicular to γ_i. We have

$$d\gamma_1 \, d\gamma_2 = d\omega_1 \, d\omega_2 \, d\mathscr{P}_1 \, d\mathscr{P}_2$$
$$= d\Omega \sin \sigma \, d\varphi_1 \, d\varphi_2 \, d\mathscr{P}_1 \, d\mathscr{P}_2.$$

Here Ω is the direction perpendicular to both ω_1 and ω_2 and φ_1 and φ_2 are one-dimensional angles determining the positions of ω_1 and ω_2 on the plane perpendicular to Ω (see (3.14.2)). We can represent each $d\mathscr{P}_i$ as a product of two linear elements, one of which (dx_i) is directed along Ω, while the other (dp_i) lies in the ω_1, ω_2 plane and is perpendicular to ω_i.

The product

$$dp_i \, d\varphi_i = dg_i$$

yields the element of invariant measure of lines on a plane (see §3.6). Therefore,

$$d\gamma_1 \, d\gamma_2 = d\Omega \sin \sigma \, dx_1 \, dx_2 \, dg_1 \, dg_2.$$

We can consider both g_1 and g_2 as lying in one plane, say in ω_1, ω_2. Using the relations

$$dx_1 \, dx_2 = dt \, dh,$$

$$dg_1 \, dg_2 = \sin \sigma \, dQ \, d\sigma \, d\varphi,$$

where Q is the intersection point of the perpendicular projections of the lines γ_1 and γ_2 onto the ω_1, ω_2 plane, dQ is the corresponding planar Lebesgue measure, t denotes a shift along the line through Q in the direction Ω, and φ is a rotation around this line. Since in $\mathbb{R}^3$ (see §3.5)

$$dM = d\Omega \, dQ \, dt \, d\varphi$$

we obtain the desired factorization

$$d\gamma_1 \, d\gamma_2 = dM \, dh \sin^2 \sigma \, d\sigma, \tag{3.16.1}$$

i.e. the distribution of the angle σ is proportional to its $\sin^2$, the measure in h is Lebesgue.

II Triads of lines in $\mathbb{R}^2$ If the case of parallel lines is ignored, then each ordered triad of lines (g_1, g_2, g_3) generates a *labeled triangle* in $\mathbb{R}^2$ (in a labeled triangle the sides (or the vertices) are given the numbers 1, 2, 3). Our purpose now is to obtain a position–size–shape factorization of $dg_1 \, dg_2 \, dg_3$. First we look at the space Σ of (labeled) triangular shapes.

Suitable coordinates in the space Σ can be as follows. We fix a 'base' a of unit length on the plane and put two lines through the endpoints of the base under angles ξ_1 and ξ_2, as shown in fig. 3.16.1. Except for the case of parallel lines, our construction always defines a labeled triangle. However, ξ_1 and ξ_2 do not always coincide with the interior angles α_1 and α_2 of the triangle.

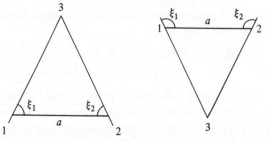

Figure 3.16.1 The numbering of the angles corresponds to the numbering of the vertices. In the usual numbering of sides $a = a_3$

We have

$$\xi_1 = a_1, \qquad \xi_2 = \alpha_2 \qquad \text{if } \xi_1 + \xi_2 < \pi,$$

and

$$\xi_1 = \pi - \alpha_1, \qquad \xi_2 = \pi - \alpha_2 \qquad \text{if } \xi_1 + \xi_2 > \pi.$$

The two cases are related via reflection. Anyhow, for our purposes, the space of triangular shapes can be identified with the product

$$\Sigma \approx (0, \pi) \times (0, \pi).$$

For a formal definition of a triangular shape and a complete description of the space Σ, see §3.18. However, the above product description will be acceptable until we consider measures on Σ given by densities

$$f(\xi_1, \xi_2) \, d\xi_1 \, d\xi_2.$$

Returning to our problem, we take g_3 for the reference axis and use (3.7.3) to get

$$dg_1 \, dg_2 \, dg_3 = dg_3 \sin \alpha_1 \sin \alpha_2 \, d\xi_1 \, d\xi_2 \, dx_1 \, dx_2.$$

Clearly

$$dx_1 \, dx_2 = dt \, da_3,$$

where t stands for a shift along the g_3 line, a_3 is the length of the side of the triangle formed by the lines g_1, g_2 and g_3 which lies on g_3 and dt and da_3 are the elements of Lebesgue measure in the corresponding spaces. By (3.13.3)

$$dg_1 \, dg_2 \, dg_3 = dM \, da_3 \, m_{a_3}(d\sigma),$$

where

$$m_{a_3}(d\sigma) = \sin \alpha_1 \sin \alpha_2 \, d\xi_1 \, d\xi_2.$$

Here and in the following *the index on m shows which size parameter has been used in the factorization*. Notice the asymmetry of m_{a_3} with respect to the angles α_1, α_2 and α_3. We obtain a symmetrical measure in Σ if we factorize using h, the *perimeter length*, instead of a_3. We obtain

$$dg_1 \, dg_2 \, dg_3 = J \, dM \, dh \, m_{a_3}(d\sigma) = dM \, dh \, m_h(d\sigma), \qquad (3.16.2)$$

where the Jacobian J is written in terms of the interior angles (see fig. 3.16.1) as

$$J = \frac{\sin \alpha_3}{\sin \alpha_1 + \sin \alpha_2 + \sin \alpha_3}$$

and therefore

$$m_h(d\sigma) = \frac{\sin \alpha_1 \sin \alpha_2 \sin \alpha_3}{\sin \alpha_1 + \sin \alpha_2 + \sin \alpha_3} \, d\xi_1 \, d\xi_2. \qquad (3.16.3)$$

III Triads of points in $\mathbb{R}^2$ We denote by dQ_i the Lebesgue measure elements in $\mathbb{R}^2$. The previous results can be used to obtain position–size–shape factorizations of $dQ_1 \, dQ_2 \, dQ_3$.

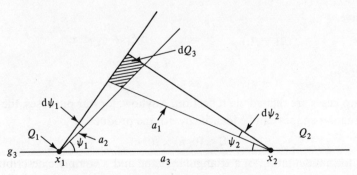

Figure 3.16.2

It is known that [2]

$$dQ_1 \, dQ_2 \, dQ_3 = 8r^3 \, dg_1 \, dg_2 \, dg_3, \qquad (3.16.4)$$

where r is the circumradius of the triangle $Q_1 Q_2 Q_3$ while g_1, g_2 and g_3 are (continuations of the) sides of the same triangle.

Let us give a simple derivation of this relation.

Using (3.14.1) we find

$$dg_1 \, dg_2 \, dg_3 = \sin \psi_1 \, dx_1 \, d\psi_1 \sin \psi_2 \, dx_2 \, d\psi_2 \, dg_3$$
$$= \sin \psi_1 \sin \psi_2 \, d\psi_1 \, d\psi_2 (a_3)^{-1} \, dQ_1 \, dQ_2.$$

As shown in fig. 3.16.2,

$$dQ_3 = \sin \psi_3 (\sin \psi_3)^{-1} a_1 \, d\psi_2 (\sin \psi_3)^{-1} a_2 \, d\psi_1,$$

therefore

$$d\psi_1 \, d\psi_2 = (a_1 a_2)^{-1} \sin \psi_3 \, dQ_3.$$

Now (3.16.4) follows by substitution since

$$\sin \psi_1 \sin \psi_2 \sin \psi_3 (a_1 a_2 a_3)^{-1} = (2r)^{-3}.$$

It is a matter of basic geometry to express $8r^3$ in the form

$$8r^3 = h^3 (\sin \alpha_1 + \sin \alpha_2 + \sin \alpha_3)^{-3}.$$

Hence, using (3.16.2), we find

$$dQ_1 \, dQ_2 \, dQ_3 = dM \, h^3 \, dh \, v_h(d\sigma), \qquad (3.16.5)$$

where the measure v_h in the space Σ is given by the density

$$v_h(d\sigma) = \frac{\sin \alpha_1 \sin \alpha_2 \sin \alpha_3}{(\sin \alpha_1 + \sin \alpha_2 + \sin \alpha_3)^4} \, d\xi_1 \, d\xi_2. \qquad (3.16.6)$$

From an elementary relation between h (the perimeter) and S (the area) of the triangle $Q_1 Q_2 Q_3$, which is as follows:

$$4S^2 = \frac{\sin^2 \alpha_1 \sin^2 \alpha_2 \sin^2 \alpha_3}{(\sin \alpha_1 + \sin \alpha_2 + \sin \alpha_3)^4} h^4,$$

we find that

$$v_S(d\sigma) = (\sin \alpha_1 \sin \alpha_2 \sin \alpha_3)^{-1} d\xi_1 d\xi_2, \qquad (3.16.7)$$

which appears in the position–size–shape factorization

$$\prod_{i=1}^{3} dQ_i = 2 \, dM \, S \, dS \, v_S(d\sigma). \qquad (3.16.8)$$

Lastly we write the corresponding density of shapes when A, the area of the circumcircle through Q_1, Q_2, Q_3, is chosen for the size parameter. Since

$$A = \frac{\pi}{2} (\sin \alpha_1 \sin \alpha_2 \sin \alpha_3)^{-1} S,$$

we have

$$dQ_1 \, dQ_2 \, dQ_3 = \frac{8}{\pi^2} \, dM \, A \, dA \sin \alpha_1 \sin \alpha_2 \sin \alpha_3 \, d\xi_1 \, d\xi_2$$

$$= dM \, A \, dA \, v_A(d\sigma) \qquad (3.16.9)$$

with

$$v_A(d\sigma) = \frac{8}{\pi^2} \sin \alpha_1 \sin \alpha_2 \sin \alpha_3 \, d\xi_1 \, d\xi_2. \qquad (3.16.10)$$

IV Triads of points in $\mathbb{R}^3$ Three points $Q_1, Q_2, Q_3 \in \mathbb{R}^3$ can be described by means of the plane e through these points and the positions $\mathscr{P}_1, \mathscr{P}_2$ and $\mathscr{P}_3$ of the points on e (thus $\mathscr{P}_i$ are points in two dimensions).

We choose the elementary volumes dQ_1, dQ_2 and dQ_3 as shown in fig. 3.16.3, so that

$$dQ_i = dl_i \, d\mathscr{P}_i \quad (i = 1, 2, 3).$$

In this situation (see (3.12.2))

$$de = (2S^{-1}) \, dl_1 \, dl_2 \, dl_3,$$

where S is the area of the triangle $\mathscr{P}_1 \mathscr{P}_2 \mathscr{P}_3$.

Thus

$$dQ_1 \, dQ_2 \, dQ_3 = 2 \, de \, S \, d\mathscr{P}_1 \, d\mathscr{P}_2 \, d\mathscr{P}_3.$$

Factorizing $d\mathscr{P}_1 \, d\mathscr{P}_2 \, d\mathscr{P}_3$ according to (3.16.8) and making use of (3.13.4), we find that

$$dQ_1 \, dQ_2 \, dQ_3 = dM \, S^2 \, dS \, v_S(d\sigma),$$

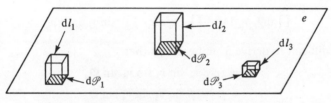

Figure 3.16.3 $d\mathscr{P}_i$ lies on e, dl_i is perpendicular to e, $i = 1, 2, 3$

where dM is the Haar measure on $\mathbb{M}_3$ and the measure v_S on Σ (the space of triangular shapes) is the same as described in subsection III. Note that the measure in the space of areas differs from that in (3.16.8).

V Quadruples of planes in $\mathbb{R}^3$ The problem here is to factorize out the Haar measure on $\mathbb{M}_3$ from $de_1\, de_2\, de_3\, de_4$; we use the same notation as in §3.15, III. Choosing e_1 for the reference plane and applying (3.12.4) we get

$$\prod_1^4 de_i = de_1 \prod_{2,3,4} \sin^2 \psi_i\, d\psi_i\, dg_i.$$

Further application of (3.16.2) and (3.13.4) yields

$$\prod_1^4 de_i = dM\, dh_1\, m_{h_1}(d\sigma_1) \prod_{2,3,4} \sin^2 \psi_i\, d\psi_i, \qquad (3.16.11)$$

where h_1 and σ_1 denote the perimeter and the shape, respectively, of the (triangular) face of the tetrahedron built by the four planes which lies on the e_1 plane. Equation (3.16.11) solves our problem; its asymmetry is due to the choice of the size parameter h_1. In principle this can be remedied by moving to symmetrical size parameters (like the volume of the tetrahedron, etc.).

VI Quadruples of points in $\mathbb{R}^3$ Of more importance in the context of the present book is the shape measure which is obtained from $dQ_1\, dQ_2\, dQ_3\, dQ_4$ (the product of Lebesgue measures in $\mathbb{R}^3$) by separation of the kinematic measure and a measure in the space of size parameter V, the volume of the tetrahedron $Q_1Q_2Q_3Q_4$.

In $\mathbb{R}^3$ we consider the usual polar coordinates with origin at Q_1. In particular let $r_2, \Omega \in \mathbb{S}_2$ be the polar coordinates of Q_2 so that

$$dQ_2 = r_2^2\, dr_2\, d\Omega.$$

We denote by v_3 and v_4 the angles between Ω and the directions of Q_1Q_3 and Q_1Q_4; we denote by Φ the rotation angle of the plane $Q_1Q_2Q_3$ around Ω; and we denote by Φ_4 the flat angle between $Q_1Q_2Q_3$ and $Q_1Q_2Q_4$. In this notation

$$dQ_3 = r_3^2\, dr_3 \sin v_3\, dv_3\, d\Phi$$

and

$$dQ_4 = r_4^2\, dr_4 \sin v_4\, dv_4\, d\Phi_4.$$

Using (3.5.2) and (3.3.2) we find

$$\prod_{i=1}^4 dQ_i = dM \prod_{i=2,3,4} r_i^2\, dr_i \prod_{i=3,4} \sin v_i\, dv_i\, d\Phi_4. \qquad (3.16.12)$$

It is not difficult to derive from fig. 3.16.4 that

$$V = \tfrac{1}{6} r_2 r_3 r_4 \sin v_3 \sin v_4 \sin \Phi_4$$
$$= \tfrac{1}{6} r_2^3 \alpha \cdot \beta \sin v_3 \sin v_4 \sin \Phi_4,$$

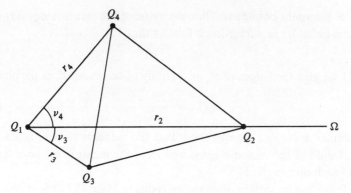

Figure 3.16.4

where

$$\alpha = r_3/r_2, \quad \beta = r_4/r_2.$$

The replacement in (3.16.12) of the variables r_2, r_3 and r_4 by V, α and β yields

$$\prod_{i=1}^{4} dQ_i = dM \, V^2 \, dV \, m_V(d\sigma_1), \tag{3.16.13}$$

where the measure m_V in the space of tetrahedral shapes σ_1 has the form

$$m_V(d\sigma_1) = 6^3 \frac{d\alpha \, d\beta \, dv_3 \, dv_4 \, d\Phi_4}{\alpha \cdot \beta \cdot \sin^2 v_3 \sin^2 v_4 \sin^3 \Phi_4}. \tag{3.16.14}$$

3.17 On measures in shape spaces

All measures in the shape space Σ derived in the previous two sections, except for the measures (3.16.7) and (3.16.14) are finite (and therefore can be normalized to become probability measures). A less obvious case is the measure m_h given by (3.16.3). Let us prove that it is finite.

We take a product set

$$A \times B \times \Sigma,$$

where A is a bounded set in the space of positions of triangles; for instance,

$A = \{$the center of gravity of the triangle formed by g_1, g_2, g_3 lies in the unit disc$\}$

B is a bounded set in the space of perimeter lengths, say

$$B = (0, 1)$$

In terms of the product cylinder topology (see §2.1) the set $A \times B \times \Sigma$ has compact closure (because the distances from O of the lines in $(g_1, g_2, g_3) \in$

$A \times B \times \Sigma$ remain bounded). Thus the value of the measure $dg_1 \, dg_2 \, dg_3$ on this set is finite. By factorization it follows that

$$m_h(\Sigma) < \infty.$$

In §6.11 we give the values of m_h on a family of subsets of Σ; in particular

$$m_h(\Sigma) = \frac{12 - \pi^2}{2}. \tag{3.17.1}$$

By a similar argument we can show that the measure (3.16.6) is also finite. Again, values of the measure v_h (as well as of the measure v_A) on a family of sets can be found in §6.11.

The following important observation is due to Miles [57]: the shape density (3.16.10) survives for triads of points *on a sphere* in $\mathbb{R}^3$.

Let Q_1, Q_2 and Q_3 be three points and let dQ_i $(i = 1, 2, 3)$ be area measure elements on the surface of a sphere of radius R. The problem is to express the product measure $dQ_1 \, dQ_2 \, dQ_3$ in terms of the parameters

w-rotation of the sphere,
r-the radius of the circular trace of the sphere on the plane through Q_1, Q_2 and Q_3,
σ-the planar triangular shape generated by Q_1, Q_2, Q_3.

We can choose

$$dQ_i = r \, d\beta_i \, dx_i$$

where $d\beta_i$ are angles which correspond to elementary arcs on the circle through Q_1, Q_2, Q_3, the linear element dx_i is perpendicular to $d\beta_i$ and lies in the tangent plane at Q_i. Using (3.12.2), (3.11.2) and (3.3.2) we obtain

$$dQ_1 \, dQ_2 \, dQ_3 = r^3 2|\Delta|(\cos v)^{-3} \, de \, d\beta_1 \, d\beta_2 \, d\beta_3$$
$$= r^3 2|\Delta|(\cos v)^{-3} \, dw \, dp \, d\beta_2 \, d\beta_3$$

Observing from (3.16.10) that

$$|\Delta| \, d\beta_2 \, d\beta_3 = \pi^2 r^2 v_A(d\sigma)$$

we easily transform the above to the result of Miles

$$dQ_1 \, dQ_2 \, dQ_3 = 2\pi^2 \, dw \, R^3 r^3 [1/\sqrt{(R^2 - r^2)}] \, dr \, v_A(d\sigma).$$

Apart from the result we present in §6.11 this can be used in calculation of $v_A(\Sigma)$. We integrate both sides over the cube of the sphere and obtain

$$(4\pi R^2)^3 = 2\pi^2 \cdot 4\pi \cdot 2\pi R^3 \int_0^R r^3 [1/\sqrt{(R^2 - r^2)}] \, dr \cdot v_A(\Sigma).$$

Since

$$\int_0^R r^3 \frac{1}{\sqrt{(R^2 - r^2)}} \, dr = -\frac{2R^2 + r^2}{3} \sqrt{(R^2 - r^2)}|_0^R = \frac{2}{3} R^3$$

we find

$$v_A(\Sigma) = 6/\pi. \tag{3.17.2}$$

As for the measure (3.16.7) we have

$$v_S(\Sigma) = \infty. \tag{3.17.3}$$

This can be seen from (6.11.1) which gives the values of this measure for a family of sets. An independent confirmation follows from the considerations of §4.6 where v_S is viewed as a Haar measure on a certain non-compact group, see (4.6.4).

Clearly the measure in the space of tetrahedral shapes which appears as a factor in (3.16.11), namely

$$m_{h_1}(d\sigma_1) \prod_{2,3,4} \sin^2 \psi_i \, d\psi_i,$$

is finite. Yet another measure in the same space, namely (3.16.14), is totally infinite. Again, this measure coincides with a Haar measure on a group (see §4.12), and this observation enables us to find explicitly its density in appropriate coordinates.

We end with a remark concerning the density

$$\sin \alpha \sin^2 \psi_2 \sin^2 \psi_3 \, d\alpha \, d\psi_2 \, d\psi_3 \tag{3.17.4}$$

which appeared in (3.15.3). It yields a natural geometrical example of three random variables which are not completely independent, yet there is independence for each pair. Indeed, the flat angles ψ_1, ψ_2 and ψ_3 of the random trihedron described by (normalized) density (3.17.4) have joint densities proportional to

$$\sin \psi_1 \sin \psi_2 \sin \psi_3 \, d\psi_1 \, d\psi_2 \, d\psi_3. \tag{3.17.5}$$

This result is obtained from (3.17.4) by applying

$$\cos \psi_1 = -\cos \psi_2 \cos \psi_3 + \sin \psi_2 \sin \psi_3 \cos \alpha,$$

a standard formula from spherical trigonometry. Since the variables ψ_2 and ψ_3 remain intact, we have

$$\sin \psi_1 \, d\psi_1 = \sin \psi_2 \sin \psi_3 \sin \alpha \, d\alpha$$

and it remains to substitute this into (3.17.4). The variables ψ_1, ψ_2 and ψ_3 *are not independent* since the domain of ψ_1, ψ_2 and ψ_3 is not a product set. Nevertheless, pairwise independence is there, as proved by integrating out the parameter α in (3.17.4).

We end the section with some remarks concerning the spaces Σ_n of (labeled) shapes of sequences $(\mathscr{P}_1, \ldots, \mathscr{P}_{n+1})$ of points in $\mathbb{R}^n$ ('practical' cases are $n = 2$, 3). We will give (in the next section) a detailed description only for $\Sigma_2 = \Sigma$ but a rough idea about the possible construction of labeled tetrahedral shapes was given in §3.16, V and VI, see also §4.12.

In the context of point processes (chapter 9) we will consider shapes of sets $\{\mathscr{P}_1, \ldots, \mathscr{P}_{n+1}\}$ of points in $\mathbb{R}^n$ (of simplexes). For them simplexial shapes without labels can be defined and the corresponding space we will denote by $\Sigma_n/(n + 1)!$. This notation reflects the fact that in Σ_n we have $(n + 1)!$ labeled shapes which correspond to different labeling of a single shape without labels.

The measures on Σ_2 and Σ_3 we considered except for (3.16.14) are all *symmetrical* i.e. invariant with respect to the permutations of numbers attached to the vertices. Any symmetrical measure m on Σ_n uniquely determines a measure m^* on $\Sigma_n/(n + 1)!$. One can think that $\Sigma_n/(n + 1)!$ is a subset of Σ_n. Then m^* can be defined to be the *restriction* of m on $\Sigma_n/(n + 1)!$.

In particular,

$$m^*(\Sigma_n/(n + 1)!) = ((n + 1)!)^{-1}m(\Sigma_n).$$ (3.17.6)

3.18 The spherical topology of Σ

The topological nature of the spaces of shapes of k-tuples of points in $\mathbb{R}^n$ has been disclosed by Kendall in [41]; see also [63] and [68] for the presentation in the setting of other work on random shapes carried out in England in recent years. In the special case of triangular shapes, Kendall's result says that, topologically, the corresponding shape space (Σ in our notation) is the usual sphere $\mathbb{S}_2$. We show now that this can be deduced from the symmetry and continuity properties of the space Σ.

First we stress that we treat triads as ordered sequences rather then sets. Thus we could call $(\mathcal{P}_1, \mathcal{P}_2, \mathcal{P}_3)$ 'labeled' triangles and Σ is the space of labeled triangular shapes.

We call two triads equivalent if one can be transformed into another by applying a motion from $\mathbb{M}_2$ and then a homothety from the group $\mathbb{H}$ of homotheties of the plane (equivalently a scale change in $\mathbb{R}^2$).

By definition, the triangular shape σ is a class of equivalent triads and Σ is the factor space

$$(\mathbb{R}^2 \times \mathbb{R}^2 \times \mathbb{R}^2 \backslash \{\mathcal{P}_1 = \mathcal{P}_2 = \mathcal{P}_3\})/\mathbb{H}\mathbb{M}_2.$$

where excluded are the triads with totally coinciding points (such triads have no shape), $\mathbb{H}\mathbb{M}_2$ is the corresponding group. We denote by p permutations of the indexes $(1, 2, 3)$. We denote by $p(\mathcal{P}_1, \mathcal{P}_2, \mathcal{P}_3)$ the accordingly transformed triad $(\mathcal{P}_1, \mathcal{P}_2, \mathcal{P}_3)$. For instance, if p sends $(1, 2, 3)$ into $(1, 3, 2)$ then

$$p(\mathcal{P}_1, \mathcal{P}_2, \mathcal{P}_3) = (\mathcal{P}_1, \mathcal{P}_3, \mathcal{P}_2).$$ (3.18.1)

If σ is the shape of $(\mathcal{P}_1, \mathcal{P}_2, \mathcal{P}_3)$ then $p\sigma$ will denote the shape of $p(\mathcal{P}_1, \mathcal{P}_2, \mathcal{P}_3)$.

Topologically the shapes of the triads from the set

$$\mathbb{R}^2 \times \mathbb{R}^2 \times \mathbb{R}^2 \backslash \{\text{triads with } \mathcal{P}_1 \neq \mathcal{P}_2\} = A$$

form the Euclidean plane. This can be seen from the map which is well-defined on the above set:

$$\sigma = ((0, 0), (0, 1), \mathcal{P}_3)$$ (3.18.2)

(under this map each $\sigma \in A$ is identified with certain representative of the class of triads with the same shape σ and eventually with $\mathscr{P}_3 \in \mathbb{R}^2$).

The triads of the set complementary to A, i.e. from the set

$$\{(\mathscr{P}_1, \mathscr{P}_2, \mathscr{P}_3) : \mathscr{P}_1 = \mathscr{P}_2 \neq \mathscr{P}_3\},$$

all have the same shape, σ_0 say. This σ_0 can be obtained as the limit of the shapes (3.18.2) under the single requirement that the distance of $\mathscr{P}_3$ from the origin tends to ∞.

Proof We can identify σ_0 with the triad

$$\sigma_0 = ((0, 0), (0, 0), (0, 1)),$$

therefore $p\sigma_0 \in A$. In the following p is as in (3.18.1)). We consider the map (3.18.2) as applied to the shapes of $p\sigma_0$ and $p\sigma$, where σ is given by (3.18.2). We observe that as $\mathscr{P}_3$ goes to ∞ then

$$\lim p\sigma = p\sigma_0 \quad \text{(in the Euclidean topology of } A\text{)}.$$

By the properties of the factor topology we can remove p on both sides. Hence the assertion.

We see that to obtain the whole Σ we have to compactify A (the plane) by a single point σ_0 *at infinity*. Such a compactification converts (topologically) the plane into a sphere. The latter fact is well known and can be visualized from the usual stereographical projection.

4

Haar measures on groups of affine transformations

In §1.3 we mentioned the possibility of finding Haar measures on a group, based upon knowledge of Haar measures on its subgroups. In this chapter this method is applied to the groups of Lebesgue measure-preserving affine transformations on $\mathbb{R}^n$ ($n = 2, 3$). As a by-product we derive the Haar measures on what we call 'affine deformations groups' and make connections with the position–size–shape factorizations considered in §3.16.

These new considerations permit the expansion of the list of explicit factorizations. Their application leads to the solution of the *modified* J. J. Sylvester problem for four points in $\mathbb{R}^2$ and five points in $\mathbb{R}^3$. This development culminates in the derivation (by means of factorization) of natural probability distributions in the spaces of 'affine shapes' of m-point sets in $\mathbb{R}^2$.

4.1 The group $\mathbb{A}_2^0$ and its subgroups

We denote by $\mathbb{A}_2^0$ the group of 2×2 matrices with determinant equal to $+1$. Equivalently, each $A^0 \in \mathbb{A}_2^0$ is an area-, origin- and 'sense'-preserving affine transformation of $\mathbb{R}^2$. ('Sense-preserving' means that for any oriented line $g \in \overline{\mathbb{G}}$ its *left* halfplane is mapped by A^0 into the left halfplane of $A^0 g$.)

Let two orthonormal vectors e_1 and e_2 emerging from the origin O be fixed (they determine the x and y coordinates on $\mathbb{R}^2$). Each $A^0 \in \mathbb{A}_2^0$ is completely determined by the vectors $A^0 e_1$ and $A^0 e_2$ ($A^0 e_i$ is the image of e_i under A^0), see fig. 4.1.1.

In fact, the coordinates of $A^0 e_1$ and $A^0 e_2$ when written columnwise constitute the matrix of A^0, and the matrix elements (albeit non-independent) can serve for coordinates in $\mathbb{A}_2^0$. However, we will use a different parametrization. Given $A^0 \in \mathbb{A}_2^0$, its inverse $(A^0)^{-1}$ can be obtained as a product

$$(A^0)^{-1} = CHw, \tag{4.1.1}$$

where the transformations $C, H, w \in \mathbb{A}_2^0$ are of a very special type. Namely: w

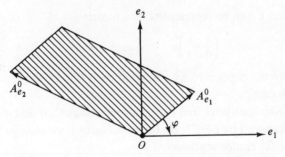

Figure 4.1.1 The area of the shaded parallelogram equals 1

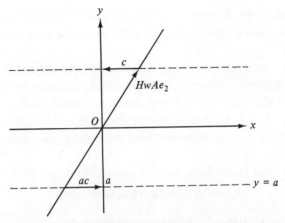

Figure 4.1.2 C maps each horizontal line onto itself. On each horizontal line C acts as a rigid shift. For the line $y = a$ the shift is ac, where c is the shift for the line $y = 1$

is a rotation by the angle φ shown in fig. 4.1.1; H corresponds to scale changes on the x and y axes, where the homothety along the x axis is chosen from the condition

$$HwA^0e_1 = e_1,$$

while the homothety along the y axis is chosen from the condition that H be area-preserving. Thus H can be represented by the matrix

$$H = \begin{pmatrix} h & 0 \\ 0 & h^{-1} \end{pmatrix}, \qquad h > 0;$$

C is a linear transformation which is explained as follows. Because of the area-preserving property of Hw, the endpoint of HwA^0e_2 will lie on the horizontal line $y = 1$. C is a transformation from $\mathbb{A}_2^0$ which leaves the Ox axis intact and maps HwA^0e_2 on e_2. This transformation is shown in fig. 4.1.2.

Clearly, each C can be represented by a matrix

$$\begin{pmatrix} 1 & c \\ 0 & 1 \end{pmatrix}, \quad -\infty < c < \infty.$$

It follows from the caption to fig. 4.1.2 that each C is a Cavalieri transformation of the plane.

It is clear from the above that the transformations C, H and w in (4.1.1) are uniquely determined by $(A^0)^{-1}$. By the usual *inversion argument* we conclude that any $A^0 \in \mathbb{A}_2^0$ can be represented as

$$A^0 = CHw$$

as well as   (4.1.2)

$$A^0 = w'H'C'$$

(representations with different orders of w, H and C elements also exist). Both representations in (4.1.2) are uniquely determined (because of uniqueness in (4.1.1)).

The transformations w belong to the group $\mathbb{W}_2$ of rotations of $\mathbb{R}^2$ (see §3.1). The transformations C constitute a group which we denote by $\mathbb{C}_1$. The transformations H also constitute a group, which we denote by $\mathbb{H}_1$.

The group $\mathbb{H}_1$ can be identified with the multiplicative group of positive numbers, while $\mathbb{C}_1$ can be identified with the additive group of real numbers (i.e. with $\mathbb{T}_1$).

On $\mathbb{W}_2$, $\mathbb{H}_1$ and $\mathbb{C}_1$ we have the bi-invariant Haar measures

$$dw = d\varphi \quad \text{(the arc length measure)},$$

$$dH = \frac{dh}{h} \quad \text{(the logarithmic measure)},$$

$$dC = dc \quad \text{(the Lebesgue measure)}.$$

4.2 Affine deformations of $\mathbb{R}^2$

The products

$$CH \qquad C \in \mathbb{C}_1, \qquad H \in \mathbb{H}_1$$

form a group which we denote by $\mathbf{V}_2$ and call 'the group of affine deformations of $\mathbb{R}^2$'. The elements of $\mathbf{V}_2$ are represented by matrices of the form

$$V = \begin{pmatrix} h & ch^{-1} \\ 0 & h^{-1} \end{pmatrix},$$

which we call 'affine deformations' of $\mathbb{R}^2$. We note that both representations

$$V = C_l H_r$$

and   (4.2.1)

$$V = H_1 C_r$$

exist and are unique. This follows from the solvability of the matrix equation

$$\begin{pmatrix} h_1 & 0 \\ 0 & h_1^{-1} \end{pmatrix}\begin{pmatrix} 1 & c_r \\ 0 & 1 \end{pmatrix} = \begin{pmatrix} 1 & c_1 \\ 0 & 1 \end{pmatrix}\begin{pmatrix} h_r & 0 \\ 0 & h_r^{-1} \end{pmatrix},$$

which has the solution

$$\begin{align} h_1 &= h_r \\ c_r &= c_1 \cdot h_r^{-2}. \end{align} \tag{4.2.2}$$

Therefore, in order to find the left $d^{(1)}V$ and the right $d^{(r)}V$ Haar measures on V_2 we can use (1.3.6) and (1.3.7), putting

$$\mathbb{X} = V_2, \qquad \mathbb{U} = C_1, \qquad \mathbb{V} = H_1.$$

Let us find $d^{(1)}V$ both in terms of the coordinates c_1, h_r and c_r, h_1. We assume that the unknown measures m_1 and m_2 in (1.3.6) have densities, i.e.

$$m_1(dH) = \rho_1(h)\, dh, \qquad m_2(dC) = \rho_2(c)\, dc.$$

Equating the right-hand sides in (1.3.6) yields

$$dc_1\, \rho_1(h_r)\, dh_r = \rho_2(c_r)\, dc_r\, h_1^{-1}\, dh_1. \tag{4.2.3}$$

Using the solution (4.2.2) we express the measure on the right-hand side of (4.2.3) in the coordinates c_1 and h_r. This yields

$$dc_1\, \rho_1(h_r)\, dh_r = \rho_2(c_1 h_r^{-2})h_r^{-3}\, dc_1\, dh_r$$

or

$$\rho_1(h_r) = \rho_2(c_1 h_r^{-2})h_r^{-3}.$$

Here the variables c_1 and h_r are independent (in the usual analytical sense). Therefore we conclude that

$$\rho_2 = \text{const},$$
$$\rho_1(h) = \text{const} \cdot h^{-3}.$$

We choose const $= 1$ and finally get

$$d^{(1)}V = h_r^{-3}\, dc_1\, dh_r = h_1^{-1}\, dc_r\, dh_1. \tag{4.2.4}$$

Similarly, we find

$$d^{(r)}V = h_r^{-1}\, dc_1\, dh_r = h_1\, dc_r\, dh_1. \tag{4.2.5}$$

We note that our expressions for $d^{(1)}V$ and $d^{(r)}V$ depend on the combination of the subscripts l and r. Clearly this is because the two maps

$$V \to (c_1, h_r)$$
$$V \to (c_r, h_1)$$

which correspond to (4.2.1) are essentially different and produce on $\mathbb{R} \times (0, \infty)$ different *images* of the same measure on V_2. Similar situations occur in other sections.

4.3 The Haar measure on $\mathbb{A}_2^0$

From (4.1.2) and (4.2.1) we derive the existence and uniqueness of the representations

$$A^0 = \nabla_l w_r$$
$$A^0 = w_l \nabla_r,$$

where

$$A^0 \in \mathbb{A}_2^0, \qquad w_l, w_r \in \mathbb{W}_2, \qquad \nabla_l, \nabla_r \in \mathbf{V}_2.$$

Therefore the approach of §1.3 can again be used. By Haar factorization, any measure on $\mathbb{A}_2^0$ which is invariant with respect to the transformations

$$A^0 \to \nabla A^0 w$$

is necessarily proportional to (see (4.2.4))

$$d^{(l)}\nabla_l\, dw_r. \tag{4.3.1}$$

Similarly, invariance with respect to the transformations

$$A^0 \to w A^0 \nabla$$

yields proportionality to the measure

$$d^{(r)}\nabla_r\, dw_l. \tag{4.3.2}$$

It is well known (see [2]) that the group $\mathbb{A}_2^0$ possesses a *bi-invariant* Haar measure dA^0. It follows from our criterion in §1.3 that both measures (4.3.1) and (4.3.2) are proportional to dA^0. By the same criterion an independent proof of the bi-invariance of dA^0 would follow if one shows that the Jacobian relating (4.3.1) to (4.3.2) equals 1 *identically*. We do this at the end of this section. Anyhow, bi-invariance of dA^0 being established, we have by (1.3.9)

$$dA^0 = d^{(l)}\nabla_l\, dw_r = d^{(r)}\nabla_r\, dw_l.$$

To obtain explicit expressions for dA^0, we use the formula (4.2.4). For instance, if we put

$$\nabla_l = H_1 C_2, \quad \text{i.e.} \quad A^0 = H_1 C_2 w_r,$$

then we get using natural indexation

$$dA^0 = h_1^{-1}\, dc_2\, dh_1\, d\varphi.$$

Putting

$$\nabla_l = C_1 H_2, \quad \text{i.e.} \quad A^0 = C_1 H_2 w_r,$$

we get

$$dA^0 = h_2^{-3}\, dc_1\, dh_2\, d\varphi, \quad \text{etc.}$$

Further expressions can be obtained using (4.2.5).

We now outline a calculation of the Jacobian relating the measures (4.3.1) and (4.3.2) which can be a useful prototype for calculations in higher dimensions.

Let us represent A^0 by a matrix:

$$A^0 = \begin{pmatrix} a_{11} & a_{12} \\ a_{21} & a_{22} \end{pmatrix}.$$

The equation $A^0 = V_1\, w_r$ in matrix form can be written as

$$\begin{pmatrix} a_{11} & a_{12} \\ a_{21} & a_{22} \end{pmatrix} = \begin{pmatrix} h_1 & h_1 c_1 \\ 0 & h_1^{-1} \end{pmatrix} \cdot \begin{pmatrix} \cos \varphi_1 & -\sin \varphi_1 \\ \sin \varphi_1 & \cos \varphi_1 \end{pmatrix},$$

from which we find

$$a_{21} = h_1^{-1} \sin \varphi_1,$$
$$a_{11} = h_1 \cos \varphi_1 + h_1 c_1 \sin \varphi_1$$
$$a_{12} = -h_1 \sin \varphi_1 + h_1 \cos \varphi_1 \cdot c_1.$$

It follows simply from this that

$$da_{21}\, da_{11}\, da_{12} = |a_{11}| h_1^{-1}\, dh_1\, d\varphi_1\, dc_1. \tag{4.3.3}$$

Similarly the matrix equation

$$\begin{pmatrix} a_{11} & a_{12} \\ a_{21} & a_{22} \end{pmatrix} = \begin{pmatrix} \cos \varphi_2 & -\sin \varphi_2 \\ \sin \varphi_2 & \cos \varphi_2 \end{pmatrix} \begin{pmatrix} h_2 & h_2 c_2 \\ 0 & h_2^{-1} \end{pmatrix}$$

(which corresponds to $A^0 = w_1 \cdot V_r$) yields

$$da_{21}\, da_{11}\, da_{12} = h_2^2 |\cos \varphi_2|\, dh_2\, d\varphi_2\, dc_2$$
$$= |a_{11}| h_2\, dh_2\, d\varphi_2\, dc_2. \tag{4.3.4}$$

The Jacobian in question is J in the relation

$$h_1^{-1}\, dh_1\, d\varphi_1\, dc_1 = J h_2\, dh_2\, d\varphi_2\, dc_2.$$

We gather from (4.3.3) and (4.3.4) that $J \equiv 1$.

4.4 The Haar measure on $\mathbb{A}_2$

We denote by $\mathbb{A}_2$ the group of all *area-preserving* affine transformations of $\mathbb{R}^2$; generally speaking they *do not preserve the origin $O \in \mathbb{R}^2$*.

By a direct geometrical argument, resembling that used in §4.1, we can show that each $A \in \mathbb{A}_2$ can be represented either as

$$A = t_1 A_r^0, \quad \text{with} \quad t_1 \in \mathbb{T}_2,\ A_r^0 \in \mathbb{A}_2^0$$

or as

$$A = A_1^0 t_r, \quad \text{with} \quad t_r \in \mathbb{T}_2,\ A_1^0 \in \mathbb{A}_2^0$$

and both representations are unique. The products denote group multiplication in $\mathbb{A}_2$ (rather than a transformation of a vector by a matrix; for the latter operations we use the sign *).

The existence of a bi-invariant Haar measure dA on $\mathbb{A}_2$ is well known (see [2]). Therefore by (1.3.9)

$$dA = dt_1\, dA_r^0 = dt_r\, dA_1^0, \tag{4.4.1}$$

where dt is the Haar (Lebesgue) measure on $\mathbb{T}_2$, and the measure dA^0 was discussed in the previous section.

In fact, bi-invariance of dA (and consequently (4.4.1)) can be established independently without much effort. The solution of the equation

$$t_1 A_r^0 = A_1^0 t_r$$

is easily seen to be

$$A_r^0 = A_1^0$$
$$t_1 = A_1^0 * r_r.$$

This transformation is of Cavalieri type, i.e. it preserves any product measure $m(dA_1^0)\, dt_r$ (because A_1^0 by definition preserves the Lebesgue measure dt_r). Therefore bi-invariance of dA follows by the criterion of §1.3.

We get yet another factorization of dA if, for $A \in \mathbb{A}_2$, we use the representations

$$A = M_1 V_r, \quad \text{with} \quad M_1 \in \mathbb{M}_2, V_r \in V_2 \tag{4.4.2}$$

and

$$A = V_1 M_r, \quad \text{with} \quad M_r \in \mathbb{M}_2, V_1 \in V_2.$$

Since, again, both representations are unique, by the criterion of §1.3 we have

$$dA = dM_1\, d^{(r)}V_r = dM_r\, d^{(l)}V_1, \tag{4.4.3}$$

where dM is the Haar measure on $\mathbb{M}_2$, and the measures $d^{(r)}V$ and $d^{(l)}V$ have been discussed in §4.2.

4.5 Triads of points in $\mathbb{R}^2$

A triad (Q_1, Q_2, Q_3) of non-collinear points from $\mathbb{R}^2$ can be represented as

$$(Q_1, Q_2, Q_3) = (A, S, \mathcal{X}), \tag{4.5.1}$$

where S is the area of the triangle Q_1, Q_2, Q_3, $A \in \mathbb{A}_2$, and $\mathcal{X}$ is either 1 or -1.

We determine the parameters A and $\mathcal{X}$ as follows.

First we choose three non-collinear points $\mathcal{P}_1, \mathcal{P}_2, \mathcal{P}_3 \in \mathbb{R}^2$ to form a base of our map. In particular we can take

$$\mathcal{P}_1 = (0, 0), \quad \text{the origin}$$
$$\mathcal{P}_2 = (1, 0)$$
$$\mathcal{P}_3 = (0, 1)$$

By s we denote a homothety of $\mathbb{R}^2$ corresponding to the matrix

$$s = \begin{pmatrix} \sqrt{S} & 0 \\ 0 & \sqrt{S} \end{pmatrix}$$

For a given triad (Q_1, Q_2, Q_3) with $S > 0$ there are two possibilities:

I. An $A \in \mathbb{A}_2$ can be found for which

$$(Q_1, Q_2, Q_3) = As(\mathscr{P}_1, \mathscr{P}_2, \mathscr{P}_3)$$

(the right-hand side stands for the image of $(\mathscr{P}_1, \mathscr{P}_2, \mathscr{P}_3)$ under the product transformation As). In this case we put

$$(Q_1, Q_2, Q_3) = (A, S, 1)$$

II. An $A \in \mathbb{A}_2$ can be found for which

$$(Q_1, Q_2, Q_3) = As(\mathscr{P}_1, \mathscr{P}_3, \mathscr{P}_2)$$

In this case we put

$$(Q_1, Q_2, Q_3) = (A, S, -1)$$

Since in both cases A is determined uniquely and no other possibilities exist, the construction of (4.5.1) is complete.

Remark The $\mathscr{X} = -1$ case corresponds to the situations where a reflection is needed. We could avoid separate considerations of the above two cases if we allowed the value -1 for det A in §4.1.

Clearly (4.5.1) implies that (see §1.6)

$$\mathbb{R}^2 \times \mathbb{R}^2 \times \mathbb{R}^2 \backslash \{\text{triads of collinear points}\} \approx \mathbb{A}_2 \times (0, \infty) \times \{1, -1\}.$$

$$(4.5.2)$$

Any transformation $A_1 \in \mathbb{A}_2$ acts on the above product in a Cavalieri way, namely (say if $\mathscr{X} = 1$)

$$A_1(Q_1, Q_2, Q_3) = A_1 As(\mathscr{P}_1, \mathscr{P}_2, \mathscr{P}_3) = (A_1 A, S, 1)$$

Hence, on applying Haar factorization, we get the following proposition.

Any measure μ on $\mathbb{R}^2 \times \mathbb{R}^2 \times \mathbb{R}^2$, for which

$$\mu(\{\text{collinear triads}\}) = 0$$

and which is $\mathbb{A}_2$-invariant, necessarily has the product representation

$$\mu = h_{\mathbb{A}_2} \times m, \tag{4.5.3}$$

where $h_{\mathbb{A}_2}$ is the Haar measure on $\mathbb{A}_2$, while m is a measure in the space $(0, \infty) \times \{1, -1\}$.

This result does not depend on the choice of the base triad $\mathscr{P}_1, \mathscr{P}_2, \mathscr{P}_3$ with which we construct the map (4.5.1). (This follows from the Haar measure properties of $h_{\mathbb{A}_2}$.)

An obvious example of a measure μ for which the assumptions of the above proposition hold is $dQ_1 \, dQ_2 \, dQ_3$, where dQ_i are planar Lebesgue. The corresponding factor measure m can be found from homogeneity considerations (see §1.4). The measure $dQ_1 \, dQ_2 \, dQ_3$ is of order 3 in area. According to (4.4.1), dA is of order 1 in area. Therefore the corresponding measure m should be of order 2 in area. Thus, necessarily,

$$m(dS \times \mathscr{X}) = c_{\mathscr{X}} S \, dS, \qquad \mathscr{X} = 1, -1.$$

It follows from reflection invariance of $dQ_1 \, dQ_2 \, dQ_3$ that $c_1 = c_{-1} = c$. We show at the end of the next section that $c = 4$. Thus,

$$dQ_1 \, dQ_2 \, dQ_3 = 4 \, dA \, S \, dS. \qquad (4.5.4)$$

As another example we consider the measure μ in the space (4.5.2) defined to be the image of the (bi-invariant, see [2]) Haar measure on the *affine group* $\mathbb{A}_2^*$ (i.e. the group of 2×2 matrices with non-zero determinants) induced by the map

$$(Q_1, Q_2, Q_3) = A^*(\mathscr{P}_1, \mathscr{P}_2, \mathscr{P}_3), \qquad A^* \in \mathbb{A}_2^*.$$

By a direct application of the principles of §1.3 we find that the Haar measure on $\mathbb{A}_2^*$ has the form

$$dA^* = cS_1^{-1} \, dS_1 \, dA_r = cS_r^{-1} \, dS_r \, dA_1$$

where s_1, A_r and s_r, A_1 are defined by the relations

$$A^* = s_1 A_r, \qquad A^* = A_1 s_r$$

(or their versions which include reflections). Thus using natural indexation

$$\mu(dQ_1 \, dQ_2 \, dQ_3) = cS_1^{-1} \, dS_1 \, dA_r = cS_r^{-1} \, dS_r \, dA_1.$$

The constant $c > 0$ is present because the measure μ was defined up to a constant factor.

4.6 Another representation of $d^{(r)}V$

The result (4.5.4) is closely related to the factorization of $dQ_1 \, dQ_2 \, dQ_3$ given by (3.16.8).

Indeed, from (4.4.3) we find that

$$dQ_1 \, dQ_2 \, dQ_3 = 4 \, dM_1 \, d^{(r)}V_r \, S \, dS \qquad (4.6.1)$$

and

$$dQ_1 \, dQ_2 \, dQ_3 = 4 \, dM_r \, d^{(1)}V_1 \, S \, dS. \qquad (4.6.2)$$

We stress that the 'variables' M_1, V_r and M_r, V_1 above are defined by the relations

$$(Q_1, Q_2, Q_3) = M_1 V_r s(\mathscr{P}_1, \mathscr{P}_2, \mathscr{P}_3)$$

and

$$(Q_1, Q_2, Q_3) = V_1 M_r s(\mathscr{P}_1, \mathscr{P}_2, \mathscr{P}_3).$$

or their versions that include reflections. Here $\mathscr{P}_1, \mathscr{P}_2, \mathscr{P}_3$ form the base of the maps (compare with §4.5). Let us compare this with (3.16.8).

A natural question arises as to whether we can eliminate the measures $dM \, S \, dS$ and thus derive a new interpretation for the measure ν_S on Σ. The answer is that we can do this with (4.6.1) but *not* with (4.6.2). The reason is that, for any $M \in \mathbb{M}_2$,

$$M(Q_1, Q_2, Q_3) = (MM_1, \nabla_r, S, \mathcal{X}),$$

while the parameters M_r and ∇_1 lack similar Cavalieri properties. The elimination of the dM_1 S dS factor yields (see (4.2.5))

$$v_S(\mathrm{d}\sigma) = 2\,\mathrm{d}^{(r)}V = 2h_r^{-1}\,\mathrm{d}c_1\,\mathrm{d}h_r = 2h_1\,\mathrm{d}c_r\,\mathrm{d}h_1. \qquad (4.6.3)$$

It follows from the considerations of §4.5 that v_S is the image of d$^{(r)}$V under the maps

$$\nabla_2 \times \{1, -1\} \to \Sigma, \qquad (4.6.4)$$

each such map depending on the choice of the 'base' triangle $\mathcal{P}_1, \mathcal{P}_2, \mathcal{P}_3$. Thus, as soon as the latter triangle is fixed, we put

$$(\nabla, 1) \to \text{shape of the triangle } \nabla\mathcal{P}_1, \nabla\mathcal{P}_2, \nabla\mathcal{P}_3$$

and

$$(\nabla, -1) \to \text{shape of a reflection of } \nabla\mathcal{P}_1, \nabla\mathcal{P}_2, \nabla\mathcal{P}_3.$$

The independence of the image of d$^{(r)}$V from the choice of the 'base' triangle clearly follows from the Haar measure properties of d$^{(r)}$V.

Let us check (4.6.3) by a direct calculation. We choose

$$\mathcal{P}_1 = (0, 0), \qquad \mathcal{P}_2 = (1, 0), \qquad \mathcal{P}_3 = (0, 1)$$

for our 'base' triangle. From fig. 4.6.1

$$\cot \xi_1 = c_1$$
$$\cot \xi_2 = (h_r - c_1/h_r)h_r = h_r^2 - c_1.$$

Hence,

$$\mathrm{d}\xi_1\,\mathrm{d}\xi_2 = \left\| \begin{matrix} -\sin^2 \xi_1 & 0 \\ \sin^2 \xi_2 & -2h_r \sin^2 \xi_2 \end{matrix} \right\| \mathrm{d}h_r\,\mathrm{d}c_1$$
$$= 2h_r \sin^2 \xi_1 \sin^2 \xi_2 \,\mathrm{d}h_r\,\mathrm{d}c_1.$$

Using (3.16.7) we get

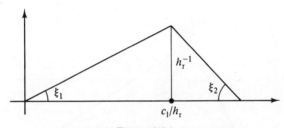

Figure 4.6.1

$$dv_S = (\sin \xi_1 \sin \xi_2 \sin(\xi_1 + \xi_2))^{-1} d\xi_1 d\xi_2$$

$$= \frac{2h_r \sin \xi_1 \sin \xi_2}{\sin(\xi_1 + \xi_2)} dh_r dc_1$$

$$= \frac{2h_r dh_r dc_1}{\cot \xi_1 + \cot \xi_2}$$

$$= 2h_r^{-1} dc_1 dh_r,$$

i.e. (4.6.3).

4.7 Quadruples of points in $\mathbb{R}^2$

We consider the space of sequences (Q_1, Q_2, Q_3, Q_4) of points in the plane with no three points on a line. This space is essentially

$$\mathbb{R}^2 \times \mathbb{R}^2 \times \mathbb{R}^2 \times \mathbb{R}^2 \setminus Z,$$

where

$$Z = \{\text{quadruples possessing collinear triples}\}.$$

Using the representation (4.5.1) we can write

$$(Q_1, Q_2, Q_3, Q_4) = (A, S, \mathcal{X}, Q_4),$$

where A, S and $\mathcal{X}$ refer to the triangle $Q_1 Q_2 Q_3$.

Let μ be any measure on $\mathbb{R}^2 \times \mathbb{R}^2 \times \mathbb{R}^2 \times \mathbb{R}^2$ which is $\mathbb{A}_2$-invariant and $\mu(Z) = 0$. In order to apply Haar factorization to μ (i.e. to separate $h_{\mathbb{A}_2}$) we need to determine Q_4 by means of $\mathbb{A}_2$-invariant parameters.

Let us consider the case shown in fig. 4.7.1. In this case the values

$$S_1 = \text{area of the triangle } Q_2 Q_3 Q_4$$

and

$$S_2 = \text{area of the triangle } Q_1 Q_4 Q_3$$

determine the position of Q_4, and we have the Cavalieri property

$$A_1(Q_1, Q_2, Q_3, Q_4) = (A_1 A, S, \mathcal{X}, S_1, S_2).$$

Hence, on the set defined by fig. 4.7.1,

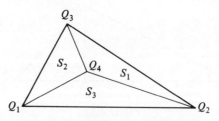

Figure 4.7.1 The triangle $Q_1 Q_2 Q_3$ encloses Q_4

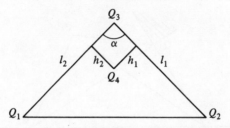

Figure 4.7.2 h_1 and h_2 are perpendicular to the sides l_1 and l_2

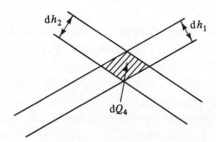

Figure 4.7.3 The angles α here and in fig. 4.7.2 coincide

$$\mu = h_{A_2} \times m, \tag{4.7.1}$$

where m is some measure in the space of sequences $(S, \mathscr{X}, S_1, S_2)$.

Let us find m in the decomposition (4.7.1) for the case

$$d\mu = dQ_1 \, dQ_2 \, dQ_3 \, dQ_4,$$

where each dQ_i is a planar Lebesgue measure. In the change of variables that follows, A, S and $\mathscr{X}$ (equivalently, $Q_1 Q_2 Q_3$) remain intact. Therefore we can treat them as fixed. We have

$$dS_1 = \tfrac{1}{2} \, dh_1 \cdot l_1, \qquad dS_2 = \tfrac{1}{2} \, dh_2 \cdot l_2;$$

see fig. 4.7.2 for notation.

The area element dQ_4 which corresponds to dh_1 and dh_2 is shown in fig. 4.7.3.

We have

$$dQ_4 = (\sin \alpha)^{-1} \, dh_1 \, dh_2 = \frac{4 \, dS_1 \, dS_2}{l_1 l_2 \sin \alpha}$$

$$= 2 \frac{dS_1 \, dS_2}{S}. \tag{4.7.2}$$

Together with (4.5.4) this yields

$$dQ_1 \, dQ_2 \, dQ_3 \, dQ_4 = 8 \, dA \, dS \, dS_1 \, dS_2. \tag{4.7.3}$$

We stress that here S is the area of the triangle $Q_1 Q_2 Q_3$ (see fig. 4.7.1). Let us

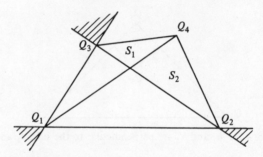

Figure 4.7.4 S_1 is the area of $Q_1Q_3Q_4$ and S_2 is the area of $Q_1Q_2Q_4$

consider the cases complementary to that shown in fig. 4.7.1.

Continuation of the sides of the triangle $Q_1Q_2Q_3$ separates the plane, as shown in fig. 4.7.4. The cases in which Q_4 lies in the angular (shaded) domains can be reduced to the above cases by a change of numeration of points.

If the points Q_1, Q_2, Q_3 and Q_4 form a *convex* quadrangle, then the position of Q_4 can be determined by the values S_1 and S_2, the areas of the triangles shown in fig. 4.7.4. A calculation in the style of (4.7.2) ensures that in this notation (4.7.2) remains valid, with S still denoting the area of $Q_1Q_2Q_3$.

4.8 The modified Sylvester problem: four points in $\mathbb{R}^2$

Let us denote by V the *area of the minimal convex hull of the points* $Q_1, Q_2,$ Q_3 and Q_4. In the case shown in fig. 4.7.1

$$V = S.$$

Therefore, when we replace S, S_1 and S_2 by the new coordinates

$$V, \qquad u = \frac{S_1}{V}, \qquad v = \frac{S_2}{V}$$

in (4.7.3) we obtain

$$dQ_1 \, dQ_2 \, dQ_3 \, dQ_4 = 8 \, dA \, V^2 \, dV \, du \, dv. \qquad (4.8.1)$$

It is clear that the range of u and v is the triangle shown in fig. 4.8.1. In the case of fig. 4.7.4 we first replace the coordinates S, S_1 and S_2 by

$$V = S_1 + S_2$$
$$S = S$$
$$S_1 = S_1$$

(the Jacobian is 1), and then

$$V = V$$
$$u = S/V$$
$$v = S_1/V$$

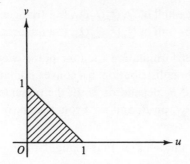

Figure 4.8.1 The area of the triangle is $\frac{1}{2}$

(the Jacobian is V^2). In these coordinates we again obtain (4.8.1). However the *range* of variables u, v is different from that in the case of fig. 4.7.1. Namely the range of u, v corresponding to fig. 4.7.4 is the *unit square*.

Summarizing, we find that the *complete range* of the u, v parameters is a union of seven components, of which four are triangles identical to that in fig. 4.8.1 and the remaining three are unit squares. From a geometrical point of view, a pair u, v defines what we call the *affine* (or $\mathbb{A}_2$ reflection and homothety-invariant) *shape* of the quadruple Q_1, Q_2, Q_3, Q_4. We denote an affine shape by τ, thus

$$\tau = (u, v)$$

In §4.16 we give a complete topological description of the *space* $\tau_{2,4}$ *of affine shapes*. If the cases where we have collinearities are discarded then what remains of $\tau_{2,4}$ can be represented by the range of the u and v parameters essentially as described above. The reason for the indexes 2 and 4 is that later on (in §4.15) we consider the spaces $\tau_{n,m}$ of affine shapes of m-tuples of points in $\mathbb{R}^n$ (thus here $n = 2$, $m = 4$).

We find from the above crude description of $\tau_{2,4}$ that the total $\mathrm{d}u\,\mathrm{d}v$-measure of this space equals 5. Therefore we define a probability measure $\mathbb{P}_{2,4}$ on $\tau_{2,4}$ putting

$$\mathbb{P}_{2,4}(\mathrm{d}\tau) = \tfrac{1}{5}\,\mathrm{d}u\,\mathrm{d}v.$$

The main result of this section can now be written as

$$\mathrm{d}Q_1\,\mathrm{d}Q_2\,\mathrm{d}Q_3\,\mathrm{d}Q_4 = 40\,\mathrm{d}A\ V^2\,\mathrm{d}V\ \mathbb{P}_{2,4}(\mathrm{d}\tau). \qquad (4.8.2)$$

The probability distribution $\mathbb{P}_{2,4}$ is of special interest because it reappears in the context of planar Poisson point processes (§9.7). However, it is of interest in its own right that certain events in $\tau_{2,4}$ have very clear geometrical interpretation and their $\mathbb{P}_{2,4}$-probabilities can now be found without difficulty.

Remarkably this is the case for the events B_1 and B_2 originally considered by J. J. Sylvester (recall his famous *Vierpunktproblem*; see [35] and [55]), where

B_1, the minimal convex hull of $Q_1Q_2Q_3Q_4$, is a triangle;
B_2, the minimal convex hull of $Q_1Q_2Q_3Q_4$, is a quadrangle.

In the J. J. Sylvester formulation the four points were dropped independently, each with uniform distribution in a convex domain $D \subset \mathbb{R}^2$; hence, the probabilities of B_1 and B_2 depend on D. In the present modified version we ask for the values of $\mathbb{P}_{2,4}$ on B_1 and B_2. From the above crude description of $\tau_{2,4}$ it is clear that

$$B_1 = \text{the union of the four triangular components,}$$

$$B_2 = \text{the union of the three squares.}$$

Thus the solution of the modified Sylvester problem is

$$\mathbb{P}_{2,4}(B_1) = \tfrac{2}{5}, \qquad \mathbb{P}_{2,4}(B_2) = \tfrac{3}{5}.$$

4.9 The group $\mathbb{A}_3^0$ and its subgroups

We denote by $\mathbb{A}_3^0$ the group of 3×3 matrices with determinants equal to $+1$. Equivalently, $\mathbb{A}_3^0$ is the group of volume, origin- and 'sense'-preserving affine transformations of $\mathbb{R}^3$.

Let three orthonormal vectors e_1, e_2 and e_3 emerging from O be fixed (they determine the x, y, z coordinates in $\mathbb{R}^3$). Given $A^0 \in \mathbb{A}_3^0$, its inverse $(A^0)^{-1}$ can be represented as a product

$$(A^0)^{-1} = HCw, \qquad (4.9.1)$$

where the transformations H, C, $w \in \mathbb{A}_3^0$ now depend on more than one parameter each. They are uniquely determined as follows: w is a rotation of $\mathbb{R}^3$ (it depends on three parameters, see §3.2). In (4.9.1) w is determined by the condition that wA^0e_1 has the direction of e_1, wA^0e_2 lies in the plane e_1e_2 (i.e. $z = 0$), and the y coordinate of wA^0e_2 is positive. The transformation C in (4.9.1) is a product of two Cavalieri transformations from $\mathbb{A}_3^0$:

$$C = C^{(2)}C^{(1)}. \qquad (4.9.2)$$

$C^{(1)}$ belongs to the group $\mathbb{C}^{(1)}$ of matrices:

$$C^{(1)} = \begin{pmatrix} 1 & a & 0 \\ 0 & 1 & 0 \\ 0 & 0 & 1 \end{pmatrix} \qquad -\infty < a < \infty. \qquad (4.9.3)$$

Under $C^{(1)}$ the plane $y = y_0$ is shifted by the vector $(y_0a, 0, 0)$ (in particular, the x axis remains intact), $-\infty < y_0 < \infty$. We choose $C^{(1)}$ from the condition

$$C^{(1)}wA^0e_2 \text{ has the direction of } e_2$$

(note that $\mathbb{C}^{(1)}$ is isomorphic to the group $\mathbb{C}$ of §4.1 and acts similarly on the $z = 0$ plane).

$C^{(2)}$ belongs to the group $\mathbb{C}^{(2)}$ of matrices:

$$C^{(2)} = \begin{pmatrix} 1 & 0 & b \\ 0 & 1 & c \\ 0 & 0 & 1 \end{pmatrix} \qquad -\infty < b < \infty,\ -\infty < c < \infty. \qquad (4.9.4)$$

Under $C^{(2)}$ the plane $z = z_0$ is shifted by the vector $z_0(b, c, 0)$ (in particular the xOy plane remains intact), $-\infty < z_0 < \infty$. We choose $C^{(2)}$ from the condition

$$C^{(2)}C^{(1)}wA^0e_3 \text{ has the direction of } e_3.$$

It follows that the product (4.9.2) sends each wA^0e_i, $i = 1, 2, 3$, onto the positive semiaxis determined by the corresponding e_i. In (4.9.1) H belongs to the group $\mathbb{H}_2$ of matrices:

$$H = \begin{pmatrix} u & 0 & 0 \\ 0 & v & 0 \\ 0 & 0 & (uv)^{-1} \end{pmatrix} \qquad u > 0, v > 0. \qquad (4.9.5)$$

Under H the scales along the x and y axes change independently by the factors u and v, and we have a compensating scale change along the z axis. We find H from the conditions

$$HCwA^0e_1 = e_1 \quad \text{(this determines } u\text{)},$$
$$HCwA^0e_2 = e_2 \quad \text{(this determines } v\text{)}.$$

All transformations considered were volume-preserving. Therefore, necessarily,

$$HCwA^0e_3 = e_3,$$

and the construction of (4.9.1) is now complete. From (4.9.1) we immediately conclude the existence of (unique) representations valid for every $A^0 \in \mathbb{A}_3^0$:

$$\begin{aligned} A^0 &= HCw \\ A^0 &= w'C'H' \end{aligned} \qquad (4.9.6)$$

(representations with different orders of elements H, C and w also exist).

The subgroups we consider possess (unique) bi-invariant Haar measures. The Haar measure on $\mathbb{C}^{(1)}$ is da (the linear Lebesgue measure); on $\mathbb{C}^{(2)}$ it is $db\, dc$ (the planar Lebesgue measure); and on $\mathbb{H}_2$ it is $(uv)^{-1}\, du\, dv$ (the product logarithmic measure). The Haar measure on $\mathbb{W}_3$ was considered in §3.3.

4.10 The group of affine deformations of $\mathbb{R}^3$

First we note that the products

$$C^{(2)}C^{(1)} \qquad C^{(2)} \in \mathbb{C}^{(2)},\ C^{(1)} \in \mathbb{C}^{(1)}$$

constitute a group which we denote by $\mathbb{C}_2$. Equivalently, an element of $\mathbb{C}_2$ can be represented by a matrix of the type

$$\begin{pmatrix} 1 & a & b \\ 0 & 1 & c \\ 0 & 0 & 1 \end{pmatrix} \qquad -\infty < a, b, c < \infty. \qquad (4.10.1)$$

Let us consider the Haar measure on $\mathbb{C}_2$. It is easy to check that the matrix equation

$$C_1^{(1)} C_r^{(2)} = C_1^{(2)} C_r^{(1)} \qquad (4.10.2)$$

is equivalent to the following equation system involving the parameters a, b and c (see (4.9.3) and (4.9.4)); the lower index will refer to the corresponding matrix in (4.10.2):

$$a_r = a_1$$
$$c_r = c_1$$
$$b_r = b_1 + a_1 c_1.$$

This is a Cavalieri-type transformation, hence

$$da_1 \, db_r \, dc_r = da_r \, db_1 \, dc_1.$$

By the criterion of §1.3 we conclude that

the group $\mathbb{C}_2$ possesses a bi-invariant Haar measure. In terms of the elements of the matrix (4.10.1), this measure is $da \, db \, dc$.

We call the products

$$\mathbf{V} = HC, \qquad H \in \mathbb{H}_2, \qquad C \in \mathbb{C}_2$$

affine deformations of $\mathbb{R}^3$. They constitute a group which we denote by $\mathbf{V}_3$ (the group of affine deformations of $\mathbb{R}^3$). It is important that both the representations

$$\mathbf{V} = H_1 C_r$$

and $\qquad\qquad\qquad\qquad\qquad\qquad\qquad\qquad\qquad\qquad\qquad$ (4.10.3)

$$\mathbf{V} = C_1 H_r$$

exist and are unique. This enables us to apply the considerations of §1.3. The matrix equation

$$C_1 H_r = H_1 C_r, \qquad C_1, C_r \in \mathbb{C}_2, \qquad H_1, H_r \in \mathbb{H}_2$$

is equivalent to the following equation system, where the elements of the matrices (4.9.5) and (4.10.1) have indexes which refer to their order in (4.10.3):

$$u_r = u_1$$
$$v_r = v_1$$
$$a_1 v_r = u_1 a_r \qquad (4.10.4)$$
$$b_1 (v_r u_r)^{-1} = u_1 b_r$$
$$c_1 (v_r u_r)^{-1} = c_r v_1.$$

Equating the right-hand side expressions in (1.3.7) (under the assumption of existence of densities ρ_1 and ρ_2 of m_1' and m_2')

$$\rho_1(a_1, b_1, c_1)\, da_1\, db_1\, dc_1(u_r v_r)^{-1}\, du_r\, dv_r = \rho_2(u_1, v_1)\, du_1\, dv_1\, da_r\, db_r\, dc_r.$$

Changing to the variable u_1, v_1, a_r, b_r, c_r on the left-hand side by using (4.10.4) yields

$$u_1^3 v_1 \rho_1(c_r u_1 v_1{}^2, a_r u_1 v_1^{-1}, b_r u_1^2 v_1) = \rho_2(u_1, v_1),$$

i.e.

$$\rho_1 = \text{const}, \qquad \rho_2(u_1, v_1) = u_1^3 v_1.$$

Thus we have found two expressions for the right-invariant Haar measure $d^{(r)}V$ on V_3:

$$d^{(r)}V = (u_r v_r)^{-1}\, du_r\, dv_r\, da_1\, db_1\, dc_1$$

and   (4.10.5)

$$d^{(r)}V = u_1^3 v_1\, du_1\, dv_1\, da_r\, db_r\, dc_r.$$

4.11 Haar measures on $\mathbb{A}_3^0$ and $\mathbb{A}_3$

The first step towards the derivation of the Haar measure on $\mathbb{A}_3^0$ (this measure *is* bi-invariant, see [2]) can be taken by using the representations

$$\begin{aligned} A^0 &= V_1 w_r \\ A^0 &= w_1 V_r \end{aligned} \qquad w_1, w_r \in W_3, \qquad V_1, V_r \in V_3$$

since their existence (and uniqueness) follow from (4.9.6).

Let us denote by dA^0 the Haar measure on the group $\mathbb{A}_3^0$. By the necessity part of the criterion in §1.3 we find that

$$dA^0 = d^{(l)}V_1\, dw_r = d^{(r)}V_r\, dw_1, \qquad (4.11.1)$$

where dw is the Haar measure on W_3.

We also consider the group $\mathbb{A}_3$ of area- and 'sense'- (but no longer origin-) preserving affine transformations of $\mathbb{R}^3$. For every $A \in \mathbb{A}_3$ we can easily derive the following unique representations:

$$\begin{aligned} A &= t_1 A_r^0 \\ A &= A_1^0 t_r \end{aligned} \qquad t_1, t_r \in \mathbb{T}_3, \qquad A_1^0, A_r^0 \in \mathbb{A}_3^0.$$

The existence of bi-invariant Haar measure on $\mathbb{A}_3$ is well known [2]. We denote this measure by dA. By the criterion of §1.3 we conclude that

$$dA = dt_1\, dA_r^0 = dt_r\, dA_1^0. \qquad (4.11.2)$$

In the next section we will use the representation

$$dA = dM_1\, d^{(r)}V_r = dM_r\, d^{(l)}V_1 \qquad (4.11.3)$$

which follows from the existence and uniqueness of representations,

$$\begin{aligned} A &= M_1 V_r \\ A &= V_1 M_r, \end{aligned} \qquad M_1, M_r \in \mathbb{M}_3, V_r, V_1 \in V_3$$

Here dM is the Haar measure on the Euclidean group $\mathbb{M}_3$ (compare with (4.4.3)).

4.12 V_3-invariant measure in the space of tetrahedral shapes

By means of a map 'based' on an appropriately fixed sequence $(\mathscr{P}_1, \mathscr{P}_2, \mathscr{P}_3, \mathscr{P}_4)$ of points in $\mathbb{R}^3$, any sequence (Q_1, Q_2, Q_3, Q_4) of non-coplanar points in $\mathbb{R}^3$ can be represented as

$$(Q_1, Q_2, Q_3, Q_4) = (A, V, \mathscr{X}),$$

where $A \in \mathbb{A}_3$, $\mathscr{X} = 1$ or -1, and V is the volume of the tetrahedron with vertices Q_1, Q_2, Q_3, Q_4 (compare with §4.5).

The variable A is determined from the relation

$$(Q_1, Q_2, Q_3, Q_4) = Av(\mathscr{P}_1, \mathscr{P}_2, \mathscr{P}_3, \mathscr{P}_4), \qquad \text{case } \mathscr{X} = 1$$

(or its version which includes a reflection when $\mathscr{X} = -1$). Here v is a homothety of $\mathbb{R}^3$ which corresponds to the matrix

$$v = \begin{pmatrix} \sqrt[3]{V} & 0 & 0 \\ 0 & \sqrt[3]{V} & 0 \\ 0 & 0 & \sqrt[3]{V} \end{pmatrix}.$$

The measure $\prod_{i=1}^4 \mathrm{d}Q_i$ (where each $\mathrm{d}Q_i$ is Lebesgue in $\mathbb{R}^3$) is invariant with respect to $\mathbb{A}_3$. Therefore, by homothety considerations,

$$\prod_{i=1}^4 \mathrm{d}Q_i = c \, \mathrm{d}A \, V^2 \, \mathrm{d}V = c \, \mathrm{d}M_1 \, \mathrm{d}^{(r)}V_r \, V^2 \, \mathrm{d}V, \tag{4.12.1}$$

below we calculate the constant c to be

$$c = 3 \cdot 6^3. \tag{4.12.2}$$

We can equate the right-hand sides of (4.12.1) and (3.16.14). Applying elimination of measures we find

$$m_V(\mathrm{d}\sigma) = 3 \cdot 6^3 \, \mathrm{d}^{(r)}V. \tag{4.12.3}$$

This reveals certain invariance properties of the measure m_V (compare with the discussion in §4.6). A direct derivation of (4.12.3) can be of interest.

It is convenient to take a general affine deformation in the form

$$V = \begin{pmatrix} 1 & a & b \\ 0 & 1 & c \\ 0 & 0 & 1 \end{pmatrix} \cdot \begin{pmatrix} u & 0 & 0 \\ 0 & v & 0 \\ 0 & 0 & (uv)^{-1} \end{pmatrix} = \begin{pmatrix} u & av & b(uv)^{-1} \\ 0 & v & c(uv)^{-1} \\ 0 & 0 & (uv)^{-1} \end{pmatrix}.$$

We take the vertices of the 'basic' tetrahedron to be

$$\mathscr{P}_1 = (0, 0, 0), \qquad \mathscr{P}_2 = (1, 0, 0), \qquad \mathscr{P}_3 = (0, 1, 0), \qquad \mathscr{P}_4 = (0, 0, 1)$$

and therefore the coordinates of the points $V\mathscr{P}_i$, $i = 2, 3, 4$, can be read from the matrix representing our V. This enables us to easily express the shape parameters of the tetrahedron

$$V(\mathscr{P}_1, \mathscr{P}_2, \mathscr{P}_3, \mathscr{P}_4)$$

which we used in §3.16, VI, in terms of a, b, c, u and v. We have (see fig. 3.16.4)

$$\cot v_3 = a,$$
$$\cot v_4 = b/\sqrt{(1 + c^2)},$$
$$\cot \Phi_4 = c,$$

as well as

$$\alpha = vu^{-1}(1 + a^2)^{-1},$$
$$\beta = (u^2 v)^{-1}\sqrt{(1 + b^2 + c^2)};$$

hence the problem is reduced to two- and three-dimensional Jacobian calculations. We easily find that

$$dv_3 \, dv_4 \, d\Phi_4 = [(1 + a^2)^{-1}(1 + b^2 + c^2)^{-1}/\sqrt{(1 + c^2)}] \, da \, db \, dc$$

and

$$d\alpha \, d\beta = 3(u^4 v)^{-1}\sqrt{(1 + a^2)}\sqrt{(1 + b^2 + c^2)} \, du \, dv.$$

Substitution of these expressions, together with

$$(\sin v_3)^{-2} = 1 + a^2,$$
$$(\sin v_4)^{-2} = (1 + b^2 + c^2)(1 + c^2)^{-1},$$
$$(\sin \Phi_4)^{-3} = (1 + c^2)^{3/2}$$

into (3.16.13) and (3.16.14) yields

$$\prod_{i=1}^{4} dQ_i = 3 \cdot 6^3 \, dM \; V^2 \, dV (uv)^{-1} \, da \, db \, dc \, du \, dv.$$

Because of (4.10.5) this is exactly the same as (4.12.3).

4.13 Quintuples of points in $\mathbb{R}^3$

We consider the sequences Q_1, Q_2, Q_3, Q_4, Q_5 of points in $\mathbb{R}^3$ which do not possess coplanar quadruples. The purpose is to factorize the product of Lebesgue measures

$$\prod_{i=1}^{5} dQ_i$$

into a product of dA (Haar measure on $\mathbb{A}_3$) and a measure in a space of $\mathbb{A}_3$-invariant parameters.

We will denote by V_i the volume of the tetrahedron

$$\{Q_i\}_{i=1}^{5} \setminus Q_i \quad \text{(thus the volume of } Q_1 Q_2 Q_3 Q_4 \text{ is } V_5\text{)}.$$

We will use (4.12.1) (with V_5 replacing V) in conjunction with the expression of dQ_5 in terms of the volumes V_1, V_2 and V_3, say.

The continuation of the faces of Q_1, Q_2, Q_3, Q_4 splits the domain of Q_5 into the following connected components:

(a) the interior of $Q_1 Q_2 Q_3 Q_4$,
(b) the four infinite trihedral regions,

(c) four infinite quadrihedral regions, each having a *face* of $Q_1 Q_2 Q_3 Q_4$ on its boundary.

(d) six infinite quadrihedral regions, each having an *edge* of $Q_1 Q_2 Q_3 Q_4$ on the boundary.

Within each of the above regions we have

$$dQ_5 = 6 \frac{dV_1 \, dV_2 \, dV_3}{V_5^2}, \tag{4.13.1}$$

thus always

$$\prod_{i=1}^{5} dQ_i = 3 \cdot 6^4 \, dA \, dV_5 \, dV_1 \, dV_2 \, dV_3. \tag{4.13.2}$$

The proof of (4.13.1) when Q_5 belongs to (a) can be as follows. Denote by

S_i – the area of the triangle $\{Q_1\}_1^4 \backslash Q_i$.

h_i – the height of Q_5 above S_i.

Then for $i = 1, 2, 3$

$$dV_i = \tfrac{1}{3} S_i \, dh_i$$

and for a parallelepipedal dQ_5

$$dQ_5 = \beta \, dh_1 \, dh_2 \, dh_3 = \beta \prod_{1,2,3} \frac{3 \, dV_i}{S_i}. \tag{4.13.3}$$

An elementary expression for the constant β (which depends on the trihedral angle at Q_4) is of no significance in this calculation. The same formula yields

$$6V_5 = \beta \prod_{1,2,3} \frac{3V_5}{S_i}$$

therefore

$$\beta \prod_{1,2,3} \frac{3}{S_i} = \frac{6}{V_5^2}$$

and we get (4.13.1) by substitution into (4.13.3).

4.14 Affine shapes of quintuples in $\mathbb{R}^3$

Following the ideas of §4.8 we rewrite here the product $dV_5 \, dV_1 \, dV_2 \, dV_3$ (see (4.13.2)) in terms of the *volume V of the minimal convex hull of* $Q_1, Q_2, Q_3, Q_4,$ Q_5 and of the ratios V_i/V, where the latter define *affine* ($\mathbb{A}_3$-invariant) shape of the quintuple. The formulae used for changing to new variables are different for different regions of variation of Q_5.

By symmetry, it is enough for our purposes to consider component (a), one component of (c), and one of (d) (see §4.13) (see Table 4.14.1).

Table 4.14.1

Region of Q_5	Domain of V_1, V_2, V_3, V_5 (all positive)	Volume of the minimal convex hull of Q_1, Q_2, Q_3, Q_4, Q_5	Intermediate variables	Domain of the intermediate variables (all positive)
(1) Type (a)	$V_1+V_2+V_3<V_5$	$V=V_5$	V_1, V_2, V_3, V	$V_1+V_2+V_3<V$
(2) Type (c) The face is $Q_1Q_2Q_3$	$V_1+V_2+V_3>V_5$	$V=V_1+V_2+V_3$	V_1, V_2, V_5, V	$V_5<V$ $V_1+V_2<V$
(3) Type (d) The edge is Q_3Q_4	$V_3<V_1+V_2+V_5$	$V=V_1+V_2+V_5$	V_1, V_2, V_3, V	$V_3<V$ $V_1+V_2<V$

In each of cases (1), (2) and (3), the change from V_1, V_2, V_3, V_5 to the intermediate variables (in the fourth column) is with Jacobian $J=1$. Now we introduce the *affine shape variables*

$$\tau = (\tau_1, \tau_2, \tau_3).$$

Namely, in cases (1) and (3) we put

$$\tau_i = V_i/V \quad (i = 1, 2, 3),$$

while in case (2)

$$\tau_1 = \frac{V_1}{V}, \qquad \tau_2 = \frac{V_2}{V}, \qquad \tau_3 = \frac{V_5}{V}.$$

From these remarks it follows that

$$\prod_{i=1}^{5} dQ_i = 3 \cdot 6^4 \, dA \, V^3 \, dV d\tau_1 \, d\tau_2 \, d\tau_3, \tag{4.14.1}$$

The range of τ can be seen from the last column of table 4.14.1 to be

case (1): $\tau_1 > 0, \tau_2 > 0, \tau_3 > 0, \tau_1 + \tau_2 + \tau_3 < 1$ (a simplex of volume $\frac{1}{6}$);
cases (2), (3): $\tau_i > 0, i = 1, 2, 3, \tau_1 + \tau_2 < 1, \tau_3 < 1$ (a prism).

Consequently, the space $\tau_{3,5}$ of affine shapes that we consider is a union of five simplices and ten prisms as above (see the regions described in §4.13). Hence the $d\tau_1 \, d\tau_2 \, d\tau_3$ volume of $\tau_{3,5}$ equals $\frac{35}{6}$.

We introduce a probability measure on $\tau_{3,5}$:

$$\mathbb{P}_{3,5}(d\tau) = \tfrac{6}{35} \, d\tau_1 \, d\tau_2 \, d\tau_3,$$

which allows (4.14.1) to be rewritten as follows:

$$\prod_{i=1}^{5} dQ_i = 105 \cdot 6^3 \, dA \, V^3 \, dV \, \mathbb{P}_{3,5}(d\tau). \tag{4.14.2}$$

This result is similar to (4.8.2); generalization of both equations is given in the next section.

With the same motivation as in §4.8, we can consider a version of J. J. Sylvester's problem for the probability distribution $\mathbb{P}_{3,5}$. Namely, we wish to find the probabilities of the events

B_3, the minimal convex hull of $Q_1Q_2Q_3Q_4Q_5$, is a tetrahedron, and of its complement

B_4, the points $Q_1Q_2Q_3Q_4Q_5$, lie on the boundary of their minimal convex hull.

Clearly, event B_3 corresponds to the union of the regions labeled (a) and (b) in §4.13, or, more properly, to the union of the five *simplex* components of $\tau_{3,5}$. Therefore

$$\mathbb{P}_{3,5}(B_3) = \frac{5/6}{35/6} = \frac{1}{7}$$

and

$$\mathbb{P}_{3,5}(B_4) = \frac{6}{7}.$$

4.15 A general theorem

Equations (4.8.2) and (4.14.2) are special cases of a general result [20] concerning the spaces $\tau_{n,m}$ of *m-point affine shapes* in $\mathbb{R}^n$ and certain probabilities $P_{n,m}$ on these spaces $\tau_{n,m}$ (see §4.8) For simplicitly we formulate and prove this theorem for the planar case.

Most of the notation used here is the same as before. We consider sequences $(Q_1, \ldots, Q_m)$, $Q_i \in \mathbb{R}^2$ where Q_1, Q_2, Q_3 are non-collinear. From each class of $\mathbb{A}_2$-equivalent sequences we choose an element of the form

$$(\mathscr{P}_1, \mathscr{P}_2, \mathscr{P}_3, Q_4^0, \ldots, Q_m^0),$$

where

$$\mathscr{P}_1 = O \text{ (the origin)}$$
$$\mathscr{P}_2 = (1, 0)$$
$$\mathscr{P}_3 = (0, y) \text{ with some } y \neq 0.$$

We put

$$\mathscr{X} = 1 \text{ if } y > 0$$
$$-1 \text{ if } y < 0.$$

We denote by V the area of the minimal convex hull of the set $\{\mathscr{P}_1, \mathscr{P}_2, \mathscr{P}_3, Q_4^0, \ldots, Q_m^0\}$ (this value is constant for all $\mathbb{A}_2$-equivalent sequences).

In the case $\mathscr{X} = 1$ we denote by τ the image of the sequence $((0, y), Q_4^0, \ldots, Q_m^0)$ after a homothety of $\mathbb{R}^2$ which reduces the area of the minimal convex hull of $\{\mathscr{P}_1, \mathscr{P}_2, \mathscr{P}_3, Q_4^0, \ldots, Q_m^0\}$ to 1. To obtain τ when $\mathscr{X} = -1$ we also make a reflection with respect to the x axis.

The equivalence classes in question consist of elements

$$(Q_1, \ldots, Q_m) = A(\mathscr{P}_1, \mathscr{P}_2, \mathscr{P}_3, Q_4^0, \ldots, Q_m^0), \qquad A \in \mathbb{A}_2$$

(A acts on a sequence). Taken together with the above this defines a map

$$(Q_1, \ldots, Q_m) = (A, \mathscr{X}, V, \tau).$$

It has the properties that for any $A_1 \in \mathbb{A}_2$

$$A_1(Q_1, \ldots, Q_m) = (A_1 A, \mathscr{X}, V, \tau) \qquad (4.15.1)$$

and

$$(\mathbb{1}, \mathscr{X}, V, \tau) = (\mathscr{P}_1, \mathscr{P}_2, \mathscr{P}_3, Q_4^0, \ldots, Q_m^0)$$

where $\mathbb{1}$ is the unit element of the group $\mathbb{A}_2$.

The $(2m - 6)$-dimensional parameter τ thus defined is invariant with respect to $\mathbb{A}_2$, reflections and homotheties. We call τ the *affine shape* of $(Q_1, \ldots, Q_m)$.

We now consider the product of planar Lebesgue measures dQ_i.

For every $m = 3, 4, \ldots$

$$\prod_1^m dQ_i = c_m \, dA \, V^{m-2} \, dV \, P_{2,m}(d\tau) \qquad (4.15.2)$$

where c_m is a constant, dA is the Haar measure on $\mathbb{A}_2$ and $P_{2,m}$ is a probability measure in the space $\tau_{2,m}$.

Proof The separation of the measures dA, $V^{m-2} \, dV$ and a measure m on the space $\tau_{2,m}$ follows from general principles of Haar factorization and homothety considerations (see §§1.3 and 1.6). Thus it remains to prove that the measure m is totally finite.

Let K_r be the disc of radius r centered at O. We have

$$(4\pi)^m = \int_{K_2} \cdots \int_{K_2} \prod_1^m dQ_i$$

$$= 2 \int_{\tau_{2,m}} m(d\tau) \iint_{B_\tau} dA \, V^{m-2} \, dV \qquad (4.15.3)$$

where

$$B_\tau = \{(A, V): \text{the set corresponding to } A, V, \tau \text{ lies within } K_2\},$$

and we assume that $\mathscr{X} = 1$. Let $(Q_1, \ldots, Q_m)$ be an arbitrary sequence for which $V \leqslant 1$ and whose affine shape is τ. We can assume that Q_1 and Q_2 lie at maximal distance, i.e.

$$|Q_1, Q_2| \geqslant |Q_i, Q_j| \quad \text{for all } i, j.$$

It is elementary to show that, by choosing appropriate transformations $M \in \mathbb{M}_2$ and $H \in \mathbb{H}_1$, we can reach the effect that both the points HMQ_1 and HMQ_2 lie on the x axis and have the abscissae -1 and 1 while all other $Q_i^* = HMQ_i$ lie within the square $(-1, 1) \times (-1, 1)$. This is illustrated by fig. 4.15.1.

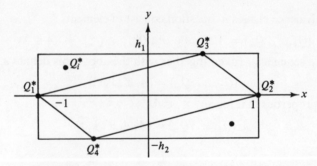

Figure 4.15.1 Q_3^* and Q_4^* are the points which have maximal (among Q_i^*) distance from the x axis

The minimal convex hull of $\{Q_i^*\}$ covers the interior of $Q_1^* Q_2^* Q_3^* Q_4^*$; hence

$$h_1 + h_2 \leqslant V \leqslant 1, \qquad \text{i.e. } h_1 \leqslant 1, h_2 \leqslant 1.$$

We now denote by $A(V, \tau)$ an element of $\mathbb{A}_2$ which transforms $(1, 1, V, \tau)$ in the way described in fig. 4.15.1. We also put

$$\alpha = \{A \in \mathbb{A}_2 : A K_{\sqrt{2}} \subset K_2\}.$$

Clearly because of (4.15.1), for every τ,

$$\iint_{B_\tau} \mathrm{d}A \ V^{m-2} \, \mathrm{d}V \geqslant \int_0^1 V^{m-2} \, \mathrm{d}V \int_{\alpha A(V, \tau)} \mathrm{d}A,$$

where

$$\alpha A(V, \tau) = \{A A(V, \tau) : A \in \alpha\}.$$

Because $\mathrm{d}A$ is *bi-invariant Haar*, the inner integral does not depend on V, τ and equals

$$h_{\mathbb{A}_2}(\alpha) > 0.$$

We conclude that

$$\iint_{B_\tau} \mathrm{d}A \ V^{m-2} \, \mathrm{d}V \geqslant h_{\mathbb{A}_2}(\alpha).$$

Now it follows from (4.15.3) that

$$(4\pi)^m \geqslant 2 h_{\mathbb{A}_2}(\alpha) m(\tau_{2, m}),$$

i.e. the measure in question is totally finite. This result can be generalized to points in $\mathbb{R}^3$ and higher dimensional spaces as suggested by the results of §4.14. Values of $P_{2, 5}$ for certain events in $\tau_{2, 5}$ have been found in [21].

4.16 The elliptical plane as a space of affine shapes

In preceding sections we considered slitted versions of the spaces $\tau_{n, m}$. Now our aim will be to give the topological description of the space $\tau_{2, 4}$. Similar results for general $\tau_{n, m}$ are seemingly unknown.

The general course of our reasoning resembles that of §3.18, i.e. we use the symmetry and continuity properties of the space $\tau_{2,4}$ which follow from its definition as a factor space.

First we stress that we consider ordered quadruples $(\mathscr{P}_1, \mathscr{P}_2, \mathscr{P}_3, \mathscr{P}_4)$ of points in $\mathbb{R}^2$ (rather than sets). Also here we prefer to consider the affine group $\mathbb{A}_2^*$ of 2×2 matrices with non-zero determinants. Two quadruples are declared equivalent if one can be transformed into another by an affine action. An affine shape τ can be formally defined to be a class of equivalent quadruples. The space $\tau_{2,4}$ is defined to be the factor space

$$[(\mathbb{R}^2 \times \mathbb{R}^2 \times \mathbb{R}^2 \times \mathbb{R}^2)\backslash\{\text{the set of totally collinear quadruples}\}]/\mathbb{A}_2^*.$$

By definition, totally collinear quadruples do not possess any affine shape.

Let us denote by E_s the set of affine shapes of the quadruples from

$$(\mathbb{R}^2 \times \mathbb{R}^2 \times \mathbb{R}^2 \times \mathbb{R}^2)\backslash\{\text{quadruples with } \mathscr{P}_i, \mathscr{P}_j, \mathscr{P}_k \text{ collinear}\},$$

where i, j and k are all distinct and assume values from $\{1, 2, 3, 4\}$ and $s \neq i$, j or k. Each E_s is homeomorphic to the Euclidean plane as seen from the following representations (maps) which are defined on different sets E_s:

on E_4, $\tau = ((0, 0), (1, 0), (0, 1), \mathscr{P}_4)$;
on E_3, $\tau = ((0, 0), (1, 0), \mathscr{P}_3, (0, 1))$;
on E_2, $\tau = ((0, 0), \mathscr{P}_2, (1, 0), (0, 1))$;
on E_1, $\tau = (\mathscr{P}_1, (0, 0), (1, 0), (0, 1))$;

where $\tau \in E_s$ is identified with a certain element chosen from the corresponding equivalence class and eventually with $\mathscr{P}_s \in \mathbb{R}^2$. In this sense we will write $\mathscr{P}_s \in E_s$.

Clearly the same affine shape can be represented by points on different planes E_s. One can also use representations of affine shapes on affine transformed planes E_s.

Let x_s, y_s be the coordinates of $\mathscr{P}_s$ on the plane E_s (see fig. 4.16.1, where we show E_4 and E_3).

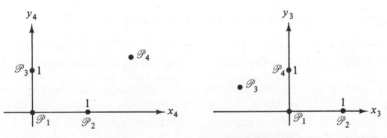

Figure 4.16.1 The points $\mathscr{P}_4$ and $\mathscr{P}_3$ on the left and right diagrams roughly correspond to the same affine shape

The relation between the affine shape coordinates (x_4, y_4) and (x_3, y_3) is as follows:

$$x_3 = \frac{-x_4}{y_4}, \qquad y_3 = \frac{1}{y_4} \qquad \text{whenever } y_4 \neq 0. \qquad (4.16.1)$$

Proof First assume that $y_4 > 0$. Then to bring the points $(0, 0)$, $(1, 0)$, $\mathscr{P}_4$, respectively, onto $(0, 0)$, $(1, 0)$, $(0, 1)$, it is enough to apply a transformation $C \in \mathbb{C}_1$ (see §4.1) with $c = -x_4/y_4$ and then to rescale the y axis by the factor $1/y_4$. Hence (4.16.1).

If $y_4 < 0$ then we first make a reflection with respect to the x axis: (x_4, y_4) goes to $(x_4, |y_4|)$ and $(0, 1)$ into $(0, -1)$. Then we apply the same transformation as above. Because under $C \in \mathbb{C}_1$ the points $(0, -1)$ and $(0, 1)$ shift in the opposite directions to equal distances, $(0, -1)$ will go to $(x_4/|y_4|, -1/|y_4|)$ which again corresponds to (4.16.1).

Let us consider the following subsets of $\tau_{2,4}$:

$B_1 = E_3$;
B_2, the affine shapes which on the plane E_4 lie on the x axis;
B_3, consists of one point (affine shape) which on both planes E_1 and E_2 is
    represented by $(0, 0)$.

The sets B_i are pairwise disjoint and $B_1 \cup B_2 \cup B_3 = \tau_{2,4}$.

Proof If for a quadruple $\mathscr{P}_1 = \mathscr{P}_2$ then $\mathscr{P}_1$, $\mathscr{P}_3$, $\mathscr{P}_4$ are necessarily non-collinear (otherwise $\mathscr{P}_1$, $\mathscr{P}_2$, $\mathscr{P}_3$, $\mathscr{P}_4$ would be totally collinear but this possibility was excluded from the start). Hence, by an affine transformation, $\mathscr{P}_1$, $\mathscr{P}_3$, $\mathscr{P}_4$ can be mapped onto $(0, 0)$, $(1, 0)$, $(0, 1)$. Therefore the corresponding τ belongs to E_2, and moreover it will be represented here by $(0, 0)$. In other words, $\mathscr{P}_1 = \mathscr{P}_2$ implies that $\tau \in B_3$.

If $\mathscr{P}_1 \neq \mathscr{P}_2$ then we have two mutually excluding subcases:

(a) $\mathscr{P}_1$, $\mathscr{P}_2$, $\mathscr{P}_4$ are non-collinear; then $\tau \in E_3 = B_1$, and
(b) $\mathscr{P}_1$, $\mathscr{P}_2$, $\mathscr{P}_4$ are collinear; then $\mathscr{P}_1$, $\mathscr{P}_2$, $\mathscr{P}_3$ are necessarily non-collinear.
    After mapping $\mathscr{P}_1$, $\mathscr{P}_2$, $\mathscr{P}_3$ on $(0, 0)$, $(1, 0)$, $(0, 1)$, $\mathscr{P}_4$ will fall on the x axis,
    i.e. $\tau \in B_2$.

Our main conclusion will be based on the following facts.

(1) In the topology of $\tau_{2,4}$ the set $B_2 \cup B_3$ is homeomorphic to a circle. Equivalently, as $|x| \to \infty$ the affine shape $(x, 0) \in E_4$ converges to $(0, 0) \in E_1$.

Proof Denote by A_0 the reflection of $\mathbb{R}^2$ with respect to the line $y = x$. Suppose $x > 0$. By rescaling the x axis by the factor $1/x$ we map $(x, 0) \in E_4$ into $(1/x, 0) \in A_0 E_2$. But a sequence which converges to $(0, 0)$ in $A_0 E_2$ also converges in E_2 to the same point.

The case $x < 0$ can be reduced to the above by a reflection.

(2) Any $\tau \in B_2 \cup B_3$ corresponds to a bundle of parallel lines in E_3 in the sense that τ is a limit (in $\tau_{2,4}$) of a point moving away in E_3 along any line of the bundle, in any of the two possible directions on the line. Different points of the circle $B_2 \cup B_3$ correspond to different bundles.

Proof Let us consider the case $\tau \in B_2$ so that $\tau = (x, 0) \in E_4$. The corresponding bundle of parallels on E_3 happens to be the image of the bundle of lines through $(x, 0)$ on E_4. This follows from (4.16.1) after substitution of the equations of the latter bundle,

$$x_4 = x + l \cos \psi, \qquad y_4 = l \sin \psi,$$

to obtain

$$x_3 = -xy_3 - \cot \psi.$$

We conclude that the direction of the line in the bundle is determined solely by x, and the horizontal shift of a line in a bundle is $\cot \psi$.

The reader may check that the point $\tau \in B_3$ corresponds to the bundle of horizontal lines on E_4.

Properties (1) and (2) imply that $\tau_{2,4}$ is obtained from E_3 in the same way that the projective (elliptical) plane is constructed by complementing a Euclidean plane with a 'line at infinity' ($B_2 \cup B_3$ in our case). We conclude that, topologically, $\tau_{2,4} = \mathbf{E}_2$.

5

Combinatorial integral geometry

In this chapter we consider in detail two topics from the vast new field mentioned in the chapter title, namely combinatorics of lines on $\mathbb{R}^2$ and of planes in $\mathbb{R}^3$. The theory has important applications in geometrical processes which we demonstrate in chapter 10 for the planar case. The combinatorics of planes is applied in this chapter in the construction of what we call *flag representations* of bounded centrally-symmetrical convex bodies in $\mathbb{R}^3$. The last section contains a brief synopsis of other ramifications of combinatorial integral geometry (a complete presentation can be found in [3]).

5.1 Radon rings in $\overline{\mathbb{G}}$ and $\mathbb{G}$

In this section we formulate the results of the theory referring to the M_2-invariant measures in the spaces $\overline{\mathbb{G}}$ and $\mathbb{G}$ (see §§2.1, 2.2 and 3.6). Two different proofs are given in §5.2 and §5.3.

We denote by m in the following a *finite set of points* on $\mathbb{R}^2$:

$$m = \{\mathscr{P}_i\}_1^n, \qquad \mathscr{P}_i \in \mathbb{R}^2.$$

Given m, let us call two lines g_1, $g_2 \in \overline{\mathbb{G}}$ equivalent if both have the same subset of m in their left halfplanes (see fig. 5.1.1).

The bundles of lines through different points of m decompose $\overline{\mathbb{G}}$ into subsets of equivalent lines which we call 'atoms'. There are only two unbounded atoms. Their lines leave the whole of m in one halfplane, which is left for one atom and right for another. The *Radon ring* $\overline{r}(m)$ is defined to be the ring generated by all *bounded* atoms. Any $B \in \overline{r}(m)$ is a union of bounded atoms. The term 'Radon ring' was introduced in this context in [3]. We stress that the atoms are pairwise disjoint.

Let g_{ij} be the line through $\mathscr{P}_i$, $\mathscr{P}_j \in m$ directed from $\mathscr{P}_i$ to $\mathscr{P}_j$. If in the set m no three points lie on a line, then each g_{ij} belongs to the boundary of exactly

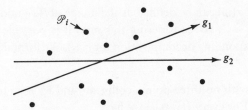

Figure 5.1.1 g_1 and g_2 are equivalent

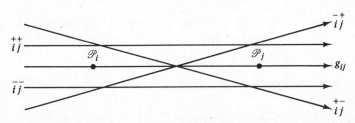

Figure 5.1.2 The sign $+$ or $-$ over an index refers to the right or left position of the corresponding point

four atoms. We can obtain lines from these four atoms by subjecting g_{ij} to sufficiently small displacements, as shown in fig. 5.1.2.

It is convenient to consider the displacements as infinitesimal. We have the following proposition.

For every set $B \in \overline{r}(m)$ its $\mathbb{M}_2$-invariant measure $\mu(B)$ can be calculated as a linear combination of the distances ρ_{ij} between points $\mathscr{P}_i, \mathscr{P}_j \in m$ with integer coefficients:

$$\mu(B) = \sum_{(i,j)} c_{ij}(B)\rho_{ij}. \qquad (5.1.1)$$

The integers $c_{ij}(B)$ can be calculated by means of the 'four indicator formula':

$$c_{ij}(B) = I_B(\overset{+}{i}, \overset{-}{j}) + I_B(\overset{-}{i}, \overset{+}{j}) - I_B(\overset{+}{i}, \overset{+}{j}) - I_B(\overset{-}{i}, \overset{-}{j}), \qquad (5.1.2)$$

where $I_B(\overset{\pm}{i}, \overset{\pm}{j})$ are the values of the indicator $I_B(g)$ on the lines (or, rather, on corresponding atoms) shown in fig. 5.1.2.

Let us now define similar rings in the space $\mathbb{G}$. Given a set m, two lines g_1, $g_2 \in \mathbb{G}$ are called equivalent if they induce the *same separation* of m. Bundles of lines through each $\mathscr{P}_i \in m$ split $\mathbb{G}$ into sets of equivalent lines, which we call *atoms*. There is only one unbounded atom, namely the one whose lines leave the whole of m in a halfplane. By definition, $r(m)$ is the minimal ring containing all bounded atoms. Every $B \in r(m)$ is a union of a finite number of atoms.

Alternatively, $r(m)$ can be defined as the image of $\bar{r}(m)$ under the direction-erasing map $\bar{G} \to G$. By considering this map it is possible to deduce from (5.1.1) the corresponding decomposition for $r(m)$. We formulate the following result.

Let m be a set with no three points collinear, and let μ be the M_2-invariant measure on G. For every $B \in r(m)$ we have

$$\mu(B) = \sum_{i<j} c_{ij}(B)\rho_{ij}, \qquad (5.1.3)$$

where $c_{ij}(B)$ are integer coefficients.

For the calculation of the values of $c_{ij}(B)$ we can again apply (5.1.2), since I_B is also naturally defined on directed lines. Since in (5.1.2) now always

$$c_{ij}(B) = c_{ji}(B)$$

the way we ascribe a direction to the line through the points $\mathscr{P}_i$ and $\mathscr{P}_j$ is irrelevant.

Examples of calculation by means of (5.1.3) are given in §5.4.

5.2 Extension of Crofton's theorem

Remarkably, the result of the previous section follows rather directly from Crofton's theorem, previously mentioned in §1.8.

The key fact is that every line g which intersects the convex hull of

$$m = \{\mathscr{P}_i\}$$

but avoids any of its points determines two non-empty subsets of m:

$$S, \text{ say, and } S^c, \qquad S \cup S^c = m.$$

The two sets are separated from each other by the line g. Therefore for every $B \in r(m)$ we can find sets $S_l \subset m$ such that B can be expressed as a disjoint union

$$B = \bigcup_l \alpha_l, \qquad (5.2.1)$$

where

$$\alpha_l = \{g \in G : g \text{ separates } S_l \text{ from } S_l^c\}.$$

We can now apply Crofton's theorem and find $\mu(\alpha_l)$ by taking D_1 and D_2 to be the (polygonal) convex hulls of S_l and S_l^c.

Using the notation of fig. 5.2.1,

$$\mu(\alpha_l) = |d_1| + |d_2| - \sum |v_i|. \qquad (5.2.2)$$

Clearly the existence of linear representations of the type (5.1.3) for general $B \in r(m)$ follows by simple addition of this expression written for all atoms participating in (5.2.1). But what about the algorithm for the coefficients?

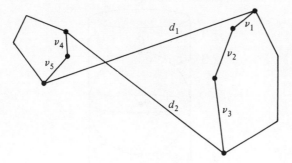

Figure 5.2.1

The simplest solution is the following: we check that (5.2.2) coincides with the expression for $\mu(\alpha_k)$ given by (5.1.2)–(5.1.3). Then we use the additive property:

$$I_B(\overset{+}{i}, \overset{+}{j}) = \sum_l I_{\alpha_l}(\overset{+}{i}, \overset{+}{j})$$

whenever B is given by (5.2.1).

The additivity follows from the definition of $I_B(\overset{+}{i}, \overset{+}{j})$. This completes the first proof of (5.1.3). We mentioned in §5.1 that (5.1.3) can be derived from (5.1.1) by means of a direction-erasing map. To prove (5.1.1) we can now use the fact that it is possible to proceed in the converse direction. In fact, (5.1.1) *follows from* (5.1.3) by invariance of the measure μ on $\overline{\mathbb{G}}$ with respect to inversion of the line directions.

5.3 Model approach and the Gauss–Bonnet theorem

The purpose of this section is to show that the decomposition (5.1.3) is based upon combinatorial relationships between certain subsets of $\mathbb{G}$. It turns out that sets of the type $[v] = \{g \in \mathbb{G} : g$ hits a linear segment $v \subset \mathbb{R}^2\}$ are of particular importance (for the reasons explained in §5.12). (In [3] the sets $[v]$ were called 'Buffon sets'.) We will use a model for $\mathbb{G}$ (i.e. a representation of $\mathbb{G}$ by a surface) in which each bundle

$$[\mathscr{P}] = \{g \in \mathbb{G} : g \text{ hits a point } \mathscr{P} \in \mathbb{R}^2\}$$

is a geodesic line. This will enable us to establish, as a by-product, connections between (5.1.3) and some well known facts of differential geometry.

First consider the cylinder $\mathbb{C} = \mathbb{S}_1 \times \mathbb{R}$ representing the space $\overline{\mathbb{G}}$ of oriented lines (see §2.1). On $\mathbb{C}$ each bundle $[\mathscr{P}]$ is represented by an ellipse, namely the trace on $\mathbb{C}$ of a plane through the point O which lies on the axis of $\mathbb{C}$ at the

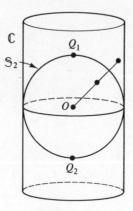

Figure 5.3.1

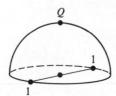

Figure 5.3.2

$p = 0$ level. We map $\mathbb{C}$ into $\mathbb{S}_2$, the unit sphere with center O in $\mathbb{R}^3$, by means of projection through O, see fig. 5.3.1.

This map is one-to-one except for the south and the north poles $Q_1, Q_2 \in \mathbb{S}_2$ which we remove. It is clear that each $[\mathscr{P}]$ is the trace on $\mathbb{S}_2$ of a plane through O, i.e. $[\mathscr{P}]$ is a *great circle*.

Now the map which reverses the direction of the lines sends each point (φ, p) of $\mathbb{C}$ to $(\varphi + \pi, -p)$. On $\mathbb{S}_2 \backslash \{Q_1, Q_2\}$ the corresponding map sends each point ω to its diametrical opposite, $-\omega$. Thus the 'direction-erasing' map $\overline{\mathbb{G}} \to \mathbb{G}$ corresponds to the *identification of opposite (antipodal) points on* $\mathbb{S}_2 \backslash \{Q_1, Q_2\}$. We come to the representation (see fig. 5.3.2)

$$\mathbb{G} = \mathbf{E}_2 \backslash Q,$$

where $\mathbf{E}_2$ is the elliptical (or projective) plane, and Q corresponds to the pair Q_1, Q_2.

Next we ask what is the form of sets from $r(m)$ in this model? We have seen that each $[\mathscr{P}]$ is a geodesic line. If $\mathscr{P}_1 \neq \mathscr{P}_2$ then $[\mathscr{P}_1]$ and $[\mathscr{P}_2]$ are two geodesics intersecting at a single 'point', namely the line through $\mathscr{P}_1$ and $\mathscr{P}_2$ (fig. 5.3.3).

These divide $\mathbf{E}_2$ into two connected components (called *lunes*): one such component is the needle set $[v]$ where $v = \mathscr{P}_1 \mathscr{P}_2$. Since $[v]$ is bounded, i.e. $[v]$

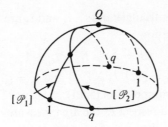

Figure 5.3.3

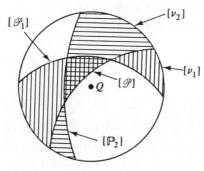

Figure 5.3.4

should have a compact closure within $\mathbf{E}_2 \setminus Q$, we identify $[v]$ as the lune which does not cover the removed point Q.

Given a set $m = \{\mathscr{P}_i\}_1^n$, we have n bundles $[\mathscr{P}_i]$ in $\mathbb{G}$, and the corresponding geodesic lines *split* $\mathbf{E}_2$ into non-overlapping geodesically convex polygons (elementary polygons). They actually represent the atoms of $r(m)$ except for the elementary polygon which covers the point Q on $\mathbf{E}_2$: it corresponds to the unbounded atom.

Consider the simplest case, involving three points $\mathscr{P}_1$, $\mathscr{P}_2$ and $\mathscr{P}$. Consider the atom $[v_1] \cap [v_2]$, where $v_1 = \mathscr{P}\mathscr{P}_1$, $v_2 = \mathscr{P}\mathscr{P}_2$; this atom corresponds to a geodesic triangle in $\mathbf{E}_2$.

The decomposition written in terms of indicator functions

$$2I_{[v_1] \cap [v_2]} = I_{[v_1]} + I_{[v_2]} - I_{[\mathscr{P}_1 \mathscr{P}_2]} \tag{5.3.1}$$

can be visualized directly from fig. 5.3.4 as a property of the demarcation of $\mathbf{E}_2$ by three geodesics. Of course in (5.3.1) we ignore points which belong to the boundaries (i.e. lines from the bundles $[\mathscr{P}]$, $[\mathscr{P}_1][\mathscr{P}_2]$). Let v be a vertex of the geodesic convex polygon $D \subset \mathbf{E}_2$. Two lunes are determined by the (two) continuations of geodesic sides meeting at v. That which contains int. D is termed the *covering lune*, while the other is the *outer lune* of D at v. Again, each of these lunes can be either unbounded or bounded, depending on whether it covers the deleted point Q or not.

If $\tau \subset \mathbf{E}_2$ is a geodesic triangle and l_1, l_2 and l_3 are the *covering lunes* of τ, then the symmetric equation

$$2I_\tau = I_{l_1} + I_{l_2} + I_{l_3} - 1 \qquad (5.3.2)$$

holds. In particular if $l_1 = [v_1]$, $l_2 = [v_2]$ and $l_3 = [\mathscr{P}_1\mathscr{P}_2]^c$ (an unbounded lune) we recover the decomposition (5.3.1).

Integrating (5.3.2) with respect to the area measure a on $\mathbf{E}_2$ yields the celebrated Euler formula for the area of a spherical triangle:

$$a(\tau) = |l_1| + |l_2| + |l_3| - \pi,$$

where $\frac{1}{2}a(l_i) = |l_i|$ is the interior angle of the lune.

At the same time, integration of (5.3.1) with respect to the M_2-invariant measure μ on G yields

$$2\mu([v_1] \cap [v_2]) = |v_1| + |v_2| - |\mathscr{P}_1\mathscr{P}_2|,$$

or in a weaker form

$$|v_1| + |v_2| - |\mathscr{P}_1\mathscr{P}_2| \geqslant 0,$$

where $|\cdot|$ now denotes length in $\mathbb{R}^2$. We see that the usual triangle inequality has combinatorial roots in common with the Euler formula.

We now return to the general case of a bounded geodesic convex polygon D on $\mathbf{E}_2$ (an atom). By the triangulation $\{\tau_i\}$ of D shown in fig. 5.3.5 we have

$$I_D(g) = \sum_j I_{\tau_j}(g), \qquad g \in \mathbf{E}_2.$$

Substituting (5.3.2) for each triangle yields

$$2I_D(g) = \sum_i I_{l_i}(g) - N,$$

where $\{l_i\}$ is the set of covering lunes of D and N is the number of triangles in the triangulation.

A covering lune l_i can be unbounded. In case it is we substitute

$$I_{l_i} = 1 - I_{(l_i)^c},$$

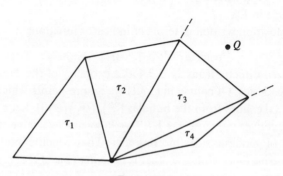

Figure 5.3.5

where c denotes the complement and obtain

$$2I_D(g) = \sum_{\substack{l_i \text{ are bounded} \\ \text{covering lunes}}} I_{l_i}(g) - \sum_{\substack{l_i \text{ are unbounded} \\ \text{covering lunes}}} I_{(l_i)^c}(g) + s - N.$$

where s is the number of unbounded covering lunes. Two remarks are important.

(a) If l_i is unbounded and, in our triangulation, $l_i = \bigcup b_k$, where b_k are covering lunes of the triangles τ_j, then exactly one of these lunes b_k is unbounded (i.e. covers Q).

(b) A bounded triangle always has exactly one unbounded covering lune.

It follows from (a) and (b) that $s = N$, and we obtain an expression for I_D, the indicator function of an atom D in terms of *bounded* lunes. Recall that only bounded lunes have the interpretation of sets $[v]$. The result is valid up to lines g which belong to a finite number of bundles:

$$2I_D = \sum_{\substack{\text{bounded} \\ \text{covering} \\ \text{lunes}}} I_{l_i} - \sum_{\substack{\text{bounded} \\ \text{outer} \\ \text{lunes}}} I_{l_i}. \tag{5.3.3}$$

Equation (5.2.2) can be obtained from (5.3.3) by integration with respect to the measure μ, recalling that

$$\mu([v]) = 2|v|.$$

The bounded covering lunes correspond to the segments d_1 and d_2 (and it follows that D always has exactly two bounded covering lunes). Thus we have given a new proof of (5.2.2) and with it of (5.1.3) and (5.1.1) (see the end of §5.2). Yet (5.3.3) is of much greater value as explained by the following two remarks.

First, (5.3.3) can be integrated with respect to any locally-finite measure m (assuming $m([\mathscr{P}]) = 0$ for every bundle of lines $[\mathscr{P}]$). After summation over atoms composing B we obtain

$$m(B) = \frac{1}{2} \sum_{i<j} c_{ij}(B) m([\mathscr{P}_i \mathscr{P}_j]), \tag{5.3.4}$$

where B is the general set from $r(m)$,

$$[\mathscr{P}_i \mathscr{P}_j] = \text{lines which separate } \mathscr{P}_i \text{ from } \mathscr{P}_j.$$

Thus (5.1.3) has a very simple generalization to non-invariant measures, and, what is important, *the coefficients* $c_{ij}(B)$ *do not depend on the choice of measure* m (in fact they are topological invariants, see [3]).

Secondly, if we rewrite (5.3.3) completely in terms of outer lunes (which can be unbounded) and integrate the result with respect to a (finite) area measure on $\mathbf{E}_2$, we obtain the classical Gauss–Bonnet formula for convex polygons on the elliptical plane. Thus (5.3.3) is a bridge which connects (5.1.3) and (5.1.1) with well-known facts of differential geometry.

5.4 Two examples

Given a finite set $m = \{\mathcal{P}_i\}$ the collection of all segments $\mathcal{P}_i\mathcal{P}_j$ is called a *companion set of needles* (needle = line segment). For brevity we denote the needles from the companion set as v_s. In (5.1.3) the range of summation is usually less than the complete companion system. This range actually defines the *skeleton* of a set $B \in r(m)$. The skeleton consists of all needles $v_s = \mathcal{P}_i\mathcal{P}_j$ for which $I_B(\overset{+}{i}, \overset{+}{j})$ depends on both of its binary arguments, $+$ and $-$. Note that if v_s belongs to the skeleton of B, then the continuation of v_s (which is a line in $\mathbb{R}^2$) necessarily belongs to ∂B, the topological boundary of B.

I As an important application of (5.1.3) and the four indicator formula, let us give the answer to J. J. Sylvester's question [23]: given a set $\{\delta_i\}$ of needles, what are the integer coefficients (5.1.2) when

$$B = \bigcap [\delta_i] \qquad \text{or} \qquad B = \bigcup [\delta_i]?$$

Let $m = \{\mathcal{P}_i\}$ be the set of endpoints of the needles δ_i with which we start. We then construct the companion set of needles $\{v_s\}$. Since we assume that no three points from m are collinear, the needles of the companion set $\{v_s\}$ may be divided into three categories:

(1) the needles δ_i of the original set;
(2) needles d_k joining two different needles δ_i and δ_j, such that δ_i and δ_j lie in *different* halfplanes with respect to d_k;
(3) needles s_k joining two different needles δ_i and δ_j which lie in the *same* halfplane with respect to s_k.

We introduce a function $I_k(v)$ defined on needles as follows:

$I_k(v) = 1$ if the continuation of θ hits the interior of exactly k
        needles from the set $\{\delta_i\}$,

      $= 0$ otherwise.

We assume here that, for the case in which $v = \delta_s$, the needle δ_s is not counted.

The coefficients $c_s(B)$ for the two cases considered by Sylvester are summarized in table 5.4.1.

Table 5.4.1

B \\ v_s	δ_i	d_k	s_k
$\bigcap [\delta_i]$	$2I_{n-1}$	I_{n-2}	$-I_{n-2}$
$\bigcup [\delta_i]$	$2I_0$	$-I_0$	I_0

Thus the representations (5.1.3) take the form

$$\mu(\bigcap [\delta_i]) = 2 \sum_{i=1}^{n} I_{n-1}(\delta_i)|\delta_i| + \sum_k I_{n-2}(d_k)|d_k| - \sum_k I_{n-2}(s_k)|s_k| \quad (5.4.1)$$

$$\mu(\bigcup [\delta_i]) = 2 \sum_{i=1}^{n} I_0(\delta_i)|\delta_i| - \sum_k I_0(d_k)|d_k| + \sum_k I_0(s_k)|s_k|. \quad (5.4.2)$$

II The second example gives a solution to another problem by J. J. Sylvester, also posed (but not solved) in his paper [23].

Let a system $D_1, \ldots, D_n$ of bounded *convex* domains be given in $\mathbb{R}^2$. We consider the sets

$$[D_i] = \{g \in \mathbb{G} : g \text{ hits } D_i\}$$

and

$$\rangle D_i, D_j \langle = \{g \in \mathbb{G} : g \text{ separates } D_i \text{ from } D_j\}.$$

Define the Sylvester ring $Sr\{D_i\}$ to be the minimal ring of subsets of $\mathbb{G}$ which contains all the sets $[D_i]$ and $\rangle D_i, D_j \langle$. The problem is to find the invariant measure of the elements of $Sr\{D_i\}$. In order to established connections with §5.1 we first assume that the domains D_i are polygonal. Let $m = \{\mathscr{P}_i\}$ be the collection of vertices of polygons D_i. We have the basic relation

$$Sr\{D_i\} \subset r(m).$$

Therefore, if the set $\{\mathscr{P}_i\}$ is non-degenerate, we may use (5.1.3) to find decompositions for every $B \in Sr\{D_i\}$. The characteristic feature of these decompositions will be the restriction of skeletons to a certain subclass of $\mathscr{P}_i\mathscr{P}_j$ needles. In particular,

the diagonals of the polygons never occur in a skeleton of a $B \in Sr\{D_i\}$ and so do the pairs $\mathscr{P}_i\mathscr{P}_j$ with $\mathscr{P}_i \in \partial D_i$, $\mathscr{P}_j \in \partial D_j$ if the continuation of $\mathscr{P}_i\mathscr{P}_j$ hits the interior of D_i or D_j.

The needles that remain are

(1) $\mathscr{P}_i$ and $\mathscr{P}_j$ are endpoints of the same polygon side;
(2) $\mathscr{P}_i$ and $\mathscr{P}_j$ are endpoints of a double support line (see fig. 5.4.1).

For brevity we call type (2) segments 'strings'.

In the case where at least some of the D_i cease to be polygons, a natural limiting procedure may lead to a decomposition of $\mu(B)$ for $B \in Sr\{D_i\}$.

Figure 5.4.1 The two types of strings

In this limiting process each (non-polygonal) D_i should be replaced by an approximating polygon $D_i^{(n)}$ (say $D_i^{(n)} \subset D_i$). The set $B \in Sr\{D_i\}$ should be replaced by the corresponding $B^{(n)} \in Sr\{D_i^{(n)}\}$. The decomposition for $\mu(B^{(n)})$ should be constructed and its limit (as $n \to \infty$) found.

In this way the following proposition can be established. Assume that for the system of convex domains $\{D_i\}$ no lines support more than two domains D_i simultaneously.

Then for every $B \in Sr\{D_i\}$ the invariant measure $\mu(B)$ can be written in the form

$$\mu(B) = \sum_i \int_{\partial D_i} c_l(B)\, dl + \sum_{\text{strings}} c_v(B)|v|, \qquad (5.4.3)$$

where

l denotes a point on ∂D,
dl is a boundary length element,
v is a generic notation for a string.

Both functions $c_l(B)$ and $c_v(B)$ can take only values 0, ± 1 or ± 2. Since we now do not exclude the fact that the boundaries of the D_i's can have curved parts, the strings in (5.4.3) can be as shown in fig. 5.4.2.

For concrete sets B the functions $c_l(B)$ and $c_v(B)$ can be written in explicit form. Let us consider the case

$$B = \{g \in \mathbb{G} : g \text{ hits at least one of the } D_i\text{'s}\}.$$

In this case

$$c_l(B) = I_0(g_l),$$
$$c_v(B) = I_0(g_v) \qquad \text{if } v \text{ is of } s \text{ type,}$$
$$\qquad = -I_0(g_v) \qquad \text{if } v \text{ is of } d \text{ type,}$$

where

$I_0(g) = 1$ if the number of the domains D_i whose interiors are hit by g is zero,
$\qquad = 0$ otherwise;

g_l is the tangent line at the point $l \in \partial D_i$; g_v is the continuation of the string v;

Figure 5.4.2

and the s and d types of a string are shown in fig. 5.4.2. The proof is by polygonal approximation.

5.5 Rings in $\mathbb{E}$

Let $m = \{\mathscr{P}_i\}$ be a finite set of points in $\mathbb{R}^3$ *with no three points on a line*. Each plane $e \in \mathbb{E}$ which does not pass through a point from m produces a separation of m in two subsets. Two planes $e_1, e_2 \in \mathbb{E}$ are called equivalent if they produce the same separation of m. Sets of equivalent planes are called *atoms*. All atoms are bounded, except for the one which corresponds to the separation $\varnothing, m$.

We denote by $r(m)$ the minimal ring of subsets of $\mathbb{E}$ which contains all bounded atoms (the Radon ring in the terminology of [3]). In this section we formulate the main combinatorial result for general *bundleless* measures m on $\mathbb{E}$, i.e. for measures satisfying the condition

$m([\mathscr{P}]) = 0$ for every $\mathscr{P} \in \mathbb{R}^3$, where $[\mathscr{P}]$ is the bundle of planes through $\mathscr{P}$.

The result is formulated in terms of 'wedges'. In the present context the term 'wedge' was first introduced in [3] although the corresponding notion has existed anonymously in integral geometry for a long time.

A *wedge* is a pair $W = (v, V)$, where v is a finite line segment in $\mathbb{R}^3$ while V is an open domain in $\mathbb{R}^3$ bounded by two planes through v, called the *faces* of W. V consists of two disjoint parts forming a flat vertical angle (fig. 5.5.1).

A wedge $W = (v, V)$ is said to belong to the *companion* system of the set $m = \{\mathscr{P}_i\}$ if

(a) v is of the form $\mathscr{P}_i\mathscr{P}_j$;
(b) the interior of V does not contain any point of $\{\mathscr{P}_i\}$;
(c) on each face of W there lies, besides the endpoints of v, at least one other point from $\{\mathscr{P}_i\}$.

Thus the companion system always consists of only finite numbers of wedges.

Given a measure m on $\mathbb{E}$, we define the following 'wedge function' $|W|$:

$$|W| = \frac{1}{2\pi} \int_{[v]} \alpha(e)m(de). \tag{5.5.1}$$

Here $[v]$ denotes the set of planes which hit v, and $\alpha(e)$ is the angle of the planar domain $e \cap V$ (the angular trace of V on the plane e).

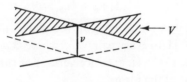

Figure 5.5.1

We have the following proposition.

Let m be a bundleless measure on $\mathbb{E}$. For every $B \in r(m)$ the value $m(B)$ can be calculated as a sum

$$m(B) = \sum c_s(B)|W_s|, \tag{5.5.2}$$

where $c_s(B)$ are integers which do not depend on the choice of the measure m, and the sum is extended over the companion set of wedges.

The algorithm for the calculation of the integers $c_s(B)$ resembles that given by (5.1.2). Let the wedge in question be $W_s = (v_s, V_s)$. Any plane e_0 containing v_s and passing within the domain V_s belongs to the boundary of four atoms, which we denote as

$$(\overset{+}{i}, \overset{+}{j}), \qquad (\overset{-}{i}, \overset{-}{j}), \qquad (\overset{+}{i}, \overset{-}{j}), \qquad (\overset{-}{i}, \overset{+}{j}).$$

These refer to the four possible positions of the endpoints $\mathscr{P}_i$ and $\mathscr{P}_j$ or v_s with respect to small displacements of e_0. That is, $+ +$ and $- -$ correspond to two cases in which both $\mathscr{P}_i$ and $\mathscr{P}_j$ remain in one halfplane; $+ -$ and $- +$ correspond to the two cases in which $\mathscr{P}_i$ and $\mathscr{P}_j$ fall in different halfspaces (compare with fig. 5.1.2). We have

$$c_s(B) = I_B(\overset{+}{i}, \overset{-}{j}) + I_B(\overset{-}{i}, \overset{+}{j}) - I_B(\overset{+}{i}, \overset{+}{j}) - I_B(\overset{-}{i}, \overset{-}{j}), \tag{5.5.3}$$

where $I_B(\overset{+}{i}, \overset{+}{j})$ are the values of the indicator $I_B(e)$ on the atoms. A complete proof of (5.5.1)–(5.5.3) can be found in [3], and we now outline its main features.

The main idea is to reduce the problem to the planar result (5.1.3). This is done in the following way.

Fix a line ω through the origin $O \in \mathbb{R}^3$ and choose a plane $e(\omega, 0)$ containing ω. For each $\Phi \in (0, \pi)$ let $e(\omega, \Phi)$ be this plane rotated about ω by the angle Φ. (Thus the initial choice of $e(\omega, 0)$ merely provides a reference direction for the Φ-parameter.)

We denote by $\mathbb{E}_\Phi \subset \mathbb{E}$ the class of planes which are perpendicular to the plane $e(\omega, \Phi)$. Obviously every $e \in \mathbb{E}$, if not from the set Π of planes perpendicular to ω, belongs to a uniquely determined class $\mathbb{E}_\Phi$.

Now each $e \in \mathbb{E}_\Phi$ is uniquely determined (within $\mathbb{E}_\Phi$) by its linear intersection with the perpendicular plane $e(\omega, \Phi)$. That is, the map

$$i(e) = e \cap e(\omega, \Phi)$$

defines a one-to-one correspondence

$$i : \mathbb{E}_\Phi \to \mathbb{G}_\Phi,$$

where $\mathbb{G}_\Phi$ is the family of lines lying in the plane $e(\omega, \Phi)$. This furnishes a parametrization of $\mathbb{E} \setminus \Pi$: to every pair (Φ, g) where $\Phi \in (0, \pi)$ and $g \in \mathbb{G}_\Phi$ there corresponds a unique $e \in \mathbb{E} \setminus \Pi$, and vice versa. For simplicity we can assume that our measure m is of the form

$$m(de) = f(\Phi, g)\, d\Phi\, dg,$$

where f is some density. Therefore, integration over any set $B \subset \mathbb{E}$ may be performed by repeated integration:

$$m(B) = \int_0^\pi d\Phi \int_{B \cap \mathbb{E}_\Phi} f(\Phi, g)\, dg.$$

The basic observation is that, given a collection m of points in $\mathbb{R}^3$,

$$B \in r(m) \qquad \text{implies} \qquad i(B \cap \mathbb{E}_\Phi) \in r(m^\Phi),$$

where the set m^Φ is the perpendicular projection of m onto the plane $e(\omega, \Phi)$, and $r(m^\Phi)$ is the ring in $\mathbb{G}_\Phi$ which corresponds to m^Φ. So for each set $B \in r(m)$ we can apply (5.3.4) to the inner integral. Another crucial step is to observe that the integer coefficients which solve the planar problem on each $e(\omega, \Phi)$ coincide with $c_s(B)$ as given by (5.5.3) (because the dependence of a planar coefficient on Φ reduces to dependence on a wedge). The final result (5.5.2)–(5.5.3) follows by eliminating the dependence on ω by means of averaging with respect to $d\omega/2\pi$ (the uniform distribution on $\mathbb{S}_2/2$).

In the following we use several times a rather special case of (5.5.2)–(5.5.3) which we consider in the next section (this case admits an elementary complete proof).

A general approach to the calculation of the function $|W|$ is considered in §5.8.

5.6 Planes cutting a convex polyhedron

Let D be a bounded convex polyhedron in $\mathbb{R}^3$, m be some bundleless measure on the space $\mathbb{E}$, and let

$$B = \{e \in \mathbb{E} : e \text{ hits } D\}.$$

Clearly this B belongs to $r(m)$, where m is defined to be the set of vertices of D. Therefore the value of $m(B)$ can be calculated by means of (5.5.1)–(5.5.3).

For this special B, however, there exists a very simple independent derivation of the same expression. We start from the observation that, for almost every plane which hits D, the intersection $D \cap e$ is a bounded convex polygon whose vertices correspond (in a one-to-one manner) to the edges of D actually hit by e. We write the elementary fact that the sum of outer angles of $D \cap e$ equals 2π in the form of an identity between indicator functions:

$$\sum I_{[a_i]}(e)\alpha_i(e) = 2\pi I_B(e). \tag{5.6.1}$$

Here $[a_i]$ is the set of planes hitting an edge a_i of D. Hence $\alpha_i(e)$ is the angle of the trace left on e by the outer wedge constructed on a_i; summation is by all edges of D. By definition, for an outer wedge the domain V (see fig. 5.5.1) does not possess points in common with the interior of D.

It remains for us to integrate (5.6.1) with respect to m:

$$m(B) = \frac{1}{2\pi} \sum \int_{[a_i]} \alpha_i(e) m(de)$$

$$= \sum_{\substack{\text{outer wedges} \\ \text{on edges}}} |W_s|. \tag{5.6.2}$$

We show in §5.8 that when m equals μ, the $\mathbb{M}_3$-invariant measure on $\mathbb{E}$, we have the following wedge function:

$$|W| = \tfrac{1}{2}|v| \cdot |V|, \tag{5.6.3}$$

where $|v|$ denotes the length of v and $|V|$ denotes the opening (flat angle) of the domain V. The results (5.6.2) and (5.6.3) together yield a classical formula:

$$\mu(B) = \frac{1}{2} \sum_{\text{edges of } D} |a_i| \cdot |V_i|. \tag{5.6.4}$$

It can be instructive to obtain (5.6.2) by a direct application of the algorithm (5.5.2)–(5.5.3). We mention in this connection that the general algorithm is capable of solving a much more difficult problem concerning polyhedrons, namely that of finding the distribution of the number of vertices in a polygon which arises when D is sectioned by a random plane, see [3].

5.7 Reconstruction of the measure from a wedge function

The wedge functions $|W|$ as introduced by (5.5.1) actually provide an alternative way of describing measures on the space $\mathbb{E}$. Indeed, (5.5.2) shows that with the knowledge of $|W|$ the problem of calculation of $m(B)$ is reduced to the linear combination of certain values of $|W|$. Of course there is a restriction that B should belong to $r(m)$ for some m. However, the class of such sets B is a measure-determining class.

Below we show how to reconstruct the density of a measure (assuming that the density exists) in terms of its function $|W|$.

Let δ_1, δ_2 and δ_3 be three parallel non-coplanar segments in $\mathbb{R}^3$. We put

$$B = \{e \in \mathbb{E} : e \text{ hits each } \delta_i, i = 1, 2, 3\}.$$

To write (5.5.2) for this B we introduce the following notation.

Let e_i be the plane containing δ_j and δ_l, $i \neq j, l$. In the plane e_i we consider the segments

$$s_i^{(1)}, \qquad s_i^{(2)}, \qquad d_i^{(1)}, \qquad d_i^{(2)}$$

shown in fig. 5.7.1. For each segment $v = s_i^{(k)}$ or $d_i^{(k)}$, $i = 1, 2, 3$, $k = 1, 2$, we define the domain V_v by the following requirements:

(a) V_v is bounded by two planes through v;
(b) these planes pass through the endpoints of δ_i;
(c) V_v contains the segment δ_i.

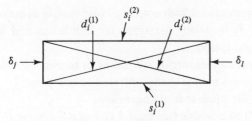

Figure 5.7.1 We assume that $\cos v_i = 1$, $i = 1, 2, 3$

We consider the wedges from the system accompanying the set

$$m = \{\text{endpoints of } \delta_i, i = 1, 2, 3\}.$$

The values of $c_s(B)$ on all wedges which are not of (v, V_v) type are zero. On the other hand, if $W_s = (v, V_v)$ we have

$$c_s(B) = 1 \qquad \text{if } v \text{ is of } d_i^{(k)} \text{ type,}$$
$$= -1 \qquad \text{if } v \text{ is of } s_i^{(k)} \text{ type.}$$

Thus the expression for $m(B)$, (5.5.2), has the form

$$m(B) = \sum_{\mathrm{I}} |W_s| - \sum_{\mathrm{II}} |W_s|, \tag{5.7.1}$$

where

I is the set of six wedges of the type $(d_i^{(k)}, V_{d_i^{(k)}})$;
II is the set of six wedges of the type $(s_i^{(k)}, V_{s_i^{(k)}})$.

Let us now assume that the measure m possesses a density u with respect to the $\mathbb{M}_3$-invariant measure de, i.e.

$$m(\mathrm{d}e) = u(e)\, \mathrm{d}e.$$

Our purpose is to find $u(e)$ for some fixed $e \in \mathbb{E}$. We now choose the segments δ_1, δ_2 and δ_3 to have a common infinitesimal length l and we let each δ_i hit the plane e. According to (3.12.2) as $l \to 0$ we have

$$m(B) = u(e) \cdot l^3 \cdot (2|\Delta|)^{-1} + o(l^3).$$

Thus

$$u(e) = 2|\Delta| \lim l^{-3} m(B).$$

This, together with (5.7.1), solves our problem. We stress that there are many degrees of freedom with which the segments δ_1, δ_2 and δ_3 satisfying the abovementioned conditions can be constructed.

5.8 The wedge function in the shift-invariant case

How can the function $|W|$ given by (5.5.1) actually be calculated? We give the answer for the $\mathbb{T}_3$-invariant measures on $\mathbb{E}$, a case that is especially important because of the connections with zonoids. Note that, because of the factorization of §2.9, all $\mathbb{T}_3$-invariant measures are bundleless and for them (5.5.2) applies without exception.

Now let μ be such a measure. It is convenient to start with the case in which the rose of directions m of μ is concentrated at one point, i.e. is of δ type. In symbols $m = \delta_\omega$ for some spatial direction ω.

Let $W = (v, V)$ be a wedge (see fig. 5.5.1). For a δ-type m, (5.5.1) yields

$$|W| = |v| \cdot F(V),$$
$$F(V) = (2\pi)^{-1} |\cos \widehat{\omega v^*}| \alpha(e_\omega),$$

where $\widehat{\omega v^*}$ is the angle between ω and the direction v^* of the segment v, and e_ω is the plane orthogonal to ω.

The formulae of spherical trigonometry enable us to write an explicit expression for the angle $\alpha(e_\omega)$. But now we will be interested in the following observation.

For a fixed direction v^* of v, each V defines a pair of arcs on the circle $\mathbb{S}_1$: they are the trace of V on $\mathbb{S}_1$ placed in a plane perpendicular to v and centered at a point of v. Therefore, for fixed v^*, $F(V)$ is a function defined on the pairs of centrally-symmetrical arcs on $\mathbb{S}_1$, or simply on arcs of $\mathbf{E}_1$ (see §1.5). Actually $F(V)$ is a measure on $\mathbf{E}_1$ (equivalently on $(0, \pi)$) which has a density

$$\frac{1}{2\pi} \sin^2 c.$$

Here $c = c(\omega)$ is the angle between v^* and the trace of e_ω on the plane e_Φ (see fig. 5.8.1): e_Φ belongs to the bundle of planes through v, and $\Phi \in \mathbf{E}_1$ denotes the angle of rotation around v^*.

Proof Applying standard formulae of spherical trigonometry we find that, if the opening of V is dv (infinitesimal), then

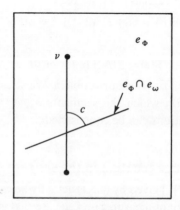

Figure 5.8.1

$$\alpha(e_\omega) = \frac{\sin c}{\sin \beta} \, dv,$$

where β is the flat angle between e_ω and the plane shown in fig. 5.8.1 The relation

$$\cos(\omega, v^*) = \sin c \sin \beta$$

completes the proof.

It is now straightforward to show that, for a general rose of direction m, the density of the corresponding F is expressed by the integral

$$\rho(v^*, \Phi) = \frac{1}{2\pi} \int \sin^2 c \, m(d\omega). \tag{5.8.1}$$

Finally, the wedge function is written in terms of ρ as follows:

$$|W| = |v| \int_V \rho(v^*, \Phi) \, d\Phi. \tag{5.8.2}$$

The space where the function $\rho(v^*, \Phi)$ is defined is the space $\mathbb{F}$ of *flags*. By definition, a *flag f is a figure consisting of a line through O and a plane through this line*. Therefore

$$f = (\Omega, \Phi), \tag{5.8.3}$$

where $\Omega \in \mathbf{E}_2, \Phi \in \mathbf{E}_1$, (see §1.5).

On $\mathbb{F}$ there is a topology whose description resembles that of $\mathbb{W}_3$ (see §3.2). Note that topologically $\mathbb{F}$ is different from the product $\mathbf{E}_2 \times \mathbf{E}_1$; however, this product can be used as a parameter space to represent $\mathbb{F}$ (in fact $\mathbf{E}_2 \times \mathbf{E}_1$ is a measure-representing model of $\mathbb{F}$).

We can represent flags in a way dual to (5.8.3)

$$f = (\omega, \varphi), \tag{5.8.4}$$

where

$\omega \in \mathbf{E}_2$ corresponds to the plane of the flag,

$\varphi \in \mathbf{E}_1$ represents the direction of the line of the flag as a direction on the plane ω.

We will see below that the possibility of dual representation of flags plays an important role in questions related to zonoids.

Example When μ is $\mathbb{M}_3$ invariant then

$$m(d\omega) = d\omega$$

and ρ reduces to a constant C. Then, according to (5.8.2),

$$|W| = C \cdot |v| \cdot |V|.$$

To calculate C we take $|V| = \pi$. By (5.5.1) $|W|$ will equal one-half the value of μ on the set of planes hitting v, i.e. $(\pi/2) \cdot |v|$. Hence, $C = 1/2$ and (5.6.3) follows.

5.9 Flag representations of convex bodies

Of special importance is the wedge density

$$\rho_\xi = \frac{1}{2\pi} \sin^2 c,$$

where we now consider the angle c (see fig. 5.8.1) as a function of flags:

$$c = c(\xi, f) = c(\xi, \Omega, \Phi) = c(\xi, \omega, \varphi).$$

We have seen above that this ρ_ξ corresponds to the measure μ_ξ in $\mathbb{E}$, which is a product:

$$\mu_\xi(de) = dp \times \delta_\xi(d\omega). \tag{5.9.1}$$

Let D be a convex domain in $\mathbb{R}^3$ and let $[D]$ be the set of planes hitting D. Then a direct integration of (5.9.1) yields

$$\mu_\xi([D]) = b(\xi), \tag{5.9.2}$$

the value of the breadth function of D in the direction ξ.

If D is a polyhedron, (5.6.4), in conjunction with the previous results, yields

$$b(\xi) = (2\pi)^{-1} \sum_{\substack{\text{outer wedges} \\ \text{of } D}} |a_i| \int_{V_i} \sin^2 c(\xi, a_i^*, \Phi) \, d\Phi$$

$$= \int \sin^2 c(\xi, f) m_D(df). \tag{5.9.3}$$

Here m_D is a measure in the space $\mathbb{F}$ concentrated on the union of the sets

$$\{(a_i^*, \Phi) : \Phi \in V_i\}.$$

is uniform in Φ on each of this sets, and

$$m_D(\{(a_i^*, \Phi) : \Phi \in V_i\}) = (2\pi)^{-1} |a_i| \cdot |V_i|.$$

Therefore the total measure is given by

$$m_D(\mathbb{F}) = \frac{1}{2\pi} \sum |a_i| \cdot |V_i| \tag{5.9.4}$$

(compare with (5.6.4.)).

The equality (5.9.3) is our flag representation for polyhedrons. *The existence of similar representations for every bounded convex body $D \subset \mathbb{R}^3$ follows now by weak convergence arguments.* Namely, for any bounded convex $D \subset \mathbb{R}^3$, we can (and do) construct a sequence of uniformly bounded convex polyhedrons D_n such that

$$\lim_{n \to \infty} b_n(\xi) = b(\xi) \text{ (uniform convergence)}$$

where b_n and b are the breadth functions of D_n and D, respectively. In the flag representations (5.9.3) for D_n's the total measures of m_{D_n}'s remain bounded (follows from (5.9.4)) and therefore a weakly convergent subsequence

$$m_{D_{n_i}} \to m$$

can be chosen. There is a complication in that the function $\sin^2 c$ is not continuous at $\xi = \omega$. Therefore at first we can state only that the equation

$$b(\xi) = \int \sin^2 c(\xi, f)m(df) \qquad (5.9.5)$$

holds for all values of ξ except for those for which

$$m(\{f = (\omega, \varphi): \omega = \xi\}) > 0,$$

i.e. except for countably many values of ξ. However, a further analysis shows that this restriction can be lifted, and in fact (5.9.5) *holds identically for every* ξ.

We call (5.9.5) a flag representation of the breadth function of a bounded convex $D \subset \mathbb{R}^3$. If D is centrally-symmetrical, then we say that (5.9.5) *is a flag representation of the body D itself*. Again we stress that in contrast with (2.10.2), which holds only for zonoids, flag representations are valid for all bounded convex centrally-symmetrical bodies in $\mathbb{R}^3$.

Given D, the measure m in (5.9.5) is not determined in a unique way. However, flag representations may shed new light on some facts of the theory of convex bodies. For instance, Blaschke's basic principle of compactness of sets of uniformly bounded convex bodies now reduces to the similar property of measures with respect to weak convergence.

In the next two sections we show that flag representations can be applied in order to obtain previously unknown facts.

5.10 Flag representations and zonoids

Zonotopes are zonoids generated (via (2.10.2)) by purely atomic measures. This means that the breadth function of a zonotope admits representation of the type

$$b(\xi) = \sum_{1}^{n} |\cos \widehat{\xi \Omega_i}| \cdot q_i, \qquad (5.10.1)$$

where $\Omega_i \in \mathbf{E}_2$ are some directions in $\mathbb{R}^3$, and q_i are some positive weights.

A zonotope is always a bounded centrally-symmetrical convex polyhedron.

A bounded centrally-symmetrical convex polyhedron D is a zonotope if and only if

(a) the collection of edges of D consists of classes such that in each class the edges are parallel and of equal length;

(b) within each class the openings of outer flat angles by the edges sum up to 2π.

The proof that (a) and (b) and (5.10.1) are equivalent is done by applying the notion of Minkowski's addition. In fact, (5.10.1) is the breadth function of

Minkowski's sum of n segments whose directions are Ω_i and whose lengths are q_i. Hence (a) and (b).

Conversely, suppose the conditions (a) and (b) are satisfied for a given polyhedron D. We write the flag representation (5.9.3) in the form

$$b(\xi) = (2\pi)^{-1} \sum |a_i| |\cos \widehat{\xi a_i^*}| \alpha_i(e_\xi)$$
$$= (2\pi)^{-1} \sum_{\substack{\text{classes}}} \sum_{\substack{\text{edges} \\ \text{from a class}}} |a_i| \cdot |\cos \widehat{\xi a_i^*}| \alpha_i(e_\xi)$$
$$= \sum_{\substack{\text{classes}}} q_j |\cos \widehat{\xi \Omega_j}|,$$

where q_j is the common length of the edges in a class, and Ω_j is their common direction.

From the point of view of flag representations the above has the following interpretations.

We have a set relation

$$\bigcup_{\substack{\text{edges} \\ \text{from a class}}} \{(\Omega, \Phi) : \Omega = a_i^*, \Phi \in V_i\} \approx a_i^* \times \mathbf{E}_1.$$

It follows from the remarks in §5.9 that on each of the above sets the measure m_D is q_j times the uniform measure on $\mathbf{E}_1$. We come to the conclusion that any zonotope D admits flag representation by means of a measure on $\mathbb{F}$ whose image under the map (5.8.3) is a factorized measure on $\mathbf{E}_2 \times \mathbf{E}_1$ with a uniform factor on $\mathbf{E}_1$.

What about flag representations for general zonoids?

It is well known that a breadth function of any zonoid admits pointwise approximation by breadth functions of zonotopes. This means that the measure m in a flag representation (5.9.5) of a zonoid can be obtained as a weak limit of measures satisfying the factorization property. The limit will necessarily be of the same factorized type.

Conversely, if the measure m in (5.9.5) is of this special type, then a direct integration reduces (5.9.5) to (2.10.2). Hence we have the following result

A bounded convex centrally-symmetrical body $D \subset \mathbb{R}^3$ is a zonoid if and only if its breadth function $b(\xi)$ admits a flag representation by means of a measure on $\mathbb{F}$ whose image under the map (5.8.3) is a factorized measure on $\mathbf{E}_2 \times \mathbf{E}_1$ with a uniform factor on $\mathbf{E}_1$; the factor on $\mathbf{E}_2$ coincides with the measure m in the representation (2.10.2) of a zonoid.

5.11 Planes hitting a smooth convex body in $\mathbb{R}^3$

In this section we derive an explicit flag representation for a general *smooth* (non-polyhedral) bounded convex body $K \subset \mathbb{R}^3$. The 'density' of the representing measure is expressed in terms of curvatures.

We start with the identity

$$\int x(e, l)\, dl = 2\pi I_{[K]}(e) \tag{5.11.1}$$

which expresses the familar fact of planar geometry that as we move along the boundary of a bounded convex domain the tangent line rotates by an angle 2π. Here $[K] \subset \mathbb{E}$ is the set of planes e which hit K, $x(e, l)$ is the curvature at the point of $\partial K \cap e$ determined by a longal coordinate l, $I_{[K]}(e)$ is the indicator function of the set $[K]$.

We integrate (5.11.1) over the space $\mathbb{E}$ with respect to the measure

$$\mu(de) = \delta_\xi(d\omega)\, dp$$

(equivalently, over the bundle of parallel planes orthogonal to the spatial direction ξ with respect to linear Lebesgue measure dp). We get the following expression for the value of the breadth function at ξ:

$$b(\xi) = \frac{1}{2\pi} \int x(\xi, p, l)\, dl\, dp$$

For a fixed ξ each pair (p, l) determines a point $\mathcal{P}$ on ∂K. By $d\mathcal{P}$ we denote an elementary rectangle at $\mathcal{P}$ which lies on $t(\mathcal{P})$ – the tangent plane at $\mathcal{P}$ ($d\mathcal{P}$ also denotes the *area* of the rectangle). Let the sides of the rectangle be dl and dh so that

$$d\mathcal{P} = dl\, dh.$$

Let γ be the angle between dh and dp (or ξ). We have

$$dp = \cos \gamma\, dh$$

and therefore

$$dl\, dp = \cos \gamma\, d\mathcal{P}.$$

The next step is to express the curvature $x(\xi, p, l)$ in terms of the main curvatures x_1 and x_2 at $\mathcal{P}$. Recall that x_1 and x_2 are two extremal values of the normal curvature $x_n(\varphi)$. The latter function is defined to be the curvature of the trace of ∂K on a plane through $\mathcal{P}$ which is perpendicular to $t(\mathcal{P})$ and leaves a linear trace in the direction φ on $t(\mathcal{P})$.

We will make use of a Euler formula [6]

$$x_n(\varphi) = x_1 \cos^2 \widehat{\beta_1 \varphi} + x_2 \cos^2 \widehat{\beta_2 \varphi}$$

where β_1 and β_2 are (mutually perpendicular) 'main' directions in $t(\mathcal{P})$.

We will also use a formula which gives the value of the curvature at $\mathcal{P}$ of the traces of ∂K on the planes which are no longer perpendicular to $t(\mathcal{P})$.

Denote by $g(\mathcal{P}, \varphi)$ a line on $t(\mathcal{P})$ which contains $\mathcal{P}$ and has (planar) direction φ. Let $e(\mathcal{P}, \varphi, \gamma)$ be the plane through $g(\mathcal{P}, \varphi)$ rotated around $g(\mathcal{P}, \varphi)$ by an angle γ. We assume that

$$e\left(\mathcal{P}, \varphi, \frac{\pi}{2}\right) = t(\mathcal{P}).$$

For the curvature $x(\mathcal{P}, \varphi, \gamma)$ of $e(\mathcal{P}, \varphi, \gamma) \cap \partial K$ at $\mathcal{P}$ we have (Meusnier's theorem)

$$x(\mathcal{P}, \varphi, \gamma) = (\cos \gamma)^{-1} x_n(\varphi).$$

The two classical results in our situation yield

$$x(\xi, p, l) = x(\mathcal{P}, \varphi_\xi, \gamma)$$
$$= (\cos \gamma)^{-1} [x_1 \cos^2 \widehat{\beta_1 \varphi_\xi} + x_2 \cos^2 \widehat{\beta_2 \varphi_\xi}]$$

where φ_ξ is the direction of the trace on $t(\mathcal{P})$ of a plane orthogonal to ξ. The expression for $b(\xi)$ now takes the form

$$b(\xi) = \frac{1}{2\pi} \int x(\xi, p, l) \cos \gamma \, d\mathcal{P}$$
$$= \frac{1}{2\pi} \int [x_1 \cos^2 \widehat{\beta_1 \varphi_\xi} + x_2 \cos^2 \widehat{\beta_2 \varphi_\xi}] \, d\mathcal{P}.$$

To obtain the desired flag representation we convert the above to integration over a sphere by means of

$$d\mathcal{P} = r_1 r_2 \, d\omega$$

where $r_i = 1/x_i$, $i = 1, 2$ are the main curvature radii, $d\omega$ is the area measure on $\mathbb{S}_2$. We get finally

$$b(\xi) = \frac{1}{2\pi} \int_{\mathbb{S}_2} [r_1 \sin^2 \widehat{\beta_1 \varphi_\xi} + r_2 \sin^2 \widehat{\beta_2 \varphi_\xi}] \, d\omega \qquad (5.11.2)$$

This is equivalent to a flàg representation of $b(\xi)$ since the integrand coincides with the result of integration of $\sin^2 c$ with respect to the measure

$$r_1 \delta_{\beta_1}(d\varphi) + r_2 \delta_{\beta_2}(d\varphi),$$

We note that (5.11.2) is a translational counterpart of the much wider known relation referring to the $\mathbb{M}_3$-invariant measure μ_0

$$\mu_0([K]) = \int_{\mathbb{E}_2} b(\xi) d\xi = \frac{1}{2} \int_{\mathbb{S}_2} (r_1 + r_2) d\omega \qquad (5.11.3)$$

The expression on the right-hand side is called the 'integral of mean curvature'. Clearly (5.11.3) follows from (5.11.2) by averaging over directions ξ. We note also that, for a general $\mathbb{T}_3$-invariant measure μ, the expression of $\mu([K])$ in terms of corresponding ρ can be obtained by integrating (5.11.2) with respect to the rose of directions of μ:

$$\mu([K]) = \frac{1}{2\pi} \int_{\mathbb{S}_2} [r_1 \rho(f_1) + r_2 \rho(f_2)] \, d\omega,$$

where the flags f_i are defined in terms of (5.8.4) as

$$f_i = (\text{orientation } \omega \text{ of the tangent plane, main direction } \beta_i).$$

5.12 Other ramifications and historical remarks

The idea of introducing measures into the space of lines in the plane was already implicit in Buffon's classical *needle problem* [36]. Let us recall its formulation. The plane is ruled by a fixed lattice of parallel lines a unit distance apart. A needle v of length $|v| < 1$ is 'thrown' at random onto the plane. What is the probability p of the event that in its final position the needle will be intersected by a line of the lattice?

In an equivalent formulation, the needle and the lattice exchange roles and one assumes that it is now the needle which is fixed in the plane with the lattice being thrown down at random.

Without loss of generality one may assume that the needle lies within some fixed open disc D of unit diameter. Then in all possible outcomes of the lattice-throwing experiment, the disc is intersected by exactly one line of the lattice if we assume that the case of tangency is 'impossible' (i.e. of probability 0). Since other lines of the lattice now play no role, we may fix our attention on this single line, g_D say, intersecting D, and Buffon's original problem is now replaced by the following one: what is the probability that a random line g_D intersecting D should also intersect the needle? We may refer to this as the *dual* problem to the classical Buffon's needle problem.

It is clear that in Buffon's original problem the solution depends on how the experiment is performed, i.e. on the distribution of the final position of the needle with respect to the lattice. Analogously in the reformulated version the result depends on the distribution P of the random line g_D.

In the classical solution to Buffon's problem it is assumed that the center of the needle and its orientation are independently and uniformly distributed. This means that the projection of the center onto a line perpendicular to the lines of the lattice is uniformly distributed on some segment of unit length, and the angle between the line containing the needle and the lines of the lattice is independent and uniformly distributed on $(0, \pi)$. With these assumptions,

$$p = \frac{2|v|}{\pi}.$$

This example is the earliest instance of the calculation of a 'geometrical probability'.

To this result corresponds the following solution of the dual problem: there is a unique distribution P of the random line g_D such that

$$P\{g_D \text{ intersects } v\} = \frac{2 \cdot |v|}{\pi}$$

for every needle v within D. This distribution P is proportional to the restriction to the set of lines intersecting D of the M_2-*invariant measure* on the space of lines in the plane (see §3.6). This direct connection with the classical Buffon problem provides justification of terming the subsets

$$[v] = \{\text{lines that hit } v\}$$

as 'Buffon sets' see [3].

The use of other distributions for g_D was first made by Bertrand [12] for the purpose of showing that the notion of 'random secant' admitted several interpretations and therefore allowed paradoxes ('Bertrand's paradoxes'). These could be avoided by adopting some principle leading to a unique choice among possible distributions, and it was considered that a 'natural' choice was one respecting the relation of Euclidean congruence, in that geometric events congruent to each other under Euclidean motions should have equal probabilities. This point of view is clearly expounded by, for example, Deltheil [22].

However, from a more modern point of view, interest in a general P has many other justifications. We mention only the simple observation that if

$$P\{g_D \text{ hits } \mathscr{P}\} = 0 \text{ for every point } \mathscr{P} \in \mathbb{R}^3$$

then the probability

$$P\{g_D \text{ intersects the needle } v\},$$

considered as a function of the needle $v \subset D$, is in fact always a continuous linearly-additive pseudometric on D (see §1.7).

Remarkably, combinatorial integral geometry has provided (see [3], chapter 6) tools for proving that the converse is also true. This amounts to a combinatorial solution of the fourth of the D. Hilbert's famous problems, see the appendix by Baddeley in [3].

The first person to realize the possibilities of integration in the space $\mathbb{G}$ (and other spaces of geometrical elements) as a useful tool for geometrical investigation was Blaschke, who proposed the title 'integral geometry' as an appropriate name for the whole topic. But it was J. J. Sylvester who first glimpsed in 1891 (see [23]) the existence of decompositions (5.1.1) and (5.1.3) in an attempt to solve a problem, which in accordance with our above remarks may be considered to be a direct generalization of the classical Buffon problem and later became known as the 'Buffon – Sylvester problem'. It is as follows (compare with §5.4).

In the plane, n needles $v_1, \ldots, v_n$ are fixed in a general position. What is the invariant measure of the sets

$$\bigcap_1^n [v_i] \qquad \text{and} \qquad \bigcup_1^n [v_i],$$

where

$$[v] = \{g \in \mathbb{G} : g \text{ separates the endpoints of } v\}?$$

Sylvester's result was that the invariant measures in question 'become Diophantine linear functions of the sides of the complete $2n$-gonal figure of which the n pairs of extremities of the needles are the angles'. Of course, this was a loose expression of (5.4.1) and (5.4.2).

Neither in [23] nor later was any practical algorithm proposed for finding the corresponding integers, and perhaps this is why the whole problem has been somewhat neglected.

Finally the problem received a very simple solution, first given in [28]. Soon it became clear that Sylvester's Diophantine decomposition principle is at the source of a potentially vast and fruitful theory. The elements for the planar case have been expounded in the preceding sections.

The theory in many dimensions has an interesting relation to a question posed by Radon. He considered n points $\mathscr{P}_i \subset \mathbb{R}^d$ in a general position and tried to calculate the number of different partition of $\{\mathscr{P}_i\}$ induced by hyperplanes.

The answer obtained by many authors (Harding [24], Schläfli [25], Watson [26]) was that this number equals

$$\sum_{K=0}^{d} \binom{n-1}{K}.$$

If d is even, the above obviously equals

$$\binom{n}{2} + \binom{n}{4} + \cdots + \binom{n}{d},$$

that is the number of odd-dimensional simplices Θ_s, which have points from $\{\mathscr{P}_i\}$ for their vertices.

Let $[\Theta]$ be the set of hyperplanes hitting the simplex Θ. Clearly for every s

$$[\Theta_s] = \bigcup \alpha_r, \tag{5.12.1}$$

where α_r are sets of equivalent hyperplanes: we call two hyperplanes equivalent if they induce the same partition of $\{\mathscr{P}_i\}$. The relation (5.12.1) ignores bundles of hyperplanes through the points $\mathscr{P}_i$. For any measure m in the space of hyperplanes for which $m(\text{any bundle}) = 0$, we have

$$m([\Theta_s]) = \sum \delta_{sr} m(\alpha_r).$$

There is the fundamental result [29]: *the matrix $\|\delta_{rs}\|$ is square if d is even, and in this case it can be inverted.* From the resulting expression

$$m(\alpha_r) = \sum c_{rs} m([\Theta_s]) \tag{5.12.2}$$

follows the existence of similar decompositions for *every* element $B \in r\{\mathscr{P}_i\}$, the 'Radon ring' generated by the bounded atoms α_r. An algorithm for calculating c_{rs} for $d = 4$ is given in [3].

The decompositions (5.12.2) have elegant implications in the geometry of the many-dimensional non-Euclidean elliptic spaces. The approach to these questions is converse to the one in §5.3: we reinterpret the facts concerning hyperplanes as facts of geometry in ellipitcal spaces.

This yields a 'combinatorial' version of the Gauss–Bonnet formula for convex polyhedrons in many dimensional elliptical spaces (see [27]). It gives the volume in terms of the openings of certain angles associated with the polyhedron (only angles having even-dimensional edges participate).

The decomposition (5.12.2) does not survive for odd-dimensional Euclidean spaces. The situation for $\mathbb{E}$ (planes in $\mathbb{R}^3$) has been considered in detail in §5.5. The basic facts of the resulting theory of flag representations of convex bodies in $\mathbb{R}^3$ were first published by the author in [58]. See also [45], [62].

We also mention that the decompositions (5.4.1) served in [3] as a starting point for the derivation of various geometrical identities and inequalities concerning both convex and non-convex domains in $\mathbb{R}^2$ and $\mathbb{R}^3$. We give an example of such a derivation in §6.9.

In the approach to the stochastic geometry problems used in chapters 9 and 10 of [3], these identities have been crucial. However, in chapter 10 of this book we treat similar problems by a direct new method of averaging the basic decompositions of (5.1.3) or (5.4.3).

6

Basic integrals

This chapter presents several integrations aimed on the one hand at complementing the preceding material and, on the other, at providing tools for the study of geometrical processes in the chapters to come.

The integrals in the first few sections refer to what can be called 'translation-invariant integral geometry': they generalize the basic formulae of classical integral geometry (as presented by Blaschke in [15], say). The section on translational analysis of discrete point sets, as well as a detailed treatment of an extremal property of the uniform distribution in the space of directions in §6.7, also belongs here.

We also include here some integrals which belong to Euclidean motion-invariant integral geometry. They are related to the analysis of sets of segments by means of moving test segments, Pleijel identities, and related isoperimetric inequalities.

We complete the chapter with the calculation of some integrals in terms of elementary functions. They refer to random chord length distribution for convex polygons as well as to triangular shapes.

6.1 Integrating the number of intersections

Let μ_β be a $\mathbb{T}_2$-invariant measure on $\mathbb{G}$ whose rose of directions is a δ-measure concentrated on a direction β, i.e.

$$\mu_\beta(\mathrm{d}g) = \mathrm{d}p \, \delta_\beta(\mathrm{d}\varphi).$$

Let $\{\delta_i\}$ be a collection of line segments in $\mathbb{R}^2$, and let $|\delta_i|$ and α_i be the length and the direction, respectively, of δ_i. It follows (see §2.10) that

$$\int I_{[\delta_i]}(g)\mu_\beta(\mathrm{d}g) = |\sin(\beta - \alpha_i)|\,|\delta_i|.$$

Summation over i yields

$$\int n(g)\mu_\beta(dg) = L \cdot \sum_i |\sin(\beta - \alpha_i)|p_i, \qquad (6.1.1)$$

where

$L = \sum |\delta_i|$ is the total length of the segments;

$p_i = |\delta_i| \cdot L^{-1}$ is the probability that a random point dropped on $\bigcup \delta_i$ (with uniform distribution on this set) will lie on a segment with orientation α_i;

$n(g) = \sum I_{\{\delta_i\}}(g)$ is the number of intersections of the line $g \in \mathbb{G}$ with the segments.

We call the integral

$$z\left(\beta + \frac{\pi}{2}\right) = \int n(g)\mu_\beta(dg) \qquad (6.1.2)$$

the multiprojection of the set $\{\delta_i\}$ on the line perpendicular to the direction β: each point p on this line is counted with multiplicity equal to the number of points from $\bigcup \delta_i$ which are projected into p.

By using (6.1.2) we extend the notion of multiprojection to that for rather general curves $\mathscr{L}$ on $\mathbb{R}^2$: it is enough to define $n(g)$ to be the number of intersections of $\mathscr{L}$ with the line g. By taking the appropriate limit, (6.1.1) can be generalized to

$$z\left(\beta + \frac{\pi}{2}\right) = L \int |\sin(\beta - \alpha)|\mathbf{p}(d\alpha) \qquad (6.1.3)$$

valid for rather general curves $\mathscr{L}$ having finite length L. Here $\mathbf{p}(d\alpha)$ is a probability distribution on $\mathbf{E}_1$, namely the distribution of the curve element direction at a point dropped at random (with uniform distribution) on $\mathscr{L}$.

Similar expressions can be obtained for multiprojections of curves or surfaces in $\mathbb{R}^3$. Let us first consider the case of a curve $\mathscr{L}$ in $\mathbb{R}^3$. We consider its perpendicular projection on a line in $\mathbb{R}^3$ which has a direction $\omega \in \mathbf{E}_2$. Let

$$n(e) = n(\omega, p)$$

be the number of intersections of the curve $\mathscr{L}$ with the plane $e \in \mathbb{E}$ having coordinates (ω, p), see §2.7. Otherwise stated, $n(\omega, p)$ is the multiplicity of the projection at p. By definition, the multiprojection of $\mathscr{L} \subset \mathbb{R}^3$ on a line with direction ω is the integral

$$z_1(\omega) = \int n(\omega, p) \, dp.$$

Applying the remarks of §2.10 and acting in the same way as before we find

$$z_1(\omega) = L \int |\cos(\omega, \xi)|\mathbf{p}(d\xi) \qquad (6.1.4)$$

first for broken lines in $\mathbb{R}^3$ and then for general curves of finite length L. Clearly

$p(d\xi)$ may be interpreted as the probability distribution of the direction of the curve element at a point dropped at random (with uniform distribution) on $\mathscr{L}$. Thus p is a probability distribution on E_2. Let $\mathscr{S}$ be a surface in $\mathbb{R}^3$. We denote by

$$n(\gamma) = n(\omega, \mathscr{P})$$

the number of intersections of $\mathscr{S}$ with a line $\gamma \in \Gamma$ having coordinates $(\omega, \mathscr{P})$, see §2.6. Equivalently $n(\omega, \mathscr{P})$ can be defined as the multiplicity of the projection at $\mathscr{P}$ on a plane having fixed normal direction ω. Now the multiprojection is given by the integral

$$z_2(\omega) = \int n(\omega, \mathscr{P}) \, d\mathscr{P}.$$

Using the remarks of §2.10 we derive the equation

$$z_2(\omega) = S \int |\cos(\omega, \xi)| \, p(d\xi) \tag{6.1.5}$$

first for finite unions of flats in $\mathbb{R}^3$ and then, by a natural passage to the limit, for rather general surfaces of finite area S. Here p is the probability distribution of the direction normal to $\mathscr{S}$ at a point dropped at random (with uniform distribution) on $\mathscr{S}$. Again p is a probability distribution on E_2.

The integrals in (6.1.3)–(6.1.5) are of the same general form as in (2.10.1) or (2.10.2) but now they are produced not by measures or convex domains but by rather arbitrary curves or surfaces. It is in the spirit of stereology to ask what can be inferred about such objects in terms of their multiprojections given as functions of directions.

The first thing that we conclude is that we can at most hope for the reconstruction of the probabilities p together with quantities L or S. Another conclusion is that multiprojections coincide with breadth functions of certain convex domains (zonoids if we are in $\mathbb{R}^3$). The solution for the planar case was described in §2.11; we discuss the spatial case in the next section. Using the formulae of §§3.6, 3.9 and 3.11 we find that

$$\int z(\beta) \, d\beta = \int n(g) \, dg = 2L, \tag{6.1.6}$$

$$\int z_1(\omega) \, d\omega = \int n(e) \, de = \pi L, \tag{6.1.7}$$

$$\int z_2(\omega) \, d\omega = \int n(\gamma) \, d\gamma = \pi S. \tag{6.1.8}$$

On the right-hand sides are classical integrals of the numbers of intersections with respect to M-invariant measures and their well known values (see [16] and [2]). Usually the above values are obtained by applying (3.7.3), (3.12.1) and (3.10.1). We have for instance

$$\int I_{[\delta_i]}(g)\, \mathrm{d}g = \int_0^{|\delta_i|} \int_0^\pi \sin\psi\, \mathrm{d}x\, \mathrm{d}\psi = 2|\delta_i|,$$

and from this (6.1.6) follows (first for broken lines by summation).

There are important corollaries of (6.1.6) and (6.1.8) referring to the case in which the curve or the surface in question happens to be the boundary ∂D of a convex domain D in $\mathbb{R}^2$ or $\mathbb{R}^3$. In such a case we have in (6.1.6)

$$n(g) = 2 \text{ if } g \text{ hits } D$$
$$= 0 \text{ otherwise,}$$

and the integral reduces to two times the measure of the set

$$[D] = \{g \in \mathbb{G} : g \text{ hits } D\}.$$

Hence for the $\mathbb{M}_2$-invariant measure μ_0 on $\mathbb{G}$ we have

$$\mu_0([D]) = |\partial D| = \text{the length of the perimeter of } D. \tag{6.1.9}$$

Similarly for the $\mathbb{M}_3$-invariant measure μ_0 in the space Γ we find

$$\mu_0(\gamma \in \overline{\Gamma} : \gamma \text{ hits a convex } D) = \frac{\pi}{2} \text{ times the surface area of } D. \tag{6.1.10}$$

In the case of a polyhedral convex $D \subset \mathbb{R}^3$ the $\mathbb{M}_3$-invariant measure of the set

$$\{e \in \mathbb{E} : e \text{ hits } D\}$$

has been found in §5.6, and for smooth D in §5.11.

6.2 The zonoid equation

It follows from the remarks of the previous section that certain problems in the field of 'reconstruction from projections' are mathematically equivalent to the problem of solving the equation

$$b(\xi) = \int |\cos \widehat{\xi\Omega}|\, m(\mathrm{d}\Omega). \tag{6.2.1}$$

This requires the recovery of the measure m defined on the space $\mathbf{E}_2$ in terms of the function $b(\xi)$ defined on the same space. The existing work on (6.2.1) and its generalization to higher dimensions is considerable and is reviewed in [7]. Here we consider some basic facts only.

I It is natural to consider the solutions of (6.2.1) in the class of *signed measures*. Within this class we have the *uniqueness result*: if $b(\xi) \equiv 0$ then (6.2.1) has only one solution, namely $m \equiv 0$.

We have seen in §2.12 that an m which has a density assuming both positive *and negative* values can produce a positive $b(\xi)$. In such cases $b(\xi)$ always happens to be a breadth function of some bounded centrally-symmetrical convex body in $\mathbb{R}^3$.

Do breadth functions always admit the representation (6.2.1) by means of signed measures? This question has an affirmative answer for sufficiently smooth bodies. If additionally we require that the solution be a measure (rather then a signed measure) then the body should be a *zonoid*, see §§1.7 and 5.10. Thus the problem of the description of zonoids in terms of their $b(\xi)$ values is essentially the problem of the description of the solvability domain of (6.2.1) in the class of measures in our usual sense. Many questions are still without answers.

II Under the assumptions of smoothness of $b(\xi)$ and the existence of the density

$$m(d\Omega) = u(\Omega) \, d\Omega$$

a *formal* solution of (6.2.1) can be based on the remarks of §2.12. Equation (2.12.9) expresses in terms of $b(\xi)$ the integrals of the density function $u(\Omega)$ extended over geodesics in the space $\mathbf{E}_2$. We face the problem first considered by Funk [5]: we know the integrals of a symmetrical function defined on $\mathbb{S}_2$ extended over great circles; the problem is to find the function. As shown by Funk, this problem is easily reduced to the classical Abel equation.

Another approach can be based upon the decomposition of $b(\xi)$ in a series using spherical functions. Here we have the advantage that spherical functions are eigenfunctions of both operators

$$L_1 H = 2H + \Delta_2 H$$

and

$$L_2 H = \int_{\langle \Omega \rangle} H \, d\Phi.$$

Thus they are eigenfunctions of the product $L_2^{-1} L_1$, see [6]. Since the corresponding eigenvalues are known (see any handbook on spherical functions), we can *formally* transform the series for $b(\xi)$ into a series for u. As a rule, the convergence rate of the series for u is worse than that for $b(\xi)$. This is the cause of the substantial complications in numerical computation. Here also we have many unanswered questions.

6.3 Integrating the Lebesgue measure of the intersection set

Taking the Lebesgue measure of the intersection set can be preferable to counting the number of points.

For instance, let $\mathscr{F}$ be a set on the plane and let $L_1(g \cap \mathscr{F})$ be the linear Lebesgue measure (total length) of the set $g \cap \mathscr{F}$. For any $\mathbb{T}_2$-invariant measure μ on the space $\mathbb{G}$ we have

$$\int L_1(g \cap \mathscr{F}) \mu(dg) = c_1 L_2(\mathscr{F}), \tag{6.3.1}$$

where $L_2(\mathscr{F})$ is the area (two-dimensional Lebesgue measure) of $\mathscr{F}$.

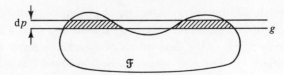

Figure 6.3.1 The shaded area is dS

To prove (6.3.1) we factorize μ as in §2.9 and use the fact that g and dp are perpendicular; therefore

$$L_1(g \cap \mathscr{F}) \cdot \mathrm{d}p = \mathrm{d}S,$$

an area element, see fig. 6.3.1.

For each direction φ we have the constant value of the integral

$$\int L_1(g \cap \mathscr{F}) \, \mathrm{d}p = L_2(\mathscr{F}),$$

hence (6.3.1). The constant c_1 depends only on the measure factor m and equals the total measure of m:

$$c_1 = m(\mathbf{E}_1).$$

For a set $\mathscr{F} \subset \mathbb{R}^3$ we have two analogs of (6.3.1). We can integrate the linear Lebesgue measure $L_1(\gamma \cap \mathscr{F})$ of the intersection set with respect to a $\mathbb{T}_3$-invariant measure μ in the space Γ. Using the factorization of §2.9 and the fact that $L_1(\gamma \cap \mathscr{F}) \, \mathrm{d}\mathscr{P}$ is a volume element, we find that

$$\int L_1(\gamma \cap \mathscr{F}) \mu(\mathrm{d}\gamma) = c_2 L_3(\mathscr{F}). \qquad (6.3.2)$$

Here

$$c_2 = m(\mathbf{E}_2)$$

where m is the measure determined by the factorization.

Let $L_2(e \cap \mathscr{F})$ be the planar Lebesgue measure of the trace on $\mathscr{F}$ on a plane $e \in \mathbb{E}$. Using the factorization of §2.9 and the fact that $L_2(e \cap \mathscr{F}) \, \mathrm{d}p$ is a volume element, we find that for every $\mathbb{T}_3$-invariant measure μ on $\mathbb{E}$

$$\int L_2(e \cap \mathscr{F}) \mu(\mathrm{d}e) = c_3 L_3(\mathscr{F}). \qquad (6.3.3)$$

Here

$$c_3 = m(\mathbf{E}_2),$$

where m is the measure determined by the factorization. The above equations generalize the classical results of integral geometry which refer to the cases of $\mathbb{M}$-invariant measures in the spaces $\mathbb{G}$, Γ and $\mathbb{E}$, where under the normalizations adopted in chapter 3 we have

$$c_1 = \pi, \qquad c_2 = 2\pi, \qquad c_3 = 2\pi.$$

In contrast to this, the following result referring to the $\mathbb{M}_3$-invariant measure μ_0 on $\mathbb{E}$ cannot be easily generalized to $\mathbb{T}_3$-invariant cases.

Let $\mathscr{S}$ be a *surface* in $\mathbb{R}^3$ with area S. Let $l(e)$ be the length of the curve $\mathscr{S} \cap e, e \in \mathbb{E}$. We have

$$\int l(e)\, \mathrm{d}e = \pi^2 S. \tag{6.3.4}$$

To prove (6.3.4) we integrate (6.1.8) over the circle $\mathbb{S}_1$ with uniform density $\mathrm{d}\varphi$:

$$\iint n(\gamma)\, \mathrm{d}\gamma\, \mathrm{d}\varphi = 2\pi^2 S.$$

From (3.13.6) and (6.1.6) we conclude that

$$\iint n(\gamma)\, \mathrm{d}\gamma\, \mathrm{d}\varphi = 2 \int l(e)\, \mathrm{d}e,$$

hence (6.3.4).

6.4 Vertical windows and shift-invariance

We will consider a pair of 'vertical windows' v_1 and v_2 as shown in fig. 6.4.1. Let a line g_0 be fixed on the plane. We consider the Haar measure of the set

$$B = \{M \in \mathbb{M}_2 : Mg_0 \in [v_1] \cap [v_2] \text{ and } MO \in b(r, 0)\},$$

where MX is the image of the set X under the notion M and $b(r, O)$ is the disc of radius r centered at O. It is natural to use the formula (3.13.3). For this purpose it is necessary to choose a directed unit segment $\delta_0 \subset \mathbb{R}^2$ and identify M with $M\delta_0$. Then g and t in (3.13.3) denote the line which carries $M\delta_0$ and the shift of $M\delta_0$ along g. We stress that by the properties of a Haar measure (3.13.3) remains valid for any choice of δ_0. We now choose δ_0 as shown in fig. 6.4.1 and then find

$$H(B) = \int_{[v_1] \cap [v_2]} \mathrm{d}g \int_{B_g} \mathrm{d}t$$

where H denotes the Haar measure and

$$B_g = \{M \in \mathbb{M}_2 : Mg_0 = g \text{ and } MO \in b(r, O)\}.$$

This expression makes clear the asymptotic behavior of $H(B)$ when we let the

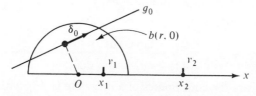

Figure 6.4.1 The segments v_1 and v_2 are perpendicular to the x axis. The source of δ_0 is the foot of the perpendicular from O onto g_0, $d = |x_1 - x_2|$, $l = |v_1| = |v_2|$.

length l of the windows tend to zero. Under this assumption the set $[v_1] \cap [v_2]$ shrinks down to the x-axis for every value of x_1, x_2.

By the integral mean theorem we get

$$H(B) = \left(\int_{B_{g^*}} dt \right) \cdot \left(\int_{[v_1] \cap [v_2]} dg \right)$$

where g^* is some line from $[v_1] \cap [v_2]$. From (3.7.4) we find that

$$\int_{[v_1] \cap [v_2]} dg = \frac{l^2}{d} + o(l^2)$$

and

$$\lim_{l \to 0} \int_{B_{g^*}} dt = \int_{B_{Ox}} dt = 2|\chi(g_0)|$$

where B_{Ox} corresponds to the x-axis, $\chi(g_0) = g_0 \cap b(r, O)$.

The last equality follows from the possibility of identifying B_{Ox} with two chords of $b(r, O)$ parallel to the x-axis whose lengths are equal to $|\chi(g_0)|$ (the locus of MO when $M \in B_{Ox}$).

We gather from the above that

$$H(B) = \frac{2|\chi(g_0)|}{d} l^2 + o(l^2),$$

where d is the distance between the windows. The important feature of this result is that

$$\lim_{l \to 0} l^{-2} H(B)$$

is invariant with respect to horizontal rigid shifts of the windows, i.e. it depends on x_1 and x_2 only through $d = |x_1 - x_2|$. We will encounter a version of this effect in (6.8.10) in the analysis of systems of lines.

6.5 Vertical windows and a pair of non-parallel lines

Let g_1 and g_2 be two lines which intersect at a point and form an angle of opening α_0. We again consider the vertical windows v_1 and v_2 of the previous section but now the problem is to find the Haar measure H of the set

$$B^* = \{ M \in \mathbb{M}_2 : Mg_1 \text{ hits } v_1, Mg_2 \text{ hits } v_2 \}.$$

There is an elegant way of calculating $H(B^*)$ which is based upon the identities (3.7.3) and (3.15.1). We write

$$dg_1 = \sin \psi_1 \, dy_1 \, d\psi_1$$
$$dg_2 = \sin \psi_2 \, dy_2 \, d\psi_2,$$

where y_i and ψ_i are the coordinates of a line as in (3.7.3), with v_i taken to be the reference line. Let

$$\alpha = \alpha(\psi_1, \psi_2)$$

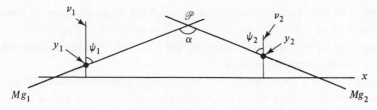

Figure 6.5.1

be the angle between the lines y_1, ψ_1 and y_2, ψ_2 (see fig. 6.5.1). By multiplication we get

$$f(\alpha)\,dg_1\,dg_2 = f(\alpha)\sin\psi_1\sin\psi_2\,dy_1\,dy_2\,d\psi_1\,d\psi_2$$

(where f is some function) or, using (3.15.1),

$$f(\alpha)\,dM\sin\alpha\,d\alpha = f(\alpha)\sin\psi_1\sin\psi_2\,dy_1\,dy_2\,d\psi_1\,d\psi_2. \qquad (6.5.1)$$

We can formally put in this equation

$$f(\alpha) = \delta_{\alpha_0}(\alpha)(\sin\alpha)^{-1}, \qquad (6.5.2)$$

where $\delta_{\alpha_0}(\alpha)$ is Dirac's δ-function concentrated on α_0. By the usual rules of operating δ-functions

$$\int_{B^*}\int \delta_{\alpha_0}(\alpha)\,dM\,d\alpha = H(B^*).$$

Therefore we can find $H(B^*)$ by integrating the left-hand side of (6.5.1) with f given by (6.5.2).

Integration by $dy_1\,dy_2$ yields

$$H(B^*) = 2l^2 \iint \delta_{\alpha_0}(\alpha)(\sin\alpha)^{-1}\sin\psi_1\sin\psi_2\,d\psi_1\,d\psi_2.$$

The coefficient 2 appeared because the domain of integration $(0, 1) \times (0, 1) \times (0, \pi) \times (0, \pi)$ corresponds to a half of B^*. This follows from the observation that after rotation by π of the lines Mg_1 and Mg_2 on fig. 6.5.1 around the point $\mathscr{P}$ we get the same values of y_1, y_2, ψ_1, ψ_2.

One more integration can be performed due to the presence of the δ-factor after we replace the variables ψ_1 and ψ_2 by ψ_1 and α. Since always

$$|\sin\psi_2| = |\sin(\psi_1 - \alpha)|$$

we get

$$H(B^*) = 2l^2(\sin\alpha_0)^{-1}\int_0^\pi \sin\psi\,|\sin(\psi - \alpha_0)|\,d\psi. \qquad (6.5.3)$$

A straightforward integration then yields the final result

$$H(B^*) = 2l^2\left[\left(\frac{\pi}{2} - \alpha_0\right)\cot\alpha_0 + 1\right]. \qquad (6.5.4)$$

Remark The above calculations remain valid for arbitrary pairs of *parallel* windows (not necessarily sides of a rectangle as in fig. 10.1.1). In particular the windows can lie on a line and even coincide (since $H(B^*)$ does not depend on the distance between the windows).

This provides a check for (6.5.4) based upon a result of Santalo quoted below in (6.8.4). Adding up Santalo's expressions for $\psi = \alpha_0$ and $\psi = \pi - \alpha_0$ yields (6.5.4). If the windows are not parallel, then (6.5.3) becomes

$$H(B^*) = 2l^2(\sin \alpha_0)^{-1} \int_0^\pi \sin \psi |\sin(\psi - \alpha_0 - \beta)| \, d\psi, \qquad (6.5.5)$$

where β is the angle between the windows. Expression of this integral in terms of elementary functions is an easy modification of (6.5.4).

There is an asymptotic counterpart of (6.5.3) for pairs of non-parallel segments. Let s_1 and s_2 be two such segments fixed in $\mathbb{R}^2$. The problem is to find the Haar measure of the set

$$B^{**} = \{M \in \mathbb{M}_2 : Ms_1 \text{ hits } v_1, Ms_2 \text{ hits } v_2\}.$$

We can think of s_1 and s_2 as being situated on the lines g_1 and g_2 in fig. 6.5.1. Then $H(B^{**})$ can be found by integrating the right-hand side of (6.5.1) with $f(\alpha)$ as in (6.5.2):

$$H(B^{**}) =$$

$$(\sin \alpha_0)^{-1} \int_0^l dy_1 \int_0^l dy_2 \int_0^\pi \sum_{i=1,2} I_{B^{**}}(y_1, y_2, \psi, i) \sin \psi |\sin(\psi - \alpha_0)| \, d\psi$$

Here the variables y_1, y_2, ψ, i determine a motion M, in particular the two values of i correspond to two rotations around the point $\mathscr{P}$ (see fig. 6.5.1) which leaves the lines Mg_1, Mg_2 intact.

From this we easily establish the existence of the limit

$$\lim_{l \to 0} l^{-2} \sin \alpha_0 H(B^{**}) = \int_0^\pi \sum_{i=1,2} I_{B^{**}}(0, 0, \psi, i) \sin \psi |\sin(\psi - \alpha_0)| \, d\psi. \qquad (6.5.6)$$

This expression is a bounded function of α_0 and other $\mathbb{M}_2$-invariant parameters which determine the segments s_1 and s_2.

Similar results can be obtained for a pair of discs of unit radius.

On the set of discs whose circumference hits a window v_i at one point we introduce the coordinates

y_i, the usual one-dimensional coordinate of the intersection point on v_i, and ψ_i, the angle at which the intersection at y_i occurs.

Because each disc is determined by its center $Q \in \mathbb{R}^2$, the planar Lebesgue measure dQ determines a measure in the space of discs. On these sets we have

$$dQ_i = \sin \psi_i \, d\psi_i \, dy_i.$$

By using (3.14.1) and applying the above method of δ-functions we come to the following result.

Given two discs b_1 and b_2 on $\mathbb{R}^2$, the Haar measure of the set

$$\{M \in \mathbb{M}_2 : Mb_1 \text{ hits } v_1, Mb_2 \text{ hits } v_2\}$$

equals $cl^2 + o(l^2)$ as $l \to 0$. The constant c depends on the distance between the windows and the distance ρ between the disc centers; ρ/l^2 times this measure is a bounded function.

Such results can be used in the study of probabilities of hitting of pairs of vertical windows in geometrical processes. The role of these probabilities will become clear from the contents of chapter 10.

6.6 Translational analysis of realizations

Integration with respect to Haar measures on groups plays a significant role in many problems of stochastic geometry. In this section we consider several integrals for the group $\mathbb{T}_n$ of translations of $\mathbb{R}^n$. We apply them in the theory of $\mathbb{T}_n$-invariant point processes in $\mathbb{R}^n$ (in chapter 8).

We denote the Haar measure on $\mathbb{T}_n$ either by h_n or, if under the sign of integral, by dt.

We begin with the following simplest result.

Assume that a point $\mathscr{P} \in \mathbb{R}^n$ and a domain $D \subset \mathbb{R}^n$ are fixed. Then

$$h_n(\{t \in \mathbb{T}_n : \mathscr{P} \in tD\}) = L_n(D), \tag{6.6.1}$$

where L_n is the Lebesgue measure in $\mathbb{R}^n$.

Proof We can assume that $\mathscr{P} = O$, the origin in $\mathbb{R}^n$. If we represent shifts by points in $\mathbb{R}^n$ according to the map

$$t \to tO$$

then the image of the set $\{t \in \mathbb{T}_n : O \in tD\}$ will be a domain $D' \subset \mathbb{R}^n$ obtained from D by means of reflection through the point O. Thus (6.6.1) follows from the fact that

$$L_n(D') = L_n(D),$$

and because the image of h_n under the above map is just L_n.

If we take

$$D = b(r, O),$$

the ball of radius r centered at O, then the image of the set $\{t \in \mathbb{T}_n : \mathscr{P} \in tD\}$ will be $b(r, \mathscr{P})$.

Now let a finite collection of points $\{\mathscr{P}_i\}$ be fixed in $\mathbb{R}^n$. We define

$$N(t) = \text{the number of points from } \{\mathscr{P}_i\} \text{ within } tD.$$

We have

$$N(t) = \sum I_{\mathscr{P}_i}(tD), \tag{6.6.2}$$

where

$$I_{\mathscr{P}}(tD) = 1 \text{ if } \mathscr{P} \in tD$$
$$= 0 \text{ otherwise.}$$

Since (6.6.1) can be written as

$$\int I_{\mathscr{P}}(tD)\, dt = L_n(D),$$

integration of (6.6.2) yields

$$\int N(t)\, dt = s \cdot L_n(D), \tag{6.6.3}$$

where s is the total number of points in $\{\mathscr{P}_i\}$.

As the number of points in $\{\mathscr{P}_i\}$ tends to infinity, so both sides of (6.6.3) tend to infinity. However, a modification of (6.6.3) yields more information than this.

We will use the following notation:

$$m = \{\mathscr{P}_i\},$$
$$N(m, D) = \text{the number of points in } m \cap D,$$
$$\bar{b}(r, O) = \{t \in \mathbb{T}_n : tO \in b(r, O)\}.$$

We will assume that for every $r > 0$

$$N(m, b(r, O)) < \infty;$$

in other words that m is a 'realization' in the sense of §7.1.

Let us consider the integral

$$\int_{\bar{b}(r,O)} N(m, tD)\, dt.$$

An asymptotic expression for this integral can be found if we assume that the domain D is infinitesimal. For simplicity let us take

$$D = b(\varepsilon, O) \tag{6.6.4}$$

and assume that

$$\varepsilon \to 0.$$

If m has no points on the circumference of $b(r, O)$ then

$$\int_{\bar{b}(r,O)} N(m, tD)\, dt = L_n(D) \cdot N(m, b(r, O)) \tag{6.6.5}$$

for all sufficiently small values of ε.

Proof Let $\varepsilon_1 > 0$ be the minimal distance of the points from m to $\partial b(r, O)$ and ε_2 be the halved minimal distance between the pairs of points from $m \cap b(r, O)$.

For values $\varepsilon < \min \varepsilon_1,\, \varepsilon_2$ the function $N(m, tD)$ has the following simple structure on the set $\bar{b}(r, O)$:

$$N(m, tD) = 1 \qquad \text{within the small circles shown in fig. 6.6.1}$$
$$= 0 \qquad \text{elsewhere in } b(r, O). \tag{6.6.6}$$

Hence (6.6.5).

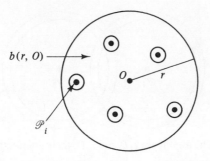

Figure 6.6.1 The radii of the small circles are ε; their centers are from the set
$$m \cap b(r, O)$$

Clearly (6.6.6) implies that for D, as in (6.6.4),

$$\lim(L_n(D))^{-1} \int_{\bar{b}(r,O)} I_1(m, tD)\, dt = N(m, b(r, O)) \qquad (6.6.7)$$

and

$$\lim(L_n(D))^{-1} \int_{\bar{b}(r,O)} \sum_{k=2}^{\infty} k I_k(m, tD)\, dt = 0. \qquad (6.6.8)$$

Here and in what follows

$$I_k(m, tD) = 1 \qquad \text{if the number of points in } m \cap tD \text{ equals } k$$
$$= 0 \qquad \text{otherwise.}$$

If we delete the condition that the domain D be infinitesimal (i.e. that D can be covered by a ball of infinitesimal radius) and retain only the requirement that

$$\lim L_n(D) = 0$$

then the last two relations do not necessarily hold. The sequence of rectangles in the plane

$$D = \{x, y : 0 < x < 1, \qquad 0 < y < \varepsilon\}, \qquad \varepsilon \to 0$$

can be an example: in this case (6.6.8) collapses if m possesses pairs of points parallel to the x-axis. Nevertheless both (6.6.7) and (6.6.8) remain valid for the sequence of annuli

$$D = b(r + \varepsilon, O) \backslash b(r, O), \qquad \varepsilon \to 0.$$

This follows from the fact that in this case for every two points $\mathscr{P}_1, \mathscr{P}_2 \in \mathbb{R}^n$

$$\{t \in \mathbb{T}_n : \mathscr{P}_1, \mathscr{P}_2 \in tD\} = t_1 D \cap t_2 D,$$

where

$$t_1 = O\vec{\mathscr{P}}_1, \qquad t_2 = O\vec{\mathscr{P}}_2,$$

since the Lebesgue measure of the above intersection set is $o(\varepsilon)$, see fig. 6.6.2.

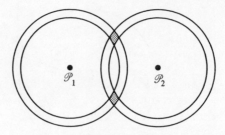

Figure 6.6.2 The annuli are $t_1 D$ and $t_2 D$; their intersections are shaded

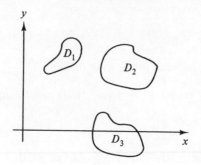

Figure 6.6.3

In the theory of point processes, yet another modification of (6.6.7) is of importance.

Let us fix a collection $D_1, \ldots, D_s$ of domains in $\mathbb{R}^n$ (see fig. 6.6.3), non-negative integers $k_1, \ldots, k_s$ and a 'realization' m. We consider the set

$$A = \bigcap_{i=1}^{s} \{t \in \mathbb{T}_n : N(m, tD_i) = k_i\}$$

and as usual

$$I_A(t) = 1 \qquad \text{if } t \in A$$
$$= 0 \qquad \text{otherwise.}$$

Let us denote by t_i the shift which sends a point $\mathscr{P}_i \in m$ into O:

$$t_i = \vec{\mathscr{P}_i O}.$$

If for a given $t > 0$ the following two conditions are satisfied:

(a) $m \cap \partial b(r, O)$ is an empty set;
(b) for every $\mathscr{P}_i \in m \cap b(r, O)$ the set $t_i m$ does not possess points on the boundaries of the domains $D_1, \ldots, D_s$;

then

$$\lim_{\varepsilon \to 0} (L_n(b(\varepsilon, O)))^{-1} \int_{\bar{b}(r,O)} I_1(m, tb(\varepsilon, O)) I_A(t) \, dt = \sum_{\mathscr{P}_i \in m \cap b(r,O)} I_A(t_i). \quad (6.6.9)$$

Proof By the remarks which led to (6.6.6), condition (a) implies

$$\int_{\bar{b}(r,O)} I_1(\textit{m}, tb(\varepsilon, O))I_A(t)\, dt = \sum_i \int_{\bar{b}(\varepsilon, -\mathscr{P}_i)} I_A(t)\, dt,$$

where the point $-\mathscr{P}_i$ corresponds to $\mathscr{P}_i$ under reflection through O.

Condition (b) implies that the function $I_A(t)$ has constant values within $b(\varepsilon, -\mathscr{P}_i)$ for sufficiently small values of ε, i.e.

$$I_A(t) \equiv I_A(t_i) \qquad \text{within} \qquad b(\varepsilon, -\mathscr{P}_i).$$

Hence,

$$\int_{\bar{b}(\varepsilon, -\mathscr{P}_i)} I_A(t)\, dt = L_n(b(\varepsilon, O)) \cdot I_A(t_i)$$

and (6.6.9) follows.

If condition (b) is violated then, in the limit of (6.6.9), terms depending on the properties of the boundaries of the domains D_i will also have to be included. We leave the details to the reader.

6.7 Integrals over product spaces

We start with several integrals from classical integral geometry, i.e. we refer to M-invariant measures and their products.

(1) The result of integration of (3.14.1) over the set $\{Q_1, Q_2 \in D\} = D \times D$, where D is a convex domain in $\mathbb{R}^2$, is as follows:

$$\int \chi^3\, dg = 3(L_2(D))^2. \tag{6.7.1}$$

Here $L_2(D)$ is the area of D, χ stands for the length of the chord $g \cap D$, and integration is over $\mathbb{G}$.

We obtain a similar result by integrating (3.14.3) and using (3.13.2), namely

$$\int \chi^4\, d\gamma = 6(L_3(D))^2, \tag{6.7.2}$$

where now D is a convex domain in $\mathbb{R}^3$, $L_3(D)$ is the volume of D, χ is the length of the chord $D \cap \gamma$, and integration is over Γ.

(2) Formula (3.15.1) is suitable for the calculation of the measure of the set

$$A_1 = \{(g_1, g_2) \in \mathbb{G} \times \mathbb{G} : \text{the point } g_1 \cap g_2 \text{ belongs to } D\},$$

where the domain D need not be convex. The result is

$$\int_{A_1} dg_1\, dg_2 = 2\pi L_2(D). \tag{6.7.3}$$

Similarly, integrating (3.15.2) we find that

$$\int_{A^2} de_1\, de_2 = (\pi^3/4)\|\partial D\|, \tag{6.7.4}$$

where the set A_2 is as follows:

$A_2 = \{(e, e_2) \in \mathbb{E} \times \mathbb{E} : \text{the line } e_1 \cap e_2 \text{ hits a convex domain } D \subset \mathbb{R}^3\}$.

The result (6.1.10) was used in (6.7.4), and $\|\partial D\|$ is the surface area of D.

Integration of (3.15.3) over the set

$A_3 = \{(e_1, e_2, e_3) \in \mathbb{E} \times \mathbb{E} \times \mathbb{E} : \text{the point } e_1 \cap e_2 \cap e_3 \text{ lies in a domain } D \subset \mathbb{R}^3\}$

yields

$$\int_{A_3} de_1 \, de_2 \, de_3 = \pi^4 L_3(D). \qquad (6.7.5)$$

The results (6.7.3)–(6.7.5) can be reinterpreted as special cases of more general relations which involve products of $\mathbb{T}$-invariant measures in the spaces $\mathbb{G}$ or $\mathbb{E}$. Let us look at this, restricting ourselves to measures in $\mathbb{G} \times \mathbb{G}$.

Let μ be a $\mathbb{T}_2$-invariant measure on $\mathbb{G}$. The image of the product measure $\mu \times \mu$ under the map

$$(g_1, g_2) \to \mathscr{P} = g_1 \cap g_2$$

is a $\mathbb{T}_2$-invariant measure in $\mathbb{R}^2$ and therefore coincides with $c \cdot L_2$, where L_2 is two-dimensional Lebesgue and c is some constant. This constant can be expressed in terms of the rose of directions $m(d\varphi)$ and of the rose of hits $\lambda(\varphi)$ which correspond to μ (see §2.10), namely

$$c = \int \lambda(\varphi) m(d\varphi).$$

Proof It is enough to find $(\mu \times \mu)(A_1)$ (see (6.7.3)). For the moment let the line $g_1 = (\varphi, p)$ be fixed and let $\chi(g_1)$ be the total length of $g_1 \cap D$. By the definition of $\lambda(\varphi)$ the μ-measure of the lines g_2 intersecting $g_1 \cap D$ equals $\lambda(\varphi) \cdot \chi(g_1)$.

In accordance with fig. 6.3.1 we find

$$(\mu \times \mu)(A_1) = \int \lambda(\varphi) \cdot \chi(g_1) \mu(dg_1)$$

$$= \int \lambda(\varphi) m(d\varphi) \int \chi(g_1) \, dp$$

$$= L_2(D) \cdot \int \lambda(\varphi) m(d\varphi). \qquad (6.7.6)$$

Because of (2.10.1) the constant c can also be written as a double integral:

$$c = \int_0^\pi \int_0^\pi |\sin(\varphi - \psi)| m(d\varphi) m(d\psi).$$

Let us restrict m to the class of *probability measures* on $(0, \pi)$. Then the above integral clearly remains bounded and it is natural to ask about its maximal value under this restriction. We show now that *for every probability measure m on $(0, \pi)$*

$$c \leqslant \frac{2}{\pi}$$

and the equality holds only for the uniform m, i.e. in the case

$$m(\mathrm{d}\varphi) = \frac{\mathrm{d}\varphi}{\pi}.$$

This result is due to Davidson [59]. The proof we give now is due to Mecke. It admits generalizations to higher dimensions (see [53], [60], [61]). Since identically

$$\int_0^\pi \sin |\varphi - \psi|\, \mathrm{d}\psi \equiv 2,$$

for every probability measure m on $(0, \pi)$ we have

$$\int_0^\pi \int_0^\pi \sin |\varphi - \psi|\, m(\mathrm{d}\varphi) l(\mathrm{d}\psi) = \frac{2}{\pi},$$

where l stands for the uniform probability measure. Hence, writing $(l - m)$ for the corresponding signed measure

$$\int_0^\pi \int_0^\pi \sin |\varphi - \psi|\, (l - m)(\mathrm{d}\psi) \cdot (l - m)(\mathrm{d}\psi) = c - \frac{2}{\pi},$$

and it remains to prove that the last double integral is non-positive. This can be done using a special integral representation for the function $\sin |\varphi - \psi|$.

Fig. 6.7.1 shows the graphs of the functions

$$t = \sin^2 z \qquad \text{and} \qquad t = \sin^2(z - a) \tag{6.7.7}$$

and we have shaded the area under the graph of the function

$$t = \min[\sin^2 z, \sin^2(z - a)]. \tag{6.7.8}$$

The shaded area above the interval $(0, \pi)$ equals

$$2\int_{-a/2}^{(\pi-a)/2} \sin^2 z\, \mathrm{d}z = \frac{\pi}{2} - \sin a,$$

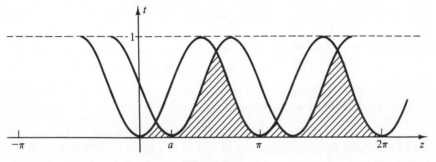

Figure 6.7.1

Because of periodicity, the shaded area above any interval of length π will have the same value. It follows that for every $x, y \in (0, \pi)$

$$\int_0^\pi \min[\sin^2(z - x), \sin^2(z - y)] \, dz \equiv \frac{\pi}{2} - \sin(|x - y|).$$

The above identity can be rewritten as follows:

$$\sin|x - y| = \frac{\pi}{2} - \int_0^1 \int_0^\pi I(t)[0, \sin^2(z - x)] I(t)[0, \sin^2(z - y)] \, dt \, dz,$$

where $I(t)[a, b]$ is the indicator function of the segment $[a, b]$. This is the desired representation for $\sin|x - y|$. Integrating it with respect to the signed measure $(l - m)$ yields

$$\int_0^\pi \int_0^\pi \sin|x - y| \cdot (l - m)(dx) \cdot (l - m)(dy)$$

$$= -\int_0^\pi \int_0^1 (l - m)^2 \{x : \sin^2(z - x) > t\} \, dz \, dt,$$

which is clearly non-positive. The uniqueness property of the uniform measure follows immediately from this.

Our last remark refers to general M_2-invariant measures on $G \times G$. Let μ be such a measure which satisfies the additional condition

μ(the set of all parallel or antiparallel pairs of lines) $= 0$.

A direct application of Haar factorization yields

$$\mu(dg_2 \, dg_2) = dM \, m(d\psi), \tag{6.7.9}$$

where we use the same notations as in §3.15.1, and m is some measure on the punctured circle

$$S_1 \setminus \{\text{the angles } \psi = 0 \text{ and } \psi = \pi\}.$$

We stress that now μ need not be a product measure, as was the case in (3.15.1).

There exist measures m on the punctured circle for which $dM \, m(d\psi)$ is not a measure on $G \times G$ in our usual sense since local finiteness can be violated. Using (6.5.5) we conclude easily that the condition

$$\int \sin^{-1} \psi m(d\psi) < \infty \tag{6.7.10}$$

is both necessary and sufficient to have

$$\mu(\{g_1, g_2\}) : g_1 \text{ hits } v_1, g_2 \text{ hits } v_2\}) < \infty$$

for any two line segments v_1 and v_2.

It is not difficult to see that the set

$$\{(g_1, g_2) : \text{both lines hit a square in } \mathbb{R}^2\}$$

can be covered by a finite number of sets of this type. Therefore (6.7.10) is a necessary and sufficient condition of local finiteness of the measures μ given by (6.7.9).

6.8 Kinematic analysis of realizations

The kinematic measure is a measure in the space of figures congruent to a given figure in $\mathbb{R}^n$. It corresponds to (and often coincides with) the Haar measure on the Euclidean group M_n (see §3.13). Integration with respect to the kinematic measure plays a significant role in integral geometry, for instance in the *Kinematische Hauptformel* of Poincaré–Blaschke; see [2] for a complete account. In this section we concentrate mainly on those results which we shall apply later on in the context of geometrical processes.

I We start with the simplest versions of the Poincaré–Blaschke formula. Let $d\delta$ be the kinematic measure in the space Δ_2^* of *directed* line segments δ of fixed length $|\delta|$ (see §2.13).

For a curve $\mathscr{L} \subset \mathbb{R}^2$ and a line segment δ we define $n(\delta)$ to be the number of points of intersection of δ with $\mathscr{L}$. Then

$$\int n(\delta)\, d\delta = 4 \cdot |\delta| \text{ times the length of } \mathscr{L}. \tag{6.8.1}$$

Proof Since we have in this case

$$d\delta = dM,$$

the Haar measure on M_2, our assertion follows from (3.13.3), (6.1.6) and the one-dimensional version of (6.6.3), that is from

$$\int n((g, t))\, dt = n(g) \cdot |\delta|,$$

where $n(g)$ is the number of points in the set $g \cap \mathscr{L}$.

Let $\mathscr{S}$ be a surface and let v be a directed line segment in $\mathbb{R}^3$. We define $n(v)$ to be the number of points of intersection of v with $\mathscr{S}$. Then

$$\int n(v)\, dv = 4\pi^2 \cdot |v| \text{ times the area of } \mathscr{S}, \tag{6.8.2}$$

where dv stands for the kinematic measure in the space Δ_3^* (the proof is by (3.13.2), (6.1.8) and (6.6.3)).

Let $\mathbb{F}$ be the space of circular *flats* of fixed area $\|f\|$ in $\mathbb{R}^3$ (see §2.13,IV). A flat can be represented as

$$f = (e, Q),$$

where e is the plane carrying f and $Q \in \mathbb{R}^3$ determines the position of the center of f on the plane e.

The kinematic measure df has the representation (compare with (3.13.4))

$$df = 4\pi \cdot de\, dQ,$$

where de is the M_3-invariant measure in $\mathbb{E}$ and dQ is the planar Lebesgue measure.

Let $n(f)$ be the number of intersections of f with a curve $\mathscr{L}$ in $\mathbb{R}^3$. By means of (6.1.7) and (6.6.3) we establish that

$$\int n(f)\, df = 4\pi^2 \|f\| \text{ times the length of } \mathscr{L}. \tag{6.8.3}$$

II The previous results can be considered to be kinematic analogs of (6.6.3). There also exist kinematic analogs of the asymptotic results presented in §6.6. These refer to the cases where $|\delta|$, $|v|$ or $\|f\|$ tend to zero.

Let us consider the planar case. Our main interest will be in collections of linear segments rather than curves.

Let m be a collection of segments on $\mathbb{R}^2$ of finite total length L. On Δ_2^* we consider the function

$I_k(\delta) = I_k(\delta, m) = 1$ if the segment δ has exactly k intersection points with m

$\qquad\qquad\quad\, = 0$ otherwise.

In contrast with (6.6.6), the function $I_k(\delta)$ can be non-zero for arbitrarily small values of $|\delta|$, even for $k \geqslant 2$. For instance, if m consists of segments situated as shown in fig. 6.8.1 then $I_k(\delta)$ can equal 1 for all values of $|\delta|$ if $k \leqslant 6$. However, there can be significant differences in the order of magnitude of the integrals

$$\alpha_k = \int I_k(\delta)\, d\delta \qquad \text{as} \qquad |\delta| \to 0$$

for the cases $k = 1$ and $k > 1$.

Recall the well known result of Santalo [2] which gives the value of α_2 in the case m is a pair of rays (infinite in one direction) emerging from a *node N*. In this case

$$\alpha_2 = \frac{1}{2}[1 + (\pi - \psi)\cot \psi]\cdot|\delta|^2, \tag{6.8.4}$$

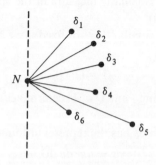

Figure 6.8.1 The segments $\delta_1, \ldots, \delta_6$ all lie in a halfplane bounded by the dotted line through N (the 'node').

where $0 < \psi < \pi$ is the angle between the rays. This result admits a number of successive generalizations.

If m consists of more than two infinite rays emerging from a node then

$$\alpha_k = C_k \cdot |\delta|^2. \tag{6.8.5}$$

The quadratic dependence on $|\delta|$ follows from purely qualitative considerations. To see this we represent the kinematic measure in the form

$$d\delta = dQ \, d\varphi,$$

where dQ is planar Lebesgue and $d\varphi$ is the angular measure on $\mathbb{S}_1$. For each value of φ, integration with respect to dQ gives the area of a domain in $\mathbb{R}^3$. By homothety this area equals $c_k(\varphi)|\delta|^2$, where $c_k(\varphi)$ does not depend on $|\delta|$. Clearly then

$$C_k = \int c_k(\varphi) \, d\varphi.$$

Also, the values of C_k can be found using (6.8.4) and applying the inclusion–exclusion rule. We leave the details to the reader who may consult [42].

The sequence

$$C_2, C_3, \ldots$$

is essentially finite; after a certain index k all C_k's are zero. The index of the last C_k which is non-zero can be called the *order* of a node. Nodes of the order two are shown in fig. 6.8.2; they appear in a stochastic context in chapter 7.

Equation (6.8.5) are no longer valid for all values of $|\delta|$ if the rays emerging from a node are of *finite length*. Yet they remain valid for sufficiently small values of $|\delta|$ (where the truncation of the rays does not affect the sets

$$\{\delta : I_k(\delta) = 1\} \qquad \text{for } k = 2, 3, \ldots.$$

From this remark and from (6.8.1) follows an asymptotic result valid for every m (a finite union of line segments of finite total length L), namely

$$\int I_k(\delta, m) \, d\delta = 4 \cdot |\delta| \cdot L + o(|\delta|) \qquad \text{if } k = 1$$

$$= |\delta|^2 \cdot \sum_i C_k(N_i) + o(|\delta|^2) \qquad \text{if } k > 1, \tag{6.8.6}$$

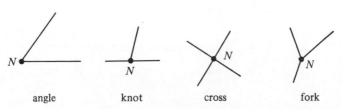

angle	knot	cross	fork

Figure 6.8.2 Nodes of order two

where the points N_i are the nodes of m (i.e. points where at least two linear segments meet at non-zero angles).

III What will happen to (6.8.6) if we allow m to possess curvilinear parts? The answer is that under appropriate smoothness conditions the nature of (6.8.6) will not change as far as in the expansion only the terms of order $|\delta|$ and $|\delta|^2$ are considered. We explain this using an example.

Let m consist of two intersecting circles m_1 and m_2 (see fig. 6.8.3). Then we have

$$I_2(\delta, m) = I_{11}(\delta, m) + I_{22}(\delta, m) + I_{12}(\delta, m),$$

where

$$I_{ii}(\delta, m) = 1 \qquad \text{if both intersections are with } m_i$$
$$= 0 \qquad \text{otherwise,}$$

and

$$I_{12}(\delta, m) = 1 \qquad \text{if one intersection is with } m_1 \text{ and the other with } m_2$$
$$= 0 \qquad \text{otherwise.}$$

In clear notation

$$\int I_{ii}(\delta, m)\, d\delta \leqslant \int I_2(\delta, m_i)\, d\delta = \frac{\pi}{6} \cdot \frac{|\delta|^3}{r_i} + o(|\delta|^3),$$

where r_i is the radius of the circle m_i. (In fact the latter integral can be calculated explicitly using (3.13.3)).

On the other hand,

$$\int I_{12}(\delta, m)\, d\delta = 2C_2(N_1) \cdot |\delta|^2 + o(|\delta|^2),$$

where $C_2(N_1)$ can be calculated using (6.8.4):

$$C_2(N_1) = 2 + (\pi - 2\psi)\cot\psi$$

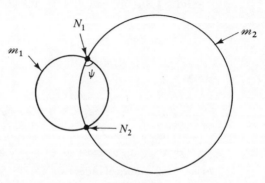

Figure 6.8.3

(the angle ψ is shown in fig. 6.8.3). Here we have used the following general principle.

If the linear rays emerging from a node N are replaced by smooth curvilinear arcs which emerge from N in the same directions and which do not create new nodes, then (6.8.5) is replaced by

$$\alpha_k = C_k|\delta|^2 + o(|\delta|^2)$$

with the same values of C_k.

The proof can be by integration of (3.13.3) over the corresponding difference sets.

From the above remarks (6.8.6) follows for $k = 2$. For $k = 3, 4$ we have

$$I_k(\delta, m) \leqslant I_{11}(\delta, m) + I_{22}(\delta, m)$$

and therefore

$$\int I_k(\delta, m) \, d\delta = o(|\delta|^2), \qquad k = 3, 4,$$

which also conforms with (6.8.6).

IV Equation (6.8.6) admits a generalization to the case where the participating set m need not be bounded. We will assume below that

the set $m \cap b(r, O)$ (as usual $b(r, O)$ is the disc of radius r centered at O) can be represented as a union of a finite number of line segments for every $r > 0$.

In other words, we assume that m is a realization of a segment process, see §7.13.

Let us choose r in such a way that there be no nodes of m on $\partial b(r, O)$ (the aim is to avoid certain 'boundary effects' which make the things more complicated).

The desired modification of (6.8.6) is as follows:

$$\lim_{|\delta| \to 0} |\delta|^{-1} \int_{\hat{b}(r, O)} I_1(\delta, m) \, d\delta = 4L(r)$$

$$\lim_{|\delta| \to 0} |\delta|^{-2} \int_{\hat{b}(r, O)} I_k(\delta, m) \, d\delta = \sum_{N_i \in b(r, O)} C_k(N_i) \qquad \text{if} \qquad k \geqslant 2,$$

(6.8.7)

where

$\hat{b}(r, O) = \{\delta \in \Delta_2^*; \text{ the origin of the directed segment } \delta \text{ belongs to } b(r, O)\}$,

$L(r) = $ the total length of $m \cap b(r, O)$.

Similar equations also hold for the sets m consisting of smooth curvilinear arcs (see the remarks of subsection III).

V Let m be from the class described in IV. Together with the number of points in $\delta \cap m$ we can consider the angles at which the intersections occur. The

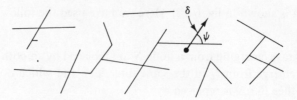

Figure 6.8.4. The segments with no arrow on them belong to m

situation is simplest if we have only one point in $\delta \cap m$, i.e. if δ belongs to the set

$$B_1 = \{\delta \in \Delta_2^* : I_1(\delta, m) = 1\}.$$

On B_1 the intersection angle $\psi = \psi(\delta)$ is a well-defined function, see fig. 6.8.4.

Let β be an arc (or, more generally, a Borel set) from $\mathbb{S}_1$ and let

$$I_\beta(\psi) = 1 \qquad \text{if } \psi \text{ belongs to } \beta$$
$$= 0 \qquad \text{otherwise.}$$

Then

$$\lim_{|\delta| \to 0} |\delta|^{-1} \int_{\hat{b}(r,O)} I_1(\delta, m) I_\beta(\psi)\, d\delta = L(r) \int_\beta |\sin \psi|\, d\psi. \qquad (6.8.8)$$

Proof We split the set $\hat{b}(r, O) \cap B_1$ into non-intersecting subsets so that in different subsets δ hits different segments from m. Let us denote the latter segments by $v_1, \ldots, v_s$. For every $\delta \in \hat{b}(r, O)$ excluding a set whose measure is $o(|\delta|)$ we have

$$I_1(\delta, m) = \sum_{i=1}^{s} I_1(\delta, v_i).$$

For $|\delta| \to 0$

$$\int_{\hat{b}(r,O)} I_1(\delta, v_i) I_\beta(\psi)\, d\delta = \int_{\Delta_2^*} I_1(\delta, v_i \cap b(r, O)) I_\beta(\psi)\, d\delta + o(|\delta|).$$

As follows from (3.13.3) and (3.7.3)

$$\int I_1(\delta, v) I_\beta(\psi)\, d\delta = |\delta| \cdot |v| \cdot \int I_\beta(\psi) |\sin \psi|\, d\psi$$

for every value of $|\delta|$ and for every segment $v \subset \mathbb{R}^2$. Equation (6.6.8) follows from these observations.

A set of segments on $\mathbb{R}^2$ is called a *mosaic* if it induces a partition of $\mathbb{R}^2$ into bounded convex polygons whose interiors do not pairwise intersect but the union of their closures gives the whole plane; the number of nodes in any disc $b(r, O)$ should be finite. A mosaic, m, cannot possess nodes of angle type, while other types shown in fig. 6.8.2 are not excluded.

Let m be a mosaic *possessing only nodes of order two.* Then on the set

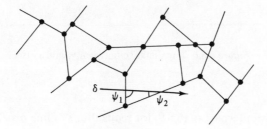

Figure 6.8.5. δ hits two segments from a mosaic which meet at a node

$$B_2 = \{\delta \in \Delta_2^* : I_2(\delta, m) = 1\}$$

we define the functions (angles) $\psi_1(\delta)$ and $\psi_2(\delta)$ as shown in fig. 6.8.5.

We choose r in such a way that there are no nodes of m on $\partial b(r, O)$ and consider the sets

$$B_{ij} = \{\delta \in \hat{b}(r, O) : \delta \text{ hits two segments } v_i, v_j \in m \text{ which meet at a node}\}.$$

If $|\delta|$ is sufficiently small then

$$B_2 \cap \hat{b}(r, O) = \bigcup B_{ij},$$

the sets in the above union are pairwise disjoint. For such values of $|\delta|$

$$|\pi - \psi_1 - \psi_2| = \Phi_{ij},$$

where Φ_{ij} is a constant on B_{ij} and equals the angle between the segments v_i and v_j. By means of (6.8.4) we find

$$\lim_{|\delta| \to 0} |\delta|^{-2} \int_{B_{ij}} \frac{2 \, d\delta}{1 + (\pi - |\pi - \psi_1 - \psi_2|) \cot |\pi - \psi_1 - \psi_2|} = 1$$

$$\lim_{|\delta| \to 0} |\delta|^{-2} \int_{B_{ij}} \frac{2 \cdot |\pi - \psi_1 - \psi_2| \, d\delta}{1 + (\pi - |\pi - \psi_1 - \psi_2|) \cot |\pi - \psi_1 - \psi_2|} = \Phi_{ij}.$$

Summation over all sets B_{ij} yields

$$\lim_{|\delta| \to 0} |\delta|^{-2} \int_{\hat{b}(r, O)} \frac{2 I_2(\delta, m) \, d\delta}{1 + (\pi - |\pi - \psi_1 - \psi_2|) \cot |\pi - \psi_1 - \psi_2|}$$
$$= 3n_f + 4n_c + 2n_k$$

$$\lim_{|\delta| \to 0} |\delta|^{-2} \int_{\hat{b}(r, O)} I_2(\delta, m) \frac{2 \cdot |\pi - \psi_1 - \psi_2| \, d\delta}{1 + (\pi - |\pi - \psi_1 - \psi_2|) \cot |\pi - \psi_1 - \psi_2|}$$
$$= 2\pi(n_f + n_c) + \pi \cdot n_k, \qquad (6.8.9)$$

where n_f, n_c and n_k are the numbers of nodes of fork, cross and knot types, respectively within $b(r, O)$. We use (6.8.9) in §7.14.

VI Now let m be a collection of countably many lines $g_i \in \mathbb{G}$. We will assume that the sequence g_i has no accumulation points in $\mathbb{G}$. In other words, m can be viewed as a realization of a line process in the plane (see §7.5).

Figure 6.8.6 The distance between v_1 and v_2 is d

There is a counterpart of (6.6.9) for realizations of line processes which we are now going to derive using the symbolism of the Haar measure on the group $\mathbb{M}_2$ rather than that of the kinematic measure on Δ_2^* (they are equivalent).

First we consider two *infinitesimal* segments v_1, v_2 situated as shown in fig. 6.8.6. We write

$$I_B(m) = 1 \qquad \text{if a line from } m \text{ hits both } v_1 \text{ and } v_2$$
$$ = 0 \qquad \text{otherwise.}$$

We consider also several *fixed* line segments $\delta_1, \ldots, \delta_s$ in $\mathbb{R}^2$ and let $k_1, \ldots, k_s$ be non-negative integers. We write

$$I_A(m) = 1 \qquad \text{if exactly } k_i \text{ lines from } m \text{ hit } \delta_i, \, i = 1, \ldots, s$$
$$ = 0 \qquad \text{otherwise.}$$

As usual, by Mm we denote the result of applying a motion M to m. For a fixed m, $I_B(Mm)$ and $I_A(Mm)$ are functions defined on $\mathbb{M}_2$. We consider the asymptotic (as $l = |v_1| = |v_2| \to 0$) behavior of the integral

$$\int_{MO \in b(r, O)} I_B(Mm) I_A(Mm) \, dM,$$

where dM is the Haar measure on $\mathbb{M}_2$.

Because in m there are only finitely many lines hitting any $b(r, O)$ for smaller values of $|v_i|$, the set $\{M : I_B(Mm) = 1\}$ can be split into *pairwise non-intersecting* subsets of the type

$C_i = \{M :$ the line from the set Mm which hits both v_1 and v_2 is the image

(under M) of some fixed line $g_i \in m\}$.

By an easy modification of the argument of §6.4 we find that

$$\int_{C_i} I_A(Mm) \, dM = l^2 \, d^{-1} \int_{\chi(g_i)} [I_A(M_{i, +, t}m) + I_A(M_{i, -, t}m)] \, dt + o(l^2).$$

It is necessary to explain $M_{i, +, t}$ (M^{-1} is the inverse of M). We consider two oppositely directed segments q_1 and q_2 both of unit length, both lying on g_i and having a common origin at

$$t \in \chi(g_i) = g_i \cap b(r, O).$$

By definition, $M_{i, +, t}$ corresponds to q_1 to $M_{i, -, t}$ corresponds to q_2 (in the sense of §2.13). The measure dt is one-dimensional Lebesgue on g_i.

By summation we get

$$\lim_{l\to 0} l^{-2} d \int_{MO \in b(r,O)} I_B(Mm) I_A(Mm) \, dM$$

$$= \sum_{g_i \text{ hits } b(r,O)} \int_{\chi(g_i)} [I_A(M_{i,+,t}^{-1} m) + I_A(M_{i,-,t}^{-1} m)] \, dt. \qquad (6.8.10)$$

We stress that here again the value of the limit does not depend on the shifts of the pair v_1 and v_2 along the x-axis. We use this result in §10.1.

6.9 Pleijel identity

A family of identities primarily associated with isoperimetric inequalities for planar convex domains was discovered by Pleijel [37], [38] in 1956. It has proved to be closely related to the questions discussed in our chapter 5. In fact below we derive them by applying a simple integration procedure to (5.4.1).

It was shown in [3] that by integration of other combinatorial formulae various generalizations of the Pleijel identities can be derived in a rather systematic way. Generalizations can be extended to more dimensions as well as to non-convex domains. Some of them can be applied in stochastic geometry, as was actually done in [3]. In this section we treat only the simplest case.

Let D be a bounded convex domain in $\mathbb{R}^2$, and let $\chi_1, \ldots, \chi_n$ be n linear chords of D. We will imagine these chords to be *directed*, therefore one endpoint of each χ_i will be called the 'head' and the other the 'tail'. In the space of sequences $(\chi_1, \ldots, \chi_n)$ we consider the product measure

$$d\mu^{(n)} = d\chi_1 \ldots d\chi_n,$$

where each $d\chi_i$ corresponds to the $\mathbb{M}_2$-invariant measure on $\overline{G}$.

First we assume that the boundary ∂D *does not possess linear parts*. Under this assumption the set of endpoints of our chords will possess triplets lying on a line for a set of sequences of measure zero. Therefore (5.4.1) (which we now consider as written for $\delta_i = \chi_i$) holds 'almost everywhere' and we can integrate it with respect to the measure $\mu^{(n)}$.

We will use the notation $|\chi|$ for the length of a chord, considered either as a function defined on $\{g \in \overline{G} : g \text{ hits } D\}$ or as a function depending on an ordered pair of points from ∂D. Several times in the integration we will use the identity

$$\int I_n(g) \, d\mu^{(n)} = (4|\chi|)^n = (4|\chi(g)|)^n,$$

where $g \in G$ is fixed and

$$I_n(g) = I_{\bigcap_1^n [\chi_i]}(g).$$

Integration of the left-hand side of (5.4.1) yields

$$\int \mu\left(\bigcap [\chi_i] \right) d\mu^{(n)} = \int d\mu^{(n)} \int_G I_n(g) \, dg$$

$$= \int_G dg \int I_n(g) \, d\mu^{(n)} = \int_G (4|\chi|)^n \, dg.$$

We turn now to the right-hand side of (5.4.1). Using symmetry we find that

$$2 \int \sum_i |\chi_i| I_{n-1}(\chi_i) \, d\mu^{(n)} = 2n \int |\chi_1| I_{n-1}(\chi_1) \, d\mu^{(n)}$$

$$= \frac{n}{2} \int_G (4|\chi|)^n \, dg = n \int_G (4|\chi(g)|)^n \, dg.$$

Again by symmetry

$$\int \left[\sum I_{n-2}(d_i)|d_i| - \sum I_{n-2}(s_i)|s_i| \right] d\mu^{(n)}$$

$$= 2n(n-1) \int |\chi_{12}| I_{n-2}(\chi_{12}) [I_d(\chi_{12}) - I_s(\chi_{12})] \, d\mu^{(n)}.$$

In this expression χ_{12} denotes the line segment joining the tails of χ_1 and χ_2, and I_d and I_s are the indicators of the events $\chi_{12} \in \{d_k\}$ and $\chi_{12} \in \{s_k\}$, respectively. After collecting similar terms and dividing by $(n-1)$ the preliminary result of integration of (5.4.1) is written as

$$- \int_G (4|\chi|)^n \, dg = 2n \int |\chi_{12}| I_{n-2}(\chi_{12}) [I_d(\chi_{12}) - I_s(\chi_{12})] \, d\mu^{(n)}. \quad (6.9.1)$$

Using (3.7.3) for dg, the right-hand side of (6.9.1) is written as

$$\frac{n}{2} \int_{G \times G} (4|\chi|)^{n-1} [I_d(\chi) - I_s(\chi)] \sin \psi_1 \sin \psi_2 \, d\psi_1 \, d\psi_2 \, dl_1 \, dl_2,$$

and here integrations over ψ_1 and ψ_2 can be performed. For fixed $l_1, l_2 \in \partial D$,

$$\int_0^\pi \int_0^\pi \sin \psi_1 \sin \psi_2 [I_d(\chi) - I_s(\chi)] \, d\psi_1 \, d\psi_2 = -4 \cos \alpha_1 \cos \alpha_2,$$

where the angles α_1 and α_2 are shown in fig. 6.9.1. Thus the final result of integration is the Pleijel identity for convex domains:

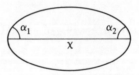

Figure 6.9.1 The angles α_1 and α_2 lie in one halfplane with respect to the inside of D

$$\int_G |\chi|^n \, dg = \frac{n}{2} \iint\limits_{(\partial D)^2} |\chi|^{n-1} \cos \alpha_1 \cos \alpha_2 \, dl_1 \, dl_2. \tag{6.9.2}$$

During derivation we have assumed that $n > 1$, but in fact this restriction is not significant.

Making use of the linearity of (6.9.2) we establish

$$\int_G f(|\chi|) \, dg = \frac{1}{2} \iint\limits_{(\partial D)^2} f'(|\chi|) \cos \alpha_1 \cos \alpha_2 \, dl_1 \, dl_2 \tag{6.9.3}$$

firstly for polynomial functions f of the form $b_2 x^2 + \cdots + b_n x^n$ (where f' denotes the derivative). Then, by Weierstrass's approximation theorem, (6.9.3) holds for every function f having continuous derivative f' and satisfying $f(0) = 0$. In particular if we take $f(x) = x$ the integral on the left-hand side reduces to $\pi \cdot \|D\|$, where $\|D\|$ is the area of D (see (6.3.1)).

On the other hand, we can use (3.7.4) to get

$$\int f(|\chi|) \, dg = \frac{1}{2} \iint\limits_{(\partial D)^2} f(|\chi|)|\chi|^{-1} \sin \alpha_1 \sin \alpha_2 \, dl_1 \, dl_2.$$

Thus, putting $f(x) = x$ we get two equations:

$$\pi \cdot \|D\| = \frac{1}{2} \iint\limits_{(\partial D)^2} \cos \alpha_1 \cos \alpha_2 \, dl_1 \, dl_2$$

and

$$\pi \|D\| = \frac{1}{2} \iint\limits_{(\partial D)^2} \sin \alpha_1 \sin \alpha_2 \, dl_1 \, dl_2$$

from which we get, by addition,

$$2\pi \|D\| = \frac{1}{2} |\partial D|^2 - \iint\limits_{(\partial D)^2} \sin^2 \frac{\alpha_1 - \alpha_2}{2} \, dl_1 \, dl_2,$$

or in a weaker form

$$4\pi \|D\| \leqslant |\partial D|^2,$$

which is the classical isoperimetric inequality. In §6.10 we will need the following corollary of (6.9.3). Using (3.7.4) we rewrite this equation in the form

$$\int_G f(|\chi|) \, dg = \int_G f'(|\chi|) \cdot |\chi| \cdot \cot \alpha_1 \cot \alpha_2 \, dg. \tag{6.9.4}$$

The above transformation is legitimate only for planar convex domains D whose boundaries do not possess line segments. If this is not the case, additional terms appear. Consider for example a *polygonal* domain D with sides a_i of length $|a_i|$. In this case, it is necessary to add the integrals

$$\frac{1}{2} \iint\limits_{a_i \times a_i} f'(\rho)\, dl_1\, dl_2,$$

where ρ is the distance between l_1 and l_2. By simple calculation we get the following modification of (6.9.4), valid for bounded convex polygons D:

$$\int_G f(|\chi|)\, dg = \int_G f'(|\chi|)\cdot|\chi| \cot \alpha_1 \cot \alpha_2\, dg + \sum \int_0^{|a_i|} f(u)\, du. \quad (6.9.5)$$

Note that in (6.9.5) the condition $f(0) = 0$ is no longer necessary.

6.10 Chords through convex polygons

In this section we assume that D is a polygon. If in (6.9.5) we formally put

$$f_y(u) = 0 \quad \text{if } u < y$$
$$= 1 \quad \text{if } u > y, \qquad (6.10.1)$$

then the left-hand integral there will equal

$$\mu\{g \in \mathbb{G} : |\chi(g)| > y\};$$

i.e. the invariant measure of the set of chords of D whose length exceeds y. We note that the distribution function of the length of a random chord χ through D is usually defined to be

$$F(y) = 1 - |\partial D|^{-1}\mu\{g \in \mathbb{G} : |\chi(g)| > y\},$$

where $|\partial D|$ is the length of the perimeter of D.

The derivative of $f_y(u)$ as given in (6.10.1) should be replaced by Dirac's δ-function concentrated at y. Thus (6.9.5) becomes

$$[1 - F(y)]\cdot|\partial D|$$

$$= \sum_{i<j} \iint\limits_{[a_i] \cap [a_j]} \delta_y(|\chi|)\cdot|\chi| \cot \alpha_1 \cot \alpha_2\, dg + \sum (|a_i| - y)^+ \quad (6.10.2)$$

where $[a_i] \cap [a_j]$ is the set of lines hitting both sides a_i and a_j of D, and $x^+ = x$ if $x > 0$, 0 otherwise. In each of the double integrals one integration can be performed by passing to the $|\chi|, \varphi$ coordinates shown in fig. 6.10.1. The $|\chi|, \varphi$ coordinates can be used if a_i and a_j are nonparallel; thus we now assume that D has no pairs of parallel sides. We have

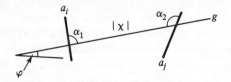

Figure 6.10.1 φ is the direction of g

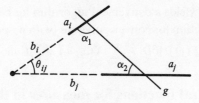

Figure 6.10.2

$$dg = \left| \frac{\sin \alpha_1 \sin \alpha_2}{\sin (\alpha_1 + \alpha_2)} \right| d|\chi| \cdot d\varphi.$$

Clearly we can put

$$\alpha_1 = \varphi, \alpha_2 = \pi - (\theta_{ij} + \alpha_1),$$

where θ_{ij} is the angle between a_i and a_j (see fig. 6.10.2). Thus we have

$$\iint\limits_{[a_i] \cap [a_j]} \delta_y(|\chi|) \cot \alpha_1 \cot \alpha_2 \, dg = -\frac{y}{\sin \theta_{ij}} \int_{\Phi_{ij}(y)} \cos \varphi \cos(\theta_{ij} + \varphi) \, d\varphi,$$

(6.10.3)

where the domain of integration is

$$\Phi_{ij}(y) = \{\psi : \text{a chord joining } a_i \text{ and } a_j \text{ exists which has direction } \varphi \text{ and its}$$
$$\text{length is } y\}.$$

The reader will check without difficulty that

if the sides a_i, a_j have a common vertex, then the set $\Phi_{ij}(y)$ is a union of disjoint intervals, their number can be one, two or at most three.

On the other hand we have

$$\int_{\varphi_1}^{\varphi_2} \cos \varphi \cdot \cos(\theta_{ij} + \varphi) \, d\varphi = \frac{1}{2}(\varphi_2 - \varphi_1) \cos \theta_{ij}$$

$$+ \frac{1}{4} \sin(\theta_{ij} + 2\varphi_2) - \frac{1}{4} \sin(\theta_{ij} + 2\varphi_1).$$

The endpoints of the intervals in question are elementary functions depending on θ_{ij}, $|a_i|$ and $|a_j|$. This means that if the sides a_i and a_j have a common vertex then the integral (6.10.3) can be expressed elementarily in terms of these quantities. The above restriction on the choice of a_i and a_j can be removed by virtue of the indicator functions relations

$$I_{[a_i] \cap [a_j]} = I_{[a_i \cup b_i] \cap [a_j \cup b_j]} - I_{[a_i \cup b_i] \cap [b_j]} - I_{[b_i] \cap [a_j \cup b_j]} + I_{[b_i] \cap [b_j]};$$

the intervals b_i and b_j are shown in fig. 6.10.2.

We conclude that a similar principle is valid for every integral appearing in (6.10.2). Thus *the integral (6.10.3) can be expressed as an elementary function* $X(a_i, a_j)$ *of the segments* a_i *and* a_j.

Clearly the above yields a convenient algorithm for calculation of random chord length distribution for convex polygons (with no pairs of sides parallel):

$$[1 - F(y)] \cdot |\partial D| = \sum_{i<j} X(a_i, a_j) + \sum_i (|a_i| - y)^+.$$

6.11 Integral functions for measures in the space of triangular shapes

In this section we describe some results referring to measures in the space Σ_2 whose densities were derived in §3.16, III. We give the values of these measures on the family of sets

$A(x) = \{$triangular shapes for which the minimal interior angle is greater than $x\}$

(see Fig. 6.11.1). More elaborate expressions for families of sets depending on more parameters have been obtained by Sukiasian in [39] and by Oganian in [40].

We start with the measure v_S which has a density given by (3.16.7). We have

$$v_S(A(x)) = 2 \int_x^{\pi-2x} d\alpha \int_x^{\pi-\alpha-x} [\sin \alpha \sin \beta \sin(\alpha + \beta)]^{-1} d\beta.$$

Making use of the identity

$$\sin \alpha [\sin \beta \sin(\alpha + \beta)]^{-1} = \cot \beta - \cot(\alpha + \beta),$$

we integrate:

$$v_S(A(x)) = 2 \int_x^{\pi-2x} \sin^{-2} \alpha \, d\alpha \int_x^{\pi-\alpha-x} [\cot \beta - \cot(\alpha + \beta)] \, d\beta$$

$$= 4 \int_x^{\pi-2x} [\ln \sin(\alpha + x) - \ln \sin x] \sin^{-2} \alpha \, d\alpha$$

$$= -4 \int_x^{\pi-2x} \ln[\sin(\alpha + x) \sin^{-1} x] \, d(\cot \alpha)$$

$$= 4 \cot x \ln(2 \cos x) + 4 \int_x^{\pi-2x} \cot \alpha \cot(\alpha + x) \, d\alpha.$$

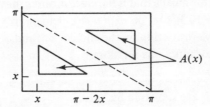

Figure 6.11.1 The two components of $A(x)$ are triangles symmetrical with respect to the diagonal

Since
$$\cot \alpha \cdot \cot(\alpha + x) = \cot x(\cot \alpha - \cot(\alpha + x)) - 1$$
we get
$$\int_x^{\pi - 2x} \cot \alpha \cot(\alpha + x) \, d\alpha = 3x - \pi + 2 \cot x \ln(2 \cos x).$$
Thus finally for $x \in (0, \pi/3)$
$$v_S(A(x)) = 4(3x - \pi + 3 \cot x \ln(2 \cos x)). \qquad (6.11.1)$$
Note that this expression implies
$$\lim_{x \to \pi/3} v_S(A(x)) = 0$$
(which can be considered as a check of (6.11.1)) and
$$\lim_{x \to 0} v_S(A(x)) = \infty$$
in concordance with (3.17.3).

Let us now consider the density (3.16.6). In terms of the quantities
$$z = tg\frac{\alpha_1}{2}, \qquad u = tg\frac{\alpha_2}{2}$$
this density can be rewritten as
$$\tfrac{1}{32}z(1 + z^2)^2 u(1 + u^2)^2 (1 - zu)(u + z)^{-2}.$$
Therefore the problem of calculation of $v_h(A(x))$ reduces to integration of this rational function. The result will be
$$v_h(A(x)) = \frac{\pi}{21} - \frac{x}{7} - \left(\frac{9}{14}b^7 + \frac{3}{2}b^5 + b^3\right) \ln[4b^2(1 + b^2)^{-1}]$$
$$+ \frac{1}{112}(99b^7 + 159b^5 + 29b^3 - 31b),$$
where
$$b = tg \, x/2.$$
The above is valid for $x \in (0, \pi/3)$; for $x \in (\pi/3, \pi)$ clearly $v_h(A(x)) = 0$. Putting $x = 0$ we get
$$v_h(\Sigma) = \frac{\pi}{21}$$
which was earlier found by Kendall in [41]. The values $m_h(A(x))$ can be found by starting from the observation that the density (3.16.3) equals the expression
$$2 \sin \frac{\alpha_1}{2} \sin \frac{\alpha_2}{2} \cos \frac{\alpha_1 + \alpha_2}{2}$$
which is very convenient for integration. The result has the form

$$m_h(A(x)) = 3 \cos x + 3 \cos 2x - 3(\pi - 3x) \sin x - \frac{(\pi - 3x)^2}{2}$$

for $x \in (0, \pi/3)$ and zero for x $(\pi/3, \pi)$. Putting $x = 0$ for the total measure we get the value (3.17.1). Similar integration for the density (3.16.10) is even more simple. We present the result: for $x \in (0, \pi/3)$

$$v_A(A(x)) = \frac{6}{\pi^2}(\pi - 3x) \cos 2x + \frac{3}{\pi^2}(\sin 4x + \sin 2x).$$

Hence (3.17.2).

7

Stochastic point processes

Random point processes in different spaces will be the main object of our study for the rest of the book. We start this chapter with a brief exposition of basic notions. The measure-theoretic ideas used here are the same for all 'carrier' spaces that we consider. However, the clearest geometrical interpretations can be given for point processes on the line, which we treat in more detail to illustrate our ideas.

We begin the study of point processes that are invariant with respect to groups with several concrete examples. Emphasis is put on models which reappear in further chapters. We also introduce the important notions of marked point processes and moment measures.

The 'analysis of realizations' of the previous chapter is put to use by means of an averaging procedure. In particular, results concerning the nodes in segment processes and random mosaics in $\mathbb{R}^2$ are obtained by this method. It can be broadly used in the study of invariant geometrical processes, and further instances of this are scattered throughout the remaining chapters.

7.1 Point processes

The notion of a point process is a formalization of the intuitive idea of a random set with countably many points. A precise definition is as follows.

Let $\mathbb{X}$ be a space (for instance a manifold or, more generally, a complete separable metric space, see [18]) which is to 'carry' the realizations of the point process.

A *realization* $m \subset \mathbb{X}$ is a subset which has no condensation points in $\mathbb{X}$. We denote by $\mathcal{M}$ the class of all realizations, $m \in \mathcal{M}$. Where necessary we will use the notation

$$\mathcal{M} = \mathcal{M}_{\mathbb{X}}$$

which stresses the basic space $\mathbb{X}$.

Remark By this definition realizations cannot possess multiple points, i.e. we consider only 'simple' point processes in the terminology of [18].

Let $B \subset \mathbb{X}$ be Borel. We put

$$N(B, m) = \text{card } B \cap m.$$

Let $\mathscr{A}$ be the minimal σ-algebra of subsets of $\mathscr{M}$ which renders all the functions $N(B, m)$ measurable. Equivalently, $\mathscr{A}$ is the σ-algebra in $\mathscr{M}$ generated by sets of the type

$$\{m \in \mathscr{M} : N(B, m) = k\}.$$

A point process in $\mathbb{X}$ is a measurable map

$$m(\omega) : \Omega \to \mathscr{M}, \tag{7.1.1}$$

where $(\Omega, \mathscr{F}, \mathbf{p})$ is a probability space. The above means that to any element $\omega \in \Omega$ corresponds a realization $m(\omega)$, with the only requirement that

$$\{\omega : m(\omega) \in A\} \in \mathscr{F}$$

as soon as A belongs to $\mathscr{A}$. (Some examples of actual maps $m(\omega)$ can be found in §§7.8–7.10.)

Every point process $m(\omega)$ induces a *probability* P on $(\mathscr{M}, \mathscr{A})$ by the formula:

$$P(A) = \mathbf{p}\{\omega : m(\omega) \in A\}, \qquad A \in \mathscr{A}.$$

This P is called the *distribution* of $m(\omega)$. Note that if a probability P on $(\mathscr{M}, \mathscr{A})$ is specified, then a point process $m(\omega)$ which has P for its distribution *always* exists. The latter can be constructed as follows. We take

$$\Omega = \mathscr{M}, \qquad \mathscr{F} = \mathscr{A}, \qquad \mathbf{p} = P$$

and define $m(\omega)$ by the identity map

$$m \to m. \tag{7.1.2}$$

Probabilities on $(\mathscr{M}, \mathscr{A})$ can be considered in their own right, apart from the notion of a point process. In such cases they are usually described by means of the theorems of continuation of probabilities (measures), starting from values of P given on some subclass $\mathscr{A}_0 \subset \mathscr{A}$. Thus arises the general problem of the description of the classes $\mathscr{A}_0$ from which continuation is possible in a unique way. Still another aspect of the same question is to describe functions which are defined on $\mathscr{A}_0$ and which permit continuation to a probability on $(\mathscr{M}, \mathscr{A})$. Below we give the answers to these two questions for a number of spaces $\mathbb{X}$ using simple geometrical considerations, which are clearest for $\mathbb{X} = \mathbb{R}$, which we treat in some detail. Other cases can be treated similarly.

7.2 *k*-subsets of a linear interval

Let $[a, b)$ be a finite semiopen interval on $\mathbb{R}$:

$$[a, b) = \{x \in \mathbb{R} : a \leqslant x < b\}.$$

Here and in §7.3 and §7.4, we will consider *only* similarly defined semiopen intervals to be termed simply as 'intervals'.

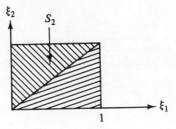

Figure 7.2.1 $a = 0, b = 1$

We will call the subsets of $[a, b)$ which contain k points *k-subsets*. The space of all *k*-subsets will be denoted by $[a, b)_k$ (in contrast to $[a, b)^k = [a, b) \times \cdots \times [a, b)$, k times). An element of $[a, b)_k$ is written as

$$m = \{x_1, \ldots, x_k\}$$

(as usual, the brackets { } denote *sets*, while the brackets ( ) denote ordered sequences).

There is only one zero-subset of $[a, b)$, namely the empty set. Thus the space $[a, b)_0$ consists of one element (is non-empty!)

Also it is clear that

$$[a, b)_1 = [a, b).$$

To construct $[a, b)_2$ we use the map

$$\{x_1, x_2\} \to (\xi_1, \xi_2),$$

where

$$\xi_1 = \min x_1, x_2, \qquad \xi_2 = \max x_1, x_2.$$

This map identifies $[a, b)_2$ with the triangle S_2 (see fig. 7.2.1). The diagonal does not belong to S_2. In general, the 'ordering' map

$$\{x_1, \ldots, x_k\} \to (\xi_1, \ldots, \xi_k),$$

where

$$\xi_1 = \min x_1, \ldots, x_k, \quad \ldots, \quad \xi_k = \max x_1, \ldots, x_k,$$

maps $[a, b)_k$ onto a *simplex* S_k in $\mathbb{R}^k$, which in terms of coordinates $\xi_1, \ldots, \xi_k$ is described by the inequalities

$$a \leqslant \xi_1 < \cdots < \xi_k < b.$$

The above maps reduce the task of the construction of *measures* on $[a, b)_k$ to the construction of measures on (subsets of) Euclidean spaces. In particular we can use usual rectangles in $\mathbb{R}^k$, noting that rectangles which lie completely within S_k make up a *semiring* of subsets of S_k and they generate the Borel subsets of S_k. We denote this semiring by $\mathscr{H}_k(a, b)$. A rectangle $r \in \mathscr{H}_k(a, b)$ has the following interpretation in terms of *k*-sets.

Given $r \in \mathscr{H}_k(a, b)$, we consider the projections

$$I_1, \ldots, I_k \tag{7.2.1}$$

of r on the coordinate axis $\xi_1, \ldots, \xi_k$, so that

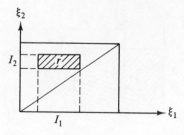

Figure 7.2.2

$$r = I_1 \times \cdots \times I_k.$$

The sequence (7.2.1) satisfies the following conditions:

(a) $I_1, \ldots, I_k$ are pairwise non-intersecting and belong to $[a, b)$;
(b) I_s lies to the right of I_l if $s > l$.

In fact (a) and (b) are also sufficient: if both (a) and (b) are satisfied then the product $I_1 \times \cdots \times I_k$ falls within S_k (see fig. 7.2.2).

By the construction of the ordering map we have

$$r = \{m \in [a, b)_k : \text{card } m \cap I_i = 1, i = 1, \ldots, k\}.$$

Below we will use the following notation. Let $B_1, \ldots, B_s$ be a system of Borel sets from $[a, b)$, and let $l_1, \ldots, l_s$ be non-negative integers. We write

$$\left\langle \begin{matrix} B_1 \\ l_1 \end{matrix}, \ldots, \begin{matrix} B_s \\ l_s \end{matrix} \right\rangle = \{m \in [a, b)_k : \text{card } m \cap B_i = l_i, i = 1, \ldots, s\}.$$

In this notation

$$r = \left\langle \begin{matrix} I_1 \\ 1 \end{matrix}, \ldots, \begin{matrix} I_k \\ 1 \end{matrix} \right\rangle.$$

We have also seen that

$$\mathscr{H}_k(a, b) = \left\{ \left\langle \begin{matrix} I_1 \\ 1 \end{matrix}, \ldots, \begin{matrix} I_k \\ 1 \end{matrix} \right\rangle : \begin{matrix} \text{intervals } I_1, \ldots, I_k \\ \text{satisfy (a) and (b)} \end{matrix} \right\}.$$

By applying the well known facts of measure theory (see [46]) we come to the following conclusions.

Every measure on $[a, b)_k$ is completely defined by its values on the events

$$\left\langle \begin{matrix} I_1, \ldots, I_k \\ 1, \ldots, 1 \end{matrix} \right\rangle$$

where the intervals satisfy (a) and (b) (i.e. on the events from $\mathscr{H}_k(a, b)$). A non-negative function F defined on $\mathscr{H}_k(a, b)$ is a measure on $[a, b)_k$ iff it is countably additive within the same class of events.

Example Let F be defined on $\mathscr{H}_k(a, b)$ as follows:

$$F\left(\left\langle \begin{matrix} I_1, \ldots, I_k \\ 1 \end{matrix} \right\rangle \right) = m(I_1) \cdots \cdot m(I_k), \qquad (7.2.2)$$

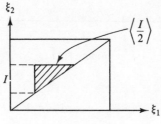

Figure 7.2.3

where m is some atomless measure on $[a, b)$. Clearly F gives the values of the product measure $m \times \cdots \times m$ (k times) on rectangles in S_k, and therefore coincides with restriction of the mentioned product measure to S_k, i.e.

$$F = m \times \cdots \times m \qquad (k \text{ times}) \qquad \text{on } S_k.$$

Clearly

$$F(S_k) = (m([a, b)))^k (k!)^{-1},$$

$$F\left(\left\langle \begin{matrix} I_1 \\ l_1 \end{matrix}, \ldots, \begin{matrix} I_s \\ l_s \end{matrix} \right\rangle\right) = \prod_{i=1}^{s} (m(I_i))^{l_i} (l_i!)^{-1} \tag{7.2.3}$$

provided $I_1, \ldots, I_s \subset [a, b)$ pairwise do not intersect, and $\sum l_i = k$. In particular, if $k = 2$, the event $\left\langle \begin{matrix} I \\ 2 \end{matrix} \right\rangle$ can be identified with a triangle in S_2 (see fig. 7.2.3) and by the last formula its $m \times m$ measure equals $\frac{1}{2}(m(I))^2$.

In the case in which $m([a, b)) < \infty$, the measure F given by (7.2.2) can be normalized to yield a probability measure

$$P = (F(S_k))^{-1} \cdot F.$$

We will have (under the same conditions as in (7.2.3))

$$P\left(\left\langle \begin{matrix} I_1 \\ l_1 \end{matrix}, \ldots, \begin{matrix} I_s \\ l_s \end{matrix} \right\rangle\right) = k! \prod_{i=1}^{s} \left(\frac{m(I_i)}{m([a, b))}\right)^{l_i} (l_i!)^{-1}.$$

We conclude that the random set described by the above probability can be obtained by an 'experiment' in which k independent random points are dropped on $[a, b)$, each point having probability distribution proportional to m.

7.3 Finite sets on $[a, b)$

We consider the space of all subsets of $[a, b)$ which possess only a finite number of points, i.e. (see §7.1)

$$\mathcal{M}_{[a, b)} = \{m \subset [a, b) : \text{card } m < \infty\}.$$

Clearly this space can be represented as a union:

$$\mathcal{M}_{[a, b)} = \bigcup_{k=0}^{\infty} [a, b)_k. \tag{7.3.1}$$

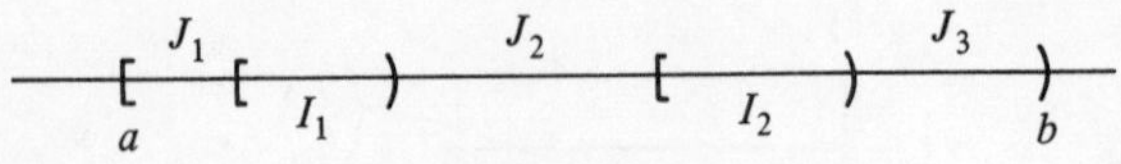

Figure 7.3.1

We will use the notation

$$\begin{bmatrix} I_1, \ldots, I_s \\ l_1 \quad\quad l_s \end{bmatrix} = \{m \in \mathcal{M}_{[a,b)} : \text{card } m \cap I_i = l_i, i = 1, \ldots, s\},$$

where $I_1, \ldots, I_s$ are intervals from $[a, b)$, and $l_1, \ldots, l_s$ are non-negative integers.

We denote by $\mathcal{H}(a, b)$ the class of subsets of $\mathcal{M}_{[a,b)}$ of the above type, which we define by the conditions

(a) $I_1, \ldots, I_s$ are disjoint and their union is $[a, b)$;

(b) each l_i is either 0 or 1.

Let us show that

$$\mathcal{H}(a, b) = \bigcup_{k=0}^{\infty} \mathcal{H}_k(a, b). \tag{7.3.2}$$

Indeed, if $A \in \mathcal{H}(a, b)$ then each element $m \in A$ has a fixed number of points, namely $k = l_1 + \cdots + l_s$, and therefore $A \subset [a, b)_k$. Conversely, if

$$A = \left\langle \begin{matrix} I_1, \ldots, I_s \\ 1 \quad\quad 1 \end{matrix} \right\rangle \in \mathcal{H}_s(a, b)$$

then we can represent this A in the form

$$A = \begin{bmatrix} I_1, \ldots, I_s, J_1, \ldots, J_j \\ 1 \quad\quad 1 \quad 0 \quad\quad 0 \end{bmatrix} \in \mathcal{H}(a, b),$$

where $J_1, \ldots, J_j$ is the collection of disjoint intervals whose union is $(\bigcup_{i=1}^s I_i)^c$ (c stands for complement, see fig. 7.3.1).

Because of (7.3.1) every measure on $\mathcal{M}_{[a,b)}$ is completely determined by its restrictions to the sets $[a, b)_k$, $k = 0, 1, \ldots$. Therefore (see §7.2) we obtain the following proposition.

Let F be a non-negative function defined on $\mathcal{H}(a, b)$ (and therefore on each $\mathcal{H}_k(a, b)$). If on every $\mathcal{H}_k(a, b)$ the function F is σ-additive then F is (equivalently, can be extended to) a measure on $\mathcal{M}_{[a,b)}$.

We are especially interested in the case in which

$$F(\mathcal{M}_{[a,b)}) = c < \infty$$

because we can put

$$P = c^{-1} \cdot F$$

and obtain a probability measure on $\mathcal{M}_{[a,b)}$.

Example Let F be defined on $\mathcal{H}(a, b)$ in the following way:

$$F\left(\begin{bmatrix} I_1, \ldots, I_s, J_1, \ldots, J_j \\ 1 \quad\quad 1 \quad 0 \quad\quad 0 \end{bmatrix}\right) = \prod_{i=1}^{s} m(I_i), \tag{7.3.3}$$

where m is a *finite* atomless measure on $[a, b]$. The restrictions of F on every $\mathcal{H}_k(a, b)$ are essentially the measures which we have considered in the Example in §7.2. Therefore F in (7.3.3) is a measure on $\mathcal{M}_{[a,b)}$ generated by the measures $m \times \cdots \times m$ (k times) on each $[a, b)_k$. In particular

$$F(\mathcal{M}_{[a,b)}) = \sum_{k=0}^{\infty} (k!)^{-1}(m([a, b))^k = \exp(m([a, b))).$$

The corresponding probability therefore is

$$P = \exp(-m([a, b))) \cdot F. \tag{7.3.4}$$

Let $I_1, \ldots, I_s$ be an arbitrary system of pairwise non-intersecting intervals on $[a, b)$. For every probability P on $\mathcal{M}_{[a,b)}$ we have

$$P\left(\begin{bmatrix} I_1 & & I_s \\ l_1 & \cdots & l_s \end{bmatrix}\right) = \sum_{r_1, \ldots r_j} P\left(\begin{bmatrix} I_1 & & I_s & J_1 & & J_j \\ l_1 & \cdots & l_s & r_1 & \cdots & r_j \end{bmatrix}\right),$$

where $J_1, \ldots, J_j$ are non-intersecting intervals which complement $\bigcup I_i$ to $[a, b)$. The event

$$\begin{bmatrix} I_1 & & I_s & J_1 & & J_j \\ l_1 & \cdots & l_s & r_1 & \cdots & r_j \end{bmatrix}$$

belongs to $[a, b)_f$, where $f = \sum l_i + \sum r_i$.

By (7.2.3) for P defined by (7.3.3) and (7.3.4) the probability of this event equals

$$\prod_{i=1}^{s} \frac{(m(I_i))^{l_i}}{l_i!} \cdot \prod_{i=1}^{j} \frac{(m(J_i))^{r_i}}{r_i!} e^{-m([a,b))}$$

$$= \prod_{i=1}^{s} \frac{(m(I_i))^{l_i}}{l_i!} e^{-m(I_i)} \cdot \prod_{i=1}^{j} \frac{(m(J_i))^{r_i}}{r_i!} e^{-m(J_i)} \tag{7.3.5}$$

(we used additivity of m). Performing the summation we find that

$$P\left(\begin{bmatrix} I_1 & & I_s \\ l_1 & \cdots & l_s \end{bmatrix}\right) = \prod_{i=1}^{s} \frac{(m(I_i))^{l_i}}{l_i!} e^{-m(I_i)}. \tag{7.3.6}$$

The probability measure $P = P_m$ defined by (7.3.3) and (7.3.4) (or, equivalently, by (7.3.6)) will be termed Poisson, governed by the measure m.

Equation (7.3.6) now can be expressed in words as follows.

For any Poisson probability on $\mathcal{M}_{[a,b)}$ the events $\begin{bmatrix} I \\ l \end{bmatrix}$ for non-overlapping intervals are independent, the distributions

$$P_m\left(\begin{bmatrix} I \\ l \end{bmatrix}\right), \qquad l = 0, 1, \ldots$$

are usual Poisson with parameter $m(I)$.

We mention also the following property.

The conditional distribution of a Poisson point process on $[a, b)$ described by (7.3.6) conditional upon the event

$$N([a, b), m) = k$$

is proportional to the k-fold product of the measure m with itself.

The proof is a direct corollary of our construction.

7.4 Consistent families

Let $[c, d)$ be a subinterval of $[a, b)$. The realizations space $\mathcal{M}_{[a,b)}$ can be mapped onto $\mathcal{M}_{[c,d)}$ by means of the truncation

$$m \to m \cap [c, d), \tag{7.4.1}$$

where $m \in \mathcal{M}_{[a,b)}$.

Assume that two probability measures, P_1 on $\mathcal{M}_{[a,b)}$ and P_2 on $\mathcal{M}_{[c,d)}$, are given. We say that P_1 and P_2 are consistent if P_2 is the image of P_1 under the map (7.4.1).

A check that P_1 and P_2 are consistent can be reduced to the verification that

$$P_2(A) = P_1(A) \qquad \text{for every } A \in \mathcal{H}(c, d). \tag{7.4.2}$$

Now let $m(\omega)$ be a point process on the line (see §7.1, $\mathbb{X} = \mathbb{R}$), and let P be the distribution of $m(\omega)$.

On every interval $[a, b)$ we construct the point process

$$m(\omega) \cap [a, b)$$

and we denote by $P_{[a,b)}$ the distribution of the latter. The *family* of distributions $\{P_{[a,b)}\}$ which arises in this way is consistent. This means that *for each $[c, d)$ and $[a, b)$ such that $[c, d) \subset [a, b)$ the probabilities $P_{[a,b)}$ and $P_{[c,d)}$ are consistent*. The last assertion follows from the relation

$$(m(\omega) \cap [a, b)) \cap [c, d) = m(\omega) \cap [c, d).$$

It is important that the converse statement is also valid. Namely, let us assume that a family $\{P_{[a,b)}\}$ of probabilities depending on $[a, b) \subset \mathbb{R}$ is given, where $P_{[a,b)}$ is a probability on $\mathcal{M}_{[a,b)}$.

If $\{P_{[a,b)}\}$ is consistent then there exists a unique probability P on $\mathcal{M}_{\mathbb{R}}$ such that each $P_{[a,b)}$ is the image of P under the map

$$m \to m \cap [a, b).$$

The proof is by standard measure-theoretic methods and is therefore omitted.

Example 1 Let m be an atomless measure on $\mathbb{R}$. Let $P_{[a,b)}$ be Poisson governed by the restriction of m to the interval $[a, b)$. By (7.3.6) this family of probabilities is consistent. We conclude that this family is generated by a probability $P = P_m$ on $\mathcal{M}_{\mathbb{R}}$. This P_m we again call Poisson, governed by m. Any point process on $\mathbb{R}$ with distribution P_m (which exists by (7.1.2)) is also Poisson, governed by m.

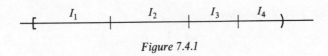

Figure 7.4.1

The above theory yields the following answer to the queries mentioned in §7.1.

Corollary Let P be a probability on $\mathcal{M}_{\mathbb{R}}$. The values of P on the class of events

$$\mathcal{H} = \bigcup \mathcal{H}(a, b),$$

where the union extends over all intervals $[a, b)$, define P uniquely.

The proof follows from the uniqueness statements of this and previous sections.

We emphasize that

the class $\mathcal{H}$ consists of elements of the form

$$\begin{pmatrix} I_1 & & I_s \\ l_1 & , \ldots, & l_s \end{pmatrix} = \{ m \in \mathcal{M}_{\mathbb{R}} : \text{card } m \cap I_i = l_i, i = 1, \ldots, s \},$$

where $I_1, \ldots, I_s$ are pairwise disjoint intervals on $\mathbb{R}$ with union again an interval (see fig. 7.4.1) and each l_i is either 0 or 1.

With the preceding theory we can state some conditions which guarantee that a non-negative function F defined on $\mathcal{H}$ is a probability measure on $\mathcal{M}_{\mathbb{R}}$. They are as follows:

(a) the restriction of F on every $\mathcal{H}_k(a, b)$ should be a measure;
(b) for every $[a, b)$ we should have

$$\sum_k F([a, b)_k) = 1,$$

i.e. on every $\mathcal{M}_{[a,b)}$ the function F generates a *probability measure*;
(c) the family of probability measures mentioned in (b) should be consistent.

For the values of F on concrete sets we use the notation

$$F\left(\begin{pmatrix} I_1 & & I_k \\ l_1 & , \ldots, & l_k \end{pmatrix} \right) = F \begin{pmatrix} I_1 & & I_k \\ l_1 & , \ldots, & l_k \end{pmatrix}$$

and similar slight abuse of notation will persist in other parts of the book.

A rather special but important way of constructing functionals F which satisfy (a), (b) and (c) is by means of so-called *relative density functions* (r.d.f.-s).

We consider non-negative functions $f_I(x_1, \ldots, x_n)$ which depend on an interval $I \subset \mathbb{R}$ as well as on points $x_1, \ldots, x_n \in I$ and which are symmetrical in the arguments x_i. In the case $n = 0$, f_I depends merely on I.

Given a system of functions

$$\{ f_I(x_1, \ldots, x_n) \}_{n=0, 1, 2, \ldots}$$

we construct an F as follows:

$$F\binom{I}{0} = f_I$$

$$F\binom{I_1, \ldots, I_s}{l_1, \ldots, l_s} = \int_{I_{i_1}} \cdots \int_{I_{i_n}} f_I(x_1, \ldots, x_n)\, dx_1 \ldots dx_n.$$

(7.4.3)

In this expression

$$n = \sum l_i > 0,$$

$$l_{i_1} = \cdots = l_{i_n} = 1, \qquad \text{all other } l_k \text{ are zero,}$$

$$I = \bigcup I_i.$$

Clearly (7.4.3) implies (a) ($f_I(x_1, \ldots, x_k)$ happens to be the density of the measure in question). (b) follows from the condition

$$\text{(b')} \sum_{n=0}^{\infty} (n!)^{-1} \int_{I^n} \cdots \int f_I(x_1, \ldots, x_n)\, dx_1 \ldots dx_n = 1.$$

Together with the consistency condition (c) this guarantees the existence of a probability P on $\mathcal{M}_{\mathbb{R}}$ (or of a point process on $\mathbb{R}$). We refer to this probability as *generated by the system* $\{f_I(x_1, \ldots, x_n)\}_{n=0,1,2\ldots}$, the latter functions are called the corresponding r.d.f.-s. The term r.d.f. is due to the interpretation

$$P\binom{dx_1, \ldots, dx_n, I\backslash(\bigcup dx_i)}{1, \ldots, 1, 0} = f_I(x_1, \ldots, x_n)\, dx_1 \ldots dx_n$$

which follows immediately from (7.4.3) if we take there $F = P$, the intervals $I_{i_1}, \ldots, I_{i_n}$ we assume centered at $x_1, \ldots, x_n$ and having infinitesimal lengths $dx_1, \ldots, dx_n$. Here the assumption of continuity is crucial.

Example 2 Let m be some measure on $\mathbb{R}$ possessing a density $\rho(x)$. We put

$$f_I = e^{-m(I)}$$

$$f_I(x_1, \ldots, x_n) = \prod \rho(x_i) e^{-m(I)}.$$

(7.4.4)

Because

$$\int_{I^n} \cdots \int \prod \rho(x_i)\, dx_1 \ldots dx_n = (m(I))^n,$$

(a) and (b') are satisfied. Moreover, we identify F generated by this system via (7.4.3) with F of the previous example. We conclude that (7.4.4) is a system of r.d.f.-s and it generates the Poisson P_m.

If continuous r.d.f.-s are subject to the condition

$$f_I(x_1, \ldots, x_n) \leqslant C^n$$

(7.4.5)

for some constant C then necessarily

$$P\binom{dx_1, \ldots, dx_n}{1, \ldots, 1} = \varphi(x_1, \ldots, x_n)\, dx_1 \ldots dx_n.$$

(7.4.6)

The functions φ are called *the densities* of P. We have

$$\varphi(x_1, \ldots, x_n) = \lim(dx_1 \ldots dx_n)^{-1} \sum_{k_1, \ldots, k_l} P\begin{pmatrix} dx_1 & & dx_n & I_1 & & I_l \\ 1 & \cdots, & 1 & k_1 & \cdots, & k_l \end{pmatrix}$$

(7.4.7)

where I_i is the interval which separates dx_i from dx_{i+1}. The condition (7.4.5) enables us to apply the dominated convergence theorem and interchange in (7.4.7) the order of limiting and summation operations. The existence of the limits of the summands will follow from the expressions of the corresponding probabilities by means of integrals which generalize (7.4.3).

The r.d.f.-s can be expressed in terms of the densities as follows

$$f_I(x_1, \ldots, x_n) = \varphi(x_1, \ldots, x_n) - \sum_{m=1}^{\infty} \frac{(-1)^m}{m!}$$

$$\times \int_{I^m} \cdots \int \varphi(x_1, \ldots, x_n, u_1, \ldots, u_m) \, du_1 \ldots du_m \qquad (7.4.8)$$

This expression can be viewed as an integro-differential counterpart of the inclusion–exclusion type formula valid for an arbitrary set of events $\{A_i\}_{i=1}^M$:

$$P\left(\bigcap_1^m A_i \cap \left(\bigcap_{m+1}^M A_i^c\right)\right) = P\left(\bigcap_1^m A_i\right) - \sum(-1)^k \sum_{J_k} P\left(\bigcap_1^m A_i \cap \left(\bigcap_{i \in J_k} A_i\right)\right)$$

(the last summation is over subsequences $J_k \subset \{m+1, \ldots, M\}$ of size k). Connections with (7.4.8) become clear when we put

$$A_i = \begin{pmatrix} \delta_i \\ 1 \end{pmatrix}$$

for an appropriately chosen set $\delta_1, \ldots, \delta_M$ of infinitesimal ($M \to \infty$) intervals for which $\delta_i \cap \delta_j = \varnothing$ if $i \neq j$, $\bigcup \delta_i = I$. There is an important question: what are the conditions which guarantee that a system of non-negative symmetrical functions

$$\varphi(x_1, \ldots, x_n), \qquad n = 1, 2, \ldots$$

actually generates a probability P via (7.4.8) and (7.4.3)? It is not difficult to prove (see [64]) that

(1) the existence of a constant C such that

$$\varphi(x_1, \ldots, x_n) \leqslant C^n, \qquad n = 1, 2, \ldots \qquad \text{and} \qquad (7.4.9)$$

(2) the right-hand side of (7.4.8) is non-negative for every I and n,

are sufficient conditions.

7.5 Situation in other spaces

The procedure of construction of probabilities on $(\mathcal{M}, \mathcal{A})$ outlined in §§7.2–7.4 for the case $\mathbb{X} = \mathbb{R}$ can be used with minimal alterations for other spaces $\mathbb{X}$ if an appropriate class of subsets of $\mathbb{X}$ with which to replace the linear intervals is at hand.

The basic and only properties of the intervals on $\mathbb{R}$ which we used in the above construction are that they constitute a *semiring* which generates the Borel sets of $\mathbb{R}$. Therefore, the class of subsets of $\mathbb{X}$ in question should also be a semiring generating the Borel sets in $\mathbb{X}$. If such a semiring S can be identified, then all the statements of §§7.2–7.4 can be transferred to point processes in $\mathbb{X}$ by simply replacing the intervals in their formulation by elements of S. Recall that S is called a semiring if (see [46])

(1) S is closed with respect to taking finite intersection, and,

(2) if $A, B \in S$, then the relative complement of A in B can be represented as a union of finite number of pairwise non-intersecting elements from S.

In the spaces

$$\mathbb{X} = \mathbb{R}^n, \qquad \overline{G}, \qquad \mathbb{M}_2(=\Delta_2^*), \qquad \mathbb{A}_2 \qquad (7.5.1)$$

and others there are natural semirings of subsets. Let us describe them for the spaces in (7.5.1) (since we later consider point processes in these spaces). These spaces have certain product representations (see chapters 2–4), therefore the following well-known result will be useful.

Let $\mathbb{X}_1$ and $\mathbb{X}_2$ be two spaces and let S_i be a semiring in $\mathbb{X}_i$ which generates the Borel sets of $\mathbb{X}_i$, $i = 1, 2$. Then the product sets $A_1 \times A_2$ with $A_1 \in S_1$, $A_2 \in S_2$ constitute a semiring which generates Borel sets in $\mathbb{X}_1 \times \mathbb{X}_2$.

The factor spaces which we need are the line $\mathbb{R}$ and the circle $\mathbb{S}_1$. On $\mathbb{R}$ we use the intervals; on $\mathbb{S}_1$ we use the arcs (under appropriate conventions concerning the boundaries). We obtain table 7.5.1 of semirings for spaces in (7.5.1).

It is clear from the remarks above that to specify a probability on the spaces from (7.5.1), say, requires knowledge of the probabilities of the events of the type

Table 7.5.1

The space $\mathbb{X}$	Product representation of $\mathbb{X}$	Elements of S
$\mathbb{R}^n$	$\mathbb{R} \times \mathbb{R} \times \cdots \mathbb{R}$ (n times)	$I_1 \times \cdots \times I_n$, where each I_i is a linear interval
$\overline{G}$	$\mathbb{S}_1 \times \mathbb{R}$	$\alpha \times I$ (a 'shield'), where α is an arc and I is a linear interval
$\mathbb{M}_2$	$\mathbb{S}_1 \times \mathbb{R}^2$	$I_1 \times I_2 \times \alpha$, where I_i, α are as above
$\mathbb{A}_2$	$\mathbb{R}^2 \times (0, \infty) \times \mathbb{R} \times \mathbb{S}_1$	$I_1 \times I_2 \times J \times I_3 \times \alpha$, where I_i, α are as above, and $J \subset (0, \infty)$ is an interval

$$\begin{pmatrix} A_1 & & A_s \\ l_1 & , \ldots, & l_s \end{pmatrix} = \{m \in \mathcal{M}_{\mathbb{X}} : \text{card } m \cap A_i = l_i, i = 1, \ldots, s\},$$

where $l_1, \ldots, l_s$ are non-negative integers. In fact it is sufficient to consider the class $\mathcal{H}_{\mathbb{X}}$ of such events defined by the following conditions:

(a) each A_i belongs to S;
(b) $A_i \cap A_j = \varnothing$ if $i \neq j$ and $\bigcup A_i \in S$;
(c) each l_i is either 0 or 1.

(Compare with §7.4.)

We can also consider Poisson probabilities (and therefore Poisson point processes) on these spaces governed by measures. The basic formula here is the same as (7.3.6); for events from $\mathcal{H}_{\mathbb{X}}$ we have

$$P\begin{pmatrix} A_1 & & A_s \\ l_1 & , \ldots, & l_s \end{pmatrix} = \prod_{i=1}^{s} \frac{(m(A_i))^{l_i}}{l_i!} e^{-m(A_i)}, \qquad (7.5.2)$$

where m is a measure on $\mathbb{X}$.

Using the extension of probability, it is possible to show that (7.5.2) remains valid if $A_i \subset \mathbb{X}$ are arbitrary Borel and l_i are arbitrary non-negative integers. The only significant requirement that remains is that the sets $A_1, \ldots, A_s$ be pairwise disjoint.

Point processes in the spaces $\overline{G}$, $\overline{\mathbb{E}}$, $\overline{\Gamma}$, Δ_2^*, etc. are also termed as *line processes on $\mathbb{R}^2$, plane and line processes in $\mathbb{R}^3$, segment processes on $\mathbb{R}^2$, etc.* In general, we call random sets of geometrical objects which correspond to point processes *geometrical processes*.

Remark A rich class of point processes arises if we allow the measure m in (7.5.2) to be *random*. A complete description of a process from this class requires the specification of the probability distribution of m.

It can be useful to think about these processes as being constructed in two stages: first we draw a realization of m, and then we generate a Poisson process governed by this realization. For this reason such processes are called *doubly stochastic Poisson* (their other name is *Cox processes*). In stochastic geometry they have been considered in connection with Davidson's hypothesis (see [10], [43]). We consider processes from this class in §10.5. The construction of point processes by means of densities as outlined at the end of §7.4 for $\mathbb{X} = \mathbb{R}$ generalizes to other spaces without substantial changes. We discuss densities for $\mathbb{X} = \mathbb{R}^n$ case in the concluding sections of chapter 8.

7.6 The example of L. Shepp

We give here an example due to Shepp of a point process in one dimension such that the number of points in any interval I is Poisson distributed with mean $\lambda|I|$ for some $\lambda > 0$, but *for disjoint intervals these random numbers are not independent*.

We restrict ourselves to a point process $m(\omega)$ on the interval $(0, 1)$ in $\mathbb{R}$. If we require a point process with similar properties on the whole line, then it is enough to construct independent realizations of $m(\omega)$ on each interval $(k, k + 1)$. If we additionally require $\overline{\mathbb{T}}$-invariance, then a random shift by a vector uniformly distributed within $(0, 1)$ will be sufficient.

We mentioned in §7.3 that a Poisson process in $(0, 1)$ can be constructed by choosing the number of points according to a Poisson distribution and dropping them independently of each other and uniformly on $(0, 1)$.

Fix (the rate) $\lambda > 0$. Choose n with probability $\lambda^n e^{-n}/n!$, $n = 0, 1, \ldots$, and let $F_n(x_1, \ldots, x_n) = x_1 \cdots x_n$ for $n \neq 3$ be the cumulative distribution function of the n points $t_1, \ldots, t_n$ of $m(\omega)$ conditional upon $N((0, 1), m(\omega)) = n$. If n happens to be 3 take

$$F_3(x_1, x_2, x_3) = x_1 x_2 x_3 + \varepsilon(x_1 - x_2)^2(x_1 - x_3)^2(x_2 - x_3)^2$$
$$\times\, x_1 x_2 x_3(1 - x_1)(1 - x_2)(1 - x_3). \qquad (7.6.1)$$

For sufficiently small $\varepsilon > 0$, F_3 is a distribution function. Note that $m(\omega)$ is not Poisson. Define

$$G_n(a, b, m) = P_n \{\text{exactly } m \text{ of } t_1, \ldots, t_n \in (a, b)\}$$
$$= C_n^m P_n \{t_1, \ldots, t_m \in (a, b) \text{ and } t_{m+1}, \ldots, t_n \bar{\in} (a, b)\}$$
$$= C_n^m E_n \prod_{j=1}^{m} (I_b(t_j) - I_a(t_j)) \prod_{j=m+1}^{n} (I_a(t_j) + I_1(t_j) - I_b(t_j)), \quad (7.6.2)$$

where the probability P_n corresponds to F_n, E_n is its expectation, and

$$I_a(t) = 1 \quad \text{if } t < a$$
$$= 0 \quad \text{if } t > a.$$

Note that

$$E_n I_{a_1}(t_1) \cdots I_{a_n}(t_n) = F_n(a_1, \ldots, a_n).$$

In the expansion of (7.6.2) only terms of the form $F_n(a_1, \ldots, a_n)$ appear, where $a_i = a$, b or 1 for all i. Thus, if

$$F_n(a_1, \ldots, a_n) = a_1 \cdots a_n \qquad (7.6.3)$$

for all such $a_1, \ldots, a_n$ then $G_n(a, b, m)$ will be just as in the Poisson case. For $n \neq 3$, F_n is chosen to be uniform; for $n = 3$ (7.6.3) follows from (7.6.1).

7.7 Invariant models

An alternative approach to the construction of probabilities P in the space $(\mathcal{M}, \mathcal{A})$ is by using *models*. A point process is called a *model* in cases where we wish to emphasize that both the probability space $(\Omega, \mathcal{F}, \mathbf{p})$ and the map $m(\omega)$ are concrete (rather than 'general abstract' as in the definition in §7.1). Often, finding the values on $\mathcal{H}$ of the distribution of a model can be too hard a problem. Then the map defining the model is the only method of description that remains. In bad cases even calculation of some expectations, one-dimensional distributions, etc., can be a problem. The models we describe in the following sections all have distributions invariant with respect to a group.

Definition: Let $\mathcal{G}$ be a group of transformations of the space $\mathbb{X}$. A probability P on the corresponding $(\mathcal{M}, \mathcal{A})$ is called invariant with respect to $\mathcal{G}$ ($\mathcal{G}$-invariant) if for every $A \in \mathcal{A}$ and every transformation $\mathfrak{G} \in \mathcal{G}$ we have

$$P(\mathfrak{G}A) = P(A), \tag{7.7.1}$$

where

$$\mathfrak{G}A = \{\mathfrak{G}m, \, m \in A\}.$$

A point process $m(\omega)$ is called $\mathcal{G}$-invariant iff its distribution is $\mathcal{G}$-invariant.

By probability continuation it is possible to prove that P is $\mathcal{G}$-*invariant whenever* (7.7.1) *holds for the events from the class* $\mathcal{H}$ (see §7.5). Cases of invariance can be found among Poisson point processes. By virtue of (7.5.2), invariance of the governing measure m with respect to $\mathcal{G}$ implies invariance of the corresponding Poisson process with respect to the same group.

For instance, if $\mathbb{X} = \mathbb{R}^n$, m is n-dimensional Lebesgue, then the corresponding Poisson point process is invariant with respect to the groups $\mathbb{T}_n$, $\mathbb{M}_n$ and $\mathbb{A}_n$.

If $\mathbb{X} = \mathbb{G}$ and m is the $\mathbb{M}_2$-invariant measure on $\mathbb{G}$ (see §3.6), then the corresponding Poisson line process in $\mathbb{M}_2$-invariant. Using the results of chapters 2–4 this list of invariant Poisson processes can be easily continued. In the three sections that follow we focus on invariant point processes whose nature is opposite to Poisson processes – those constructed by means of lattices of points.

7.8 Random shift of a lattice

Let $m_0 \subset \mathbb{R}^n$ be a lattice in $\mathbb{R}^n$, i.e. a set of points in $\mathbb{R}^n$ which in some 'affine' (non-orthogonal) system of coordinates have integer-valued coordinates (see fig. 7.8.1).

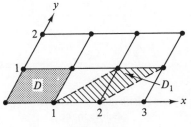

Figure 7.8.1

We consider shifts tm_0 of this lattice. The shifts t_1, $t_2 \in \mathbb{T}_n$ are termed equivalent if

$$t_1 m_0 = t_2 m_0.$$

A set $D \subset \mathbb{T}_n$ is called a fundamental region (see fig. 7.8.1) if

(a) there are no pairs of equivalent shifts in D;
(b) for every $t \in \mathbb{T}_n$ there is an equivalent shift in D.

We choose for Ω a fundamental region D, to be considered below as fixed, and let $\mathbf{p}_1$ be (the restriction of) the *Lebesgue measure* on D normalized by the condition

$$\mathbf{p}_1(D) = 1.$$

We define a point process on $\mathbb{R}^n$ by means of the map

$$m(t) = tm_0, \qquad t \in D. \tag{7.8.1}$$

In words: $m(t)$ is a lattice obtained from m_0 by a random parallel shift which is distributed uniformly within D. To see that $m(t)$ is $\mathbb{T}_n$-invariant we remark that to every event $A \subset \mathcal{M}_{\mathbb{R}^n}$ corresponds a set $A^* \subset D$, namely

$$A^* = \{t \in D : tm_0 \in A\}.$$

Let P be the distribution of $m(t)$. Clearly

$$P(tA) = \mathbf{p}_1((tA)^*) = \mathbf{p}_1(tA^*/D),$$

where tA^* denotes the usual shift of a set $A^* \subset D$ by a vector t, and tA^*/D denotes the result of replacing each point of tA^* which lies outside D by the equivalent point in D (see fig. 7.8.2).

Clearly, the transformation tA^*/D is Lebesgue measure-preserving, therefore

$$P(tA) = \mathbf{p}_1(A^*) = P(A).$$

We leave it to the reader to prove that the distribution P of $m(t)$ does not depend on the choice of the fundamental region D.

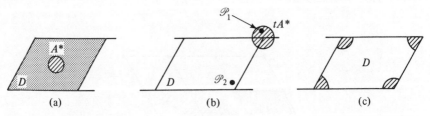

Figure 7.8.2 The point $\mathcal{P}_2$ is equivalent to $\mathcal{P}_1$. The set tA^*/D is the union of the four sectors in (c)

7.9 Random motions of a lattice

Let us first consider invariance with respect to rotations only.

For Ω we choose the group $\mathbb{W}_n$ of rotations of $\mathbb{R}^n$ around O. Let the probability $\mathbf{p}_2$ on $\mathbb{W}_n$ be proportional to the Haar measure on $\mathbb{W}_n$. The map

$$m(w) = w m_0, \tag{7.9.1}$$

where $w m_0$ is the result of rotation of m_0 by $w \in \mathbb{W}_n$, *defines a $\mathbb{W}_n$-invariant point process for any fixed $m_0 \in \mathcal{M}$* (not necessarily a lattice). Indeed, the probability of any $A \subset \mathcal{M}_{\mathbb{R}^n}$ equals

$$\mathbf{p}_2(\{w \in \mathbb{W}_n : w m_o \in A\}).$$

Our assertion follows from the relation

$$\{w \in \mathbb{W}_n : w m_0 \in w_1 A\} = w_1^{-1} \{w \in \mathbb{W}_n : w m_0 \in A\}.$$

In fact a stronger proposition is valid. Let $m(\omega)$ be an arbitrary point process in $\mathbb{R}^n$ defined on some probability space $(\Omega, \mathscr{F}, \mathbf{p})$. We construct a new point process

$$m(w, \omega) = w m(\omega) \tag{7.9.2}$$

which is defined on the product space $\mathbb{W}_n \times \Omega$ where we consider the probability measure $\mathbf{p}_2 \times \mathbf{p}$. *The point process (7.9.2) is always $\mathbb{W}_n$-invariant.* Remarkably, this follows from a kind of Cavalieri principle, according to which $\mathbf{p}_2 \times \mathbf{p}$ is always invariant with respect to the transformations

$$w_1(w, \omega) \to (w_1 w, \omega), \qquad w_1 \in \mathbb{W}_n$$

of the product space $\mathbb{W}_n \times \Omega$. Applying this remark to the process of §7.8 we conclude that $w t m_0$ (where m_0 is now a lattice, this process is defined on $\mathbb{W}_n \times D$) is $\mathbb{W}_n$-invariant.

The random lattice $w t m_0$ is also T_n-invariant. It is enough to show that for every $t_1 \in T_n$ the point process $t_1 w t m_0$ has the same distribution as $w t m_0$. This follows from the existence of $t_2 = t_2(w)$ for which

$$t_1 w = w t_2.$$

We have

$$t_1 w t m_0 = w t_2 t m_0$$

and our assertion follows from the fact that the distribution of $t_2 t m_0$ does not depend on t_2 (coincides with that of $t m_0$, see §7.8). From the above two properties follows $\mathbb{M}_n$-invariance of $w t m_0$.

7.10 Lattices of random shape and position

In this section we consider only planar lattices whose fundamental region has unit area (1-lattices). A lattice m is called *anchored* if $O \in m$.

We can represent the space of all anchored 1-lattices by an appropriate subset D of the group $\mathbb{A}_2^0$ (see §4.1); namely, each anchored m can be transformed into the standard square anchored 1-lattice by applying the following sequence of transformations of $\mathbb{R}^2$ (the order is important).

(1) A rotation w around O which brings the densest *one-dimensional* sublattice of m into a horizontal position. We can assume that O belongs to this sublattice.

(2) A transformation H from the group $\mathbb{H}_1$ (see §4.1) defined by the condition that horizontal one-dimensional sublattices in

$$Hwm$$

should be of unit space.

(3) A transformation C from the group $\mathbb{C}_1$ (see §4.1) which is defined by the condition that

$$CHwm = m_0$$

be the square lattice. Under C the shift of the line $y = 1$ should be to the right (say) and of minimal possible length.

This construction enables us to describe the space of anchored 1-lattices by means of the variables φ, h and c which correspond to the transformations w, H and C as described in §4.1. The domain D of these variables is as follows:

$$0 < \varphi < \pi$$
$$\sqrt[4]{\frac{3}{4}} < h < \infty$$
$$0 < c < 1 \qquad \text{if } h \geqslant 1 \tag{7.10.1}$$
$$\sqrt{(1 - h^4)} < c < 1 - \sqrt{(1 - h^4)} \qquad \text{if } h \leqslant 1.$$

(Note that the value $(4/3)^{1/4}$ can be obtained as the maximum of the minimal distance between the vertices of a parallelogram of area 1.)

It is important to note that $D \subset \mathbb{A}_2^0$ is a fundamental region in $\mathbb{A}_2^0$ in a sense similar to that of §7.8. Here two elements A_1^0, $A_2^0 \in \mathbb{A}_2^0$ are equivalent if

$$A_1^0 m = A_2^0 m$$

for some fixed m (the square 1-lattice say). To calculate the value of the Haar measure of D we represent dA^0 in the form (see §4.3)

$$dA^0 = h^{-3} \, dc \, dh \, d\varphi.$$

Then simple integration over the range (7.10.1) yields

$$\int_D dA^0 = \frac{\pi^2}{6} \tag{7.10.2}$$

i.e. *the Haar measure of D is finite.*

Now we construct a model of a *random* $\mathbb{A}_2^0$-invariant anchored 1-lattice. We take

$$\Omega = D,$$

$\mathbf{p} = \mathbf{p}_3$, the restriction of the Haar measure dA^0 to D, normalized by the condition

$$\mathbf{p}_3(D) = 1$$

and put

$$m(A^0) = A^0 m_0, \qquad A^0 \in D, \tag{7.10.3}$$

where m_0 is the usual square 1-lattice. The proof of invariance can be obtained by appropriate modifying of the reasoning presented by fig. 7.8.2.

An $\mathbb{A}_2$-invariant model of (non-anchored) random 1-lattices can be constructed as follows (the group $\mathbb{A}_2$ was introduced in §4.4). We take

$$\Omega = (0, 1)^2 \times D,$$

and we let the probability $\mathbf{p}_4$ on Ω be the product of the normalized Lebesgue measure on $(0,1)^2$ and the $\mathbf{p}_3$ considered above. Given an element $(\mathscr{P}, A^0) \in \Omega$ we define the shift

$$t = \overrightarrow{0, A^0\mathscr{P}},$$

i.e. t shifts O to the image of $\mathscr{P}$ under A^0. Clearly t has uniform distribution in the fundamental parallelogram $A^0(0, 1)^2$ of $A^0 m_0$, where m_0 is the standard square (anchored) 1-lattice. We now put

$$m(\mathscr{P}, A^0) = tA^0 m_0, \tag{7.10.4}$$

where $(\mathscr{P}, A^0) \in \Omega$. A proof of $\mathbb{A}_2$-invariance of this point process can be obtained by appropriate modification of the reasoning presented by fig. 7.8.2 since (t, A^0) change in a 'fundamental region' of $\mathbb{A}_2$.

Remark There is a connection between (7.10.2) and a theorem attributed to Siegel in [2]. The Siegel theorem states that Haar measures of fundamental domains in $\mathbb{A}_n$ are finite for every $n > 0$. The fundamental domains are defined with respect to lattices in $\mathbb{R}^n$. This implies the existence of $\mathbb{A}_n$-invariant random lattices in $\mathbb{R}^n$.

7.11 Kallenberg–Mecke–Kingman line processes

This example was devised by Kallenberg, Mecke and Kingman [9], [48] to provide a counterexample to a hypothesis by Davidson (see [43]). The line process we describe is $\mathbb{T}_2$-invariant, does not contain parallel lines, cannot be represented as an m-Poisson line process with *random m* (is not Cox), and its second moment measure is locally-finite. By applying a random rotation around O, we obtain an $\mathbb{M}_2$-invariant line process which retains all remaining properties. Davidson's hypothesis was that processes with such properties do not exist.

We use the representation of lines $g \in \mathbb{G}_x$ (see §2.7) with coordinates

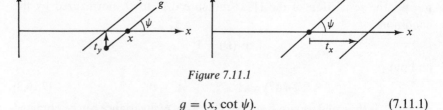

Figure 7.11.1

$$g = (x, \cot \psi). \tag{7.11.1}$$

Let us see how a parallel shift transforms the x, $\cot \psi$ plane.

We denote by t_x and t_y the projections of a shift vector $t \in \mathbb{T}_2$ on the x and y axes. Clearly, t_x and t_y act on the line (7.11.1) as follows (see fig. 7.11.1):

$$t_x g = (x + t_x, \cot \psi),$$

while

$$t_y g = (x - t_y \cdot \cot \psi, \cot \psi).$$

Therefore the action of t on the x, $\cot \psi$ plane reduces to, first, applying a transformation from the group $\mathbb{C}_1$ (see §4.1) with the $\cot \psi$ axis, now playing the role of y axis, and, second, applying a shift parallel to the x axis. The order of these two transformations is not significant. The $\mathbb{A}_2$-invariant model of the previous section, if transplated onto the x, $\cot \psi$ plane will be invariant with respect to the transformations just described (because the latter belong to subgroups of $\mathbb{A}_2$). However, we are still not finished, since there are too many lines corresponding to our random lattice. That is, with probability 1 we have infinitely many points of the lattice projecting on any finite interval on the x axis. In the line reinterpretation, this means that with probability 1 infinitely many lines hit this interval. This contradicts the basic requirement that any $m \in \mathcal{M}_G$ should have a finite number of points in a bounded region of $\mathbb{G}$. Thus what we obtain in this way is not a line process. The remedy lies in *taking only those points from the random lattice which fall in some strip parallel to the x-axis*. The (finite) width of the strip and the position of its central axis can be arbitrary (they become parameters of the line process).

Thus the truncated random lattice will no longer remain $\mathbb{A}_2$-invariant. Yet it will retain invariance with respect to any subgroup of $\mathbb{A}_2$ which maps the strip onto itself. As we have seen, the group $\mathbb{T}_2$ acting on the x, y-plane induces just such a subgroup acting on the x, $\cot \psi$-plane. Hence we have $\mathbb{T}_2$-invariance of the line process whose lines correspond to the points of the random lattice which lie in the strip. We note that further examples of $\mathbb{T}_2$-invariant line processes can be obtained by applying appropriate independent thinning to the points of the $\mathbb{A}_2$-invariant lattice (the details are left to the reader). Sukiasian has constructed still further examples by taking parabolic inversion of the random lattice [47].

7.12 Marked point processes: independent marks

To represent a point process in a space $\mathbb{X}$ we will sometimes write $\{x_i\}$ instead of $m(\omega)$. This notation appeals directly to the concept of the countable set of points $x_i \in \mathbb{X}$; the 'chance' variable ω is suppressed. According to this a point process in a product space

$$\mathbb{X} \times \mathbb{Y}$$

can be represented as

$$\{(x_i, y_i)\}, \qquad (x_i, y_i) \in \mathbb{X} \times \mathbb{Y}. \tag{7.12.1}$$

It is natural to call the random set $\{x_i\}$ the *projection* on $\mathbb{X}$ of the above process. A projection can fail to be a point process. For instance, the set obtained by projecting on the x axis the points of a planar Poisson process governed by L_2 contains probability 1 infinitely many points in each interval on this axis.

Definition: A point process $\{(x_i, y_i)\}$ in a product space $\mathbb{X} \times \mathbb{Y}$ is called a marked point process in $\mathbb{X}$ with marks in $\mathbb{Y}$ if its projection $\{x_i\}$ is a point process.

Let $\mathscr{G}$ be a group acting on $\mathbb{X}$. A marked point process $\{(x_i, y_i)\}$ is called $\mathscr{G}$-invariant if the distribution of

$$\{(\mathfrak{G}x_i, y_i)\}$$

does not depend on $\mathfrak{G} \in \mathscr{G}$.

In the group-invariant case it is possible under certain general conditions to define what is the *distribution of typical mark*; this idea plays a key role in the remaining part of the book and we give its definition in §8.1. Here we describe the situation in the simplest (but by no means most frequent) case of independent marks.

We say that in $\{(x_i, y_i)\}$ *the marks y_i are independent* if

(1) the point process $\{x_i\}$ and the sequence $\{y_i\}$ are independent;
(2) $\{y_i\}$ is a sequence of independent identically distributed random variables.

The theory of the classes $\mathscr{H}_\mathbb{X}$ developed in §7.5 enables us to equivalently reformulate the above definition in terms of the point process in the product space $\mathbb{X} \times \mathbb{Y}$ which corresponds to $\{(x_i, y_i)\}$. Namely we choose $A \in \mathscr{H}_\mathbb{X}$ of the form

$$A = \begin{pmatrix} I_1 & & I_s & J_1 & & J_r \\ 1 & ,\dots, & 1 & 0 & ,\dots, & 0 \end{pmatrix}, \qquad I_i, J_k \in S$$

and any sets $C_1, \dots, C_s \subset \mathbb{Y}$. We define

$A^* = A \cap \{\text{the point from } \{x_i\} \text{ which lies in } I_i \text{ has its mark in } C_i, i = 1, \dots, s\}$.

The marks are independent if and only if

$$P(A^*) = P_1(A) \cdot \prod_{i=1}^{s} F(C_i),$$

where P_1 is the distribution of some point process in $\mathbb{X}$, while F is some probability distribution in the space $\mathbb{Y}$. F is called the distribution of a mark and coincides with the distribution of the typical mark (whenever the latter exists, see §8.1).

Now we give some examples of marked point processes with independent marks.

I Poisson point process in a strip Let us denote by S_a the planar strip

$$\{(x, y): -\infty < x < \infty, 0 < y < a\} = \mathbb{R} \times (0, a) \subset \mathbb{R}^2.$$

Let $m(\omega)$ be a Poisson point process in $\mathbb{R}^2$ governed by the restriction of a Lebesgue measure $\lambda \cdot L_2$ to S_a (equivalently, $m(\omega)$ is the part of a planar Poisson process governed by $\lambda \cdot L_2$ which lies in S_a). We have

$$m(\omega) = \{(x_i, y_i)\},$$

where x_i and y_i are the usual Cartesian coordinates of points from $m(\omega)$.

In S_a let us consider for instance the system of rectangles shown in fig. 7.12.1. We put

$$A = \begin{pmatrix} I & J \\ 1 & 0 \end{pmatrix}$$

$A^* = A \cap$ {the mark of the point from $\{x_i\}$ which lies in I belongs to C}.

Cleary A^* coincides with the following subset of $\mathcal{M}_{S_a}$:

$$A^* = \begin{pmatrix} B_1 & B_2 & B_3 \\ 1 & 0 & 0 \end{pmatrix}.$$

According to (7.5.2) we have

$$P(A^*) = e^{-\lambda|I|a}\lambda|I||C|e^{-\lambda|J|a}$$

$$= \frac{|C|}{a} P_1(A),$$

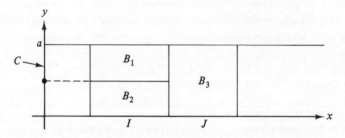

Figure 7.12.1 The interval C is the projection of B_1 on the y axis

where P_1 is the distribution of a Poisson process on the x axis governed by the measure $\lambda a \cdot L_1$. Because a similar factorization applies to every $A \in \mathscr{H}_{S_a}$, we conclude that in our example *the projection process* $\{x_i\}$ *is Poisson governed by* $\lambda a \cdot L_1$, *and the marks* y_i *are independent with uniform distribution on* $(0, a)$.

II Line processes The line process constructed in §7.11 could obviously be presented as a marked point process

$$\{(x_i, \cot \psi_i)\} \qquad \text{or} \qquad \{(x_i, \psi_i)\},$$

where the points x_i have been defined to lie on the Ox axis.

In fact similar representations exist for any $\mathbb{T}_2$-invariant line process $\{g_i\}$ and any direction θ if the condition

$$P\{\text{there are lines in } \{g_i\} \text{ having direction } \theta\} = 0$$

is satisfied. In this case we speak about θ-representation of $\{g_i\}$: by definition, in a θ-representation, $\{x_i\}$ are the points of intersection of lines from $\{g_i\}$ with an axis having direction θ, ψ_i is the angle between this axis and the corresponding line from $\{g_i\}$.

Let μ be a $\mathbb{T}_2$-invariant measure on $\mathbb{G}$ whose rose of directions m given by the factorization table 2.9.1 is atomless. What will be the θ-representations of the Poisson line process governed by μ?

The answer is as follows

For every direction θ, $\{x_i\}$ is a one-dimensional Poisson process and $\{\psi_i\}$ is a sequence of independent angles; the measure which governs $\{x_i\}$ is $\lambda(\theta)\,dx$ (Lebesgue) where $\lambda(\theta)$ is the rose of hits corresponding via (2.10.1) to the rose of directions of μ; the distribution of each angle ψ_i is proportional to

$$\sin \psi \, m_\theta(d\psi),$$

where m_θ is the rotated rose of directions.

To prove for instance Poisson property of $\{x_i\}$ it is enough to take in (7.5.2)

$$A_i = [I_i] = \{g \in \mathbb{G} : g \text{ hits the interval } I_i\},$$

where $I_1, \ldots, I_s$ are disjoint intervals on the line which carries the $\{x_i\}$ process, and to note that the sets A_i can be treated as pairwise non-intersecting.

It follows that if a Poisson $\{g_i\}$ is governed by an $\mathbb{M}_2$-invariant measure on $\mathbb{G}$ then the governing measure of $\{x_i\}$ and the distribution of ψ_i cease to depend on θ; the distribution of ψ_i always has the density

$$\tfrac{1}{2} \sin \psi \, d\psi.$$

This corresponds to the property (7.15.7) below since such a line process happens to be $\mathbb{M}_2$-invariant.

Yet another way to obtain an $\mathbb{M}_2$-invariant line process is to take Poisson $\{g_i\}$ governed by a $\mathbb{T}_2$-invariant measure μ and then to subject it to an independent random rotation with uniform distribution on W_2. The result will clearly be a doubly-stochastic Poisson point process (see the end of §7.5). It follows that on any fixed axis the process $\{x_i\}$ in this case will be doubly-stochastic Poisson governed by random measure $\lambda(\theta)\,dx$ (where θ is *random* and has uniform distribution on $(0, \pi)$). Also (generally speaking) the marks ψ_i in this process are not independent. A calculation of an expectation for this process is given in §10.5.

III Poisson processes of balls A process $\{b_i\}$ of balls in $\mathbb{R}^n$ can be described as a marked point process in $\mathbb{R}^n$, namely

$$\{b_i\} = \{(\mathscr{P}_i, r_i)\}, \tag{7.12.2}$$

where $\mathscr{P}_i$ is the center of the ball b_i and r_i is its radius. We are interested in $\mathbb{M}_n$-invariant ball processes and here the simplest case is where $\{\mathscr{P}_i\}$ is Poisson governed by the Lebesgue measure $\lambda \cdot L_n$ and the radii r_i are independent with some distribution $\mathbf{p}$ common for all centers. Such a process can be equivalently represented as a Poisson point process in the space

$$\mathbb{R}^n \times (0, \infty)$$

governed by the product measure $\lambda \cdot L_n \times \mathbf{p}$. Therefore in this case we call the ball process itself Poisson. Ball processes in $\mathbb{R}^2$ are naturally termed as discs, and in $\mathbb{R}$ as segment processes.

A natural condition usually imposed on such Poisson ball processes is that

$$\int_0^\infty r^{n-1} \mathbf{p}(dr) < \infty, \tag{7.12.3}$$

which guarantees the following properties:

(a) with probability 1 the balls of the process do not cover the whole of $\mathbb{R}^n$;
(b) on each line $g_0 \subset \mathbb{R}^n$ the 'trace' segment process

$$\{b_i \cap g_0\}$$

is Poisson governed by some measure of the form $\lambda_1 \cdot L_1 \times \mathbf{p}_1$.

The expression of λ_1 and the distribution $\mathbf{p}_1$ in terms of λ may be performed in terms of $\mathbf{p}$ – a traditional topic in geometrical probability theory (see, e.g. [2] and [12]).

IV Poisson cluster processes In a less trivial case

$$\mathbb{X} = \mathbb{R}^n, \qquad \mathbb{Y} = \mathscr{M}_{\mathbb{R}^n},$$

i.e. marks are realizations of point processes in $\mathbb{R}^n$. Let $\{x_i\}$ be Poisson governed by $\lambda \cdot L_n$, and let the marks

$$y_i = m_i$$

be independent. (This situation contrasts sharply with the one we face in our approach to Palm distribution in chapter 8 where marks from $\mathcal{M}_{\mathbb{R}^n}$ will be strongly dependent.)

We can consider the m_i's to be a sequence of independent identically distributed point processes in $\mathbb{R}^n$:

$$m_i = m_i(\omega).$$

If there exists a ball $b(r, O)$ such that

$$P(m_i(\omega) \subset b(r, O)) = 1$$

and

$$E_P N(m_i, b(r, O)) < \infty$$

then $m_i(\omega)$ is called a *cluster*. In this case the union

$$m(\omega) = \bigcup t_i m_i(\omega),$$

where $t_i = \overline{Ox_i}$, and x_i belong to the Poisson process $\{x_i\}$, is a $\mathbb{T}_n$-invariant point process; we call it a *Poisson cluster* process in $\mathbb{R}^n$. If

$$P(N(m_i(\omega), \mathbb{R}^n) = 0 \text{ or } 1) = 1$$

then $m(\omega)$ will be Poisson. But if

$$P(N(m_i(\omega), \mathbb{R}^n) > 1) > 0$$

then $m(\omega)$ is non-Poisson. This assertion follows from the theory of §8.11.

A similar construction can be used in the space G. We start from an $\mathbb{M}_2$-invariant Poisson line process $\{g_i\}$ whose lines we now call 'parent lines'. To each parent line g_i we attach a group (*cluster*) of lines parallel to g_i; let the probability law governing the number of lines in a cluster, as well as their distances from g_i, be the same for all parent lines; and let the realizations of clusters belonging to different parent lines be independent. We may assume that the lines of a cluster lie (with probability 1) within some strip of finite breadth. Then the union of all clusters will be an $\mathbb{M}_2$-invariant line process, which can also be called a Poisson cluster process. The theory of Palm distributions of line processes (see §10.1) can be an appropriate tool for their study.

7.13 Segment processes and random mosaics

In most cases the $\mathbb{T}_2$-invariant line processes in $\mathbb{R}^2$ are completely determined by their marked point process representations. But this is not the case for a similar construction applied to segment processes.

Let $m_1(\omega)$ be a *Poisson* segment process on $\mathbb{R}^2$ (a point process on Δ_2). It happens to be $\mathbb{M}_2$-invariant iff its governing measure μ is $\mathbb{M}_2$-invariant. By Haar factorization

$$\mu(d\delta) = dM\, m(dl),$$

where m is a measure in the space of segment lengths.

To guarantee that the number of intersections of the segments from $m_1(\omega)$ with any finite interval on the x axis be finite with probability 1, we require that

$$\int lm(\mathrm{d}l) < \infty. \tag{7.13.1}$$

Under this additional condition the marked point process $\{(x_i, \psi_i)\}$ is well defined, where $\{x_i\}$ is the point process of intersections of $m_1(\omega)$ with the x axis, and ψ_i is the angle of intersection at x_i. The process $\{(x_i, \psi_i)\}$ generated by $m_1(\omega)$ has a distribution of the same type as the process $\{(x_i, \psi_i)\}$ generated by the $\mathbb{M}_2$-invariant Poisson line process $\{g_i\}$ (see §7.12, II).

For brevity we call this distribution PIAsin (Poisson, independent angles, sine law).

Each line of Poisson $\{g_i\}$ is split into segments by other lines from $\{g_i\}$, and the collection of all such segments is a *segment process* $m_2(\omega)$. Clearly $m_2(\omega)$ is non-Poisson: with probability 1 $m_2(\omega)$ is a mosaic, while for $m_1(\omega)$ the probability of this event is zero (because the segments of $m_1(\omega)$ display 'loose ends' which are excluded in mosaics, see §6.8, V). Thus the segment process $m_2(\omega)$ is a *random mosaic*, and yet its $\{(x_i, \psi_i)\}$ process is of the same type as for Poisson $m_1(\omega)$, i.e. it is PIAsin.

We construct now a random mosaic $m(\omega)$ whose distribution is different from that of $m_2(\omega)$, although its $\{(x_i, \psi_i)\}$ process is again PIAsin. Let

$$m^{(0)}, \qquad m^{(1)}, \qquad m^{(2)}, \ldots$$

be a sequence of independent random mosaics with common distribution identical to that of $m_2(\omega)$.

Let $\{\pi_k\}$ be the collection of polygons generated by $m^{(0)}$. For fixed k, the polygon π_k is subdivided into an almost sure finite collection $\{\pi_{ki}\}$ by the random mosaic $m^{(k)}$. The random collection of polygons $\{\pi_{ki}\}$ generates a random mosaic $m(\omega)$ but its distribution is substantially different from that of $m_2(\omega)$. The difference can be seen in the kinds of nodes exhibited by the random mosaics in question. In fact, $m_2(\omega)$ has nodes only of X type, while $m(\omega)$ has nodes of both X and T types (see fig. 7.13.1).

By repeating this procedure we can obtain random mosaics where the 'percentage' of X-type nodes tends to zero, while that of T-type nodes tends to unity. For all these random mosaics their $\{(x_i, \psi_i)\}$ process remains PIAsin. We also make the following observation. The example of $m_1(\omega)$ shows that in

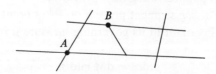

Figure 7.13.1 Part of a realization of $m(\omega)$. A is of X-type, B is of T-type

general the distribution of the process $\{(x_i, \psi_i)\}$ by no means determines the distribution of the length of the typical (see §8.1) segment in an $\mathbb{M}_2$-invariant segment process. However, if a segment process is a random mosaic with no T-type nodes the situation changes. We give the corresponding formula in chapter 10.

7.14 Moment measures

Conditions for the existence of different mathematical expectations underlie the treatment of many questions in stochastic geometry; we give some examples in the next section. Among these conditions the more frequent are the assumptions of existence of the first and the second moment measures.

Let $m(\omega)$ be a point process in a space $\mathbb{X}$, and let P be its distribution. As usual the random variable $N(B, m)$ equals the number of points of $m(\omega)$ in a set $B \subset \mathbb{X}$. We consider the expectation

$$m_1(B) = E_P N(B, m).$$

For every $m \in \mathcal{M}_{\mathbb{X}}$, $N(B, m)$ as a function of B is a measure on $\mathbb{X}$. Therefore $m_1(B)$ is also a measure on $\mathbb{X}$ but possibly not a locally-finite one. If m_1 happens to be locally-finite (i.e. a measure in our usual sense) then m_1 is called the *expectation measure* or the *first moment measure* of $m(\omega)$; the point process itself is of *first order*.

Now let $\mathcal{G}$ be a group acting on $\mathbb{X}$ and let there be a unique (up to a constant factor) measure μ on $\mathbb{X}$ which is invariant with respect of $\mathcal{G}$. Clearly if $m(\omega)$ is $\mathcal{G}$-invariant and of first order then m_1 inherits the property of $\mathcal{G}$-invariance, and therefore necessarily

$$m_1 = \lambda \cdot \mu, \qquad 0 < \lambda < \infty. \tag{7.14.1}$$

In such a case we say that (a $\mathcal{G}$-invariant) $m(\omega)$ is of *finite intensity*, and the number λ is called the *intensity* of $m(\omega)$. The intensity has the meaning of the expected number of points from $m(\omega)$ in any $B \subset \mathbb{X}$ of unit μ-measure.

The above ideas and terminology extend to random measures on $\mathbb{X}$ of a more general nature. Sometimes the random measures are generated by marked point processes in the same space $\mathbb{X}$, and then one should be careful to distinguish between different interpretations of the finite intensity condition. We illustrate this by an example.

Let $m(\omega)$ be a $\mathbb{T}_2$-invariant segment process in $\mathbb{R}^2$. Suppose that using (2.13.3) we can describe $m(\omega)$ as a marked point process

$$m(\omega) = \{(Q_i, \varphi_i, l_i)\}.$$

Here one of the possible finite intensity conditions is that the point process $\{Q_i\}$ be of finite intensity. The second is in terms of the random measure

$\mathscr{L}(m, B) = \{$the total length of the segments (or their parts) from m which
   lie within $B \subset \mathbb{R}^2\}$,

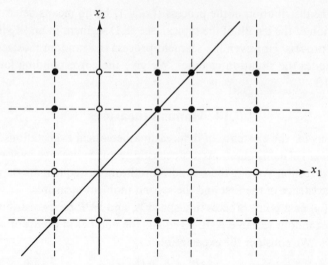

Figure 7.14.1 The circles represent the points of an $m \in \mathcal{M}_\mathbb{R}$; the solid discs represent the points of m^2. Note the presence of points from m^2 on the diagonal $\{x_1 = x_2\}$

and requires that

$$E_P \mathcal{L} = \lambda_1 \cdot L_2 \qquad \text{with some} \qquad 0 < \lambda_1 < \infty. \qquad (7.14.2)$$

Clearly the two assumptions say quite different things concerning $m(\omega)$ (neither follows from the other).

Let us turn now to second moment measures. For every $m \in \mathcal{M}_\mathbb{X}$ we construct its 'square' m^2 which is the following point set in $\mathbb{X} \times \mathbb{X}$ (see fig. 7.14.1):

$$m^2 = \{(x_i, x_j) : x_i, x_j \in m\}.$$

Clearly $m^2 \in \mathcal{M}_{\mathbb{X} \times \mathbb{X}}$; the dependence of m and ω induces dependence of m^2 on ω, thus we obtain a point process $m^2(\omega)$ on $\mathbb{X} \times \mathbb{X}$.

We say that $m(\omega)$ is of *second order* if $m^2(\omega)$ as a point process on $\mathbb{X} \times \mathbb{X}$ is of first order. The expectation measure of $m^2(\omega)$ is written for product sets $B_1 \times B_2 \subset \mathbb{X} \times \mathbb{X}$ as

$$m_2(B_1 \times B_2) = E_P N(B_1, m) N(B_2, m).$$

m_2 is a measure in $\mathbb{X} \times \mathbb{X}$ and is called the *second moment measure* of $m(\omega)$.

Let, for instance, $m(\omega)$ be a Poisson point process in $\mathbb{R}^n$ governed by $\lambda \cdot L_n$. If $B_1 \cap B_2 = \varnothing$ then $N(B_1, m)$ and $N(B_2, m)$ are independent. Therefore

$$m_2(B_1 \times B_2) = \lambda^2 \cdot L_n(B_1) L_n(B_2) \qquad \text{if } B_1 \cap B_2 = \varnothing.$$

On the other hand, for every $B \subset \mathbb{X}$,

$$m_2(\{(x_1, x_2) : x_1 = x_2 \in B\}) = E_P N^2(B, m) = \lambda L_n(B)[\lambda L_n(B) + 1].$$

Thus by continuation m_2 is a product of Lebesgue measures plus a singular component concentrated on the diagonal of $\mathbb{X} \times \mathbb{X}$. A singular component concentrated on the diagonal always appears in the second moment measure of a point process.

If a second order $m(\omega)$ is invariant with respect to a group acting on $\mathbb{X}$ then m_2 inherits this invariance property. Therefore in appropriate cases *factorizations in the style of our §2.14 or §3.15 can be applied to m_2.*

Example 1 The unit radius disc process on $\mathbb{R}^2$ which we consider in chapter 10 can be described as an $\mathbb{M}_2$-invariant second order point process (of disc centers) on $\mathbb{R}^2$. The assumption that the second moment measure m_2 of such a process possesses a density means that outside the 'diagonal' m_2 has a density with respect to $L_2 \times L_2$ or equivalently, in the notation of (3.14.1), m_2 has the form

$$f(l)l \, dl \, dM$$

(f is called the density function).

Example 2 Let $\{g_i\}$ be an $\mathbb{M}_2$-invariant second order line process on $\mathbb{R}^2$ which has with probability 1 no pairs of parallel lines. Then

$$m_2 = \mu,$$

which is an $\mathbb{M}_2$-invariant symmetrical measure on $\mathbb{G} \times \mathbb{G}$ satisfying the conditions (6.7.9) and (6.7.10). In particular for the vertical windows v_1 and v_2, as shown in fig. 6.5.1, we will have

$$m_2(\{(g_1, g_2): g_1 \neq g_2, g_1 \text{ hits } v_1, g_2 \text{ hits } v_2\}) = c \cdot l^2, \qquad (7.14.3)$$

where

$$c = \int [(\pi/2 - \alpha) \cot \alpha + 1] m(d\alpha) < \infty. \qquad (7.14.4)$$

The proof is by integration of (6.5.4); m is as in (6.7.9).

Now we show that the convergence of the integral (7.14.4) is equivalent to the existence of the expectation of a random variable (which we shall encounter in §10.3). The argument we use is universal and therefore we outline it in general terms for a point process $m(\omega)$ in a space $\mathbb{X}$.

For every set $A \subset \mathbb{X} \times \mathbb{X}$ we have

$$m_2(A) = \int I_A(x_1, x_2) m_2(dx_1 \, dx_2)$$

$$= E_P N(m^2, A) = E_P \sum_{x_i, x_j \in m} I_A(x_i, x_j).$$

Let $f(x_1, x_2)$ be a 'simple' function defined on $\mathbb{X} \times \mathbb{X}$, i.e.

$$f(x_1, x_2) = f_s \qquad \text{if } (x_1, x_2) \in A_s,$$

where

$$A_{s_1} \cap A_{s_2} = \varnothing \qquad \text{if } s_1 \neq s_2, \bigcup A_s = \mathbb{X} \times \mathbb{X};$$

f_s are any numbers.

Multiplying the above equation for A_s by f_s and summing, we get

$$\int f(x_1, x_2) m_2(dx_1\, dx_2) = E_P \sum_{x_i, x_j \in m} f(x_i, x_j). \qquad (7.14.5)$$

By a standard argument of the theory of integration we conclude that (7.14.5) *is valid for every measurable f.*

We again consider the process of Example 2 and put

$$f(g_1, g_2) = I_{b(r,O)}(g_1 \cap g_2)\left[\left(\frac{\pi}{2} - \alpha\right)\cot\alpha + 1\right].$$

(The indicator equals 1 if the intersection point belongs to the disc $b(r, O)$ and α is the angle between g_1 and g_2.) Clearly (7.14.5) now becomes

$$\pi r^2 \cdot c = E_P \sum_{g_i \cap g_j \in b(r,O)}\left[\left(\frac{\pi}{2} - \alpha_{ij}\right)\cot\alpha_{ij} + 1\right] \qquad (7.14.6)$$

where α_{ij} is the angle between g_i and g_j from a realization.

Clearly $c < \infty$ (equivalently, $\{g_i\}$ is of second order) implies that the expectation in (7.14.6) is finite.

Example 3 Let $\{s_i\}$ be an $\mathbb{M}_2$-invariant second order segment process on $\mathbb{R}^2$. Its first moment measure is an $\mathbb{M}_2$-invariant measure of the space

$$\Delta_2 = \mathbb{M}_2 \times (0, \infty)$$

and necessarily has the form

$$m_1(d\delta) = dM\, m(dl),$$

where m is a measure in the space of segment lengths $l \in (0, \infty)$. The requirement that the process $\{\mathscr{P}_i\}$ of the sources of the segments be of first order is equivalent to the condition

$$m((0, \infty)) = \int_{(0,\infty)} m(dl) < \infty.$$

The requirement that the process $\{x_i\}$ of intersections of s_i's with the x axis be of the first order is equivalent to the stronger convergence condition

$$\int_0^\infty l\, m(dl) < \infty$$

(the existence of the mean length of the typical segment in the terminology of §8.1).

Let us look at the second moment measure m_2 under the additional assumption that *with probability 1 there are no pairs of parallel or antiparallel segments in* $\{s_i\}$.

Outside the 'diagonal', m_2 is an $\mathbb{M}_2$-invariant measure in the space

$$\Delta_2 \times \Delta_2 \backslash \{\text{pairs of parallel or antiparallel segments}\}$$

$$\approx \mathbb{M}_2 \times \mathbb{S}_1 \times \mathbb{T}_1 \times \mathbb{T}_1 \times (0, \infty) \times (0, \infty).$$

We use the notation

$\mathbb{S}_1$ for the space of angles ψ between δ_1 and δ_2, and $\delta_i \in \Delta_2$; $\mathbb{T}_1$ for the space of one-dimensional shifts t_i of δ_i along the line g_i which carries δ_i, $i = 1, 2$; $(0, \infty)$ for the space of lengths l_i of the segments.

By Haar factorization

$$m_2(d\delta_1 \, d\delta_2) = dM \, m(d\psi \, dl_1 \, dl_2 \, dt_1 \, dt_2) \qquad (7.14.7)$$

with some measure m on the corresponding factor-space.

If m_2 has the form

$$m_2(d\delta_1 \, d\delta_2) = f_2 \, dg_1 \, dg_2 \, dt_1 \, dt_2 \, dl_1 \, dl_2 \qquad (7.14.8)$$

we say that m_2 has density f_2. $\mathbb{M}_2$-invariance of m_2 implies that f_2 depends solely on the $\mathbb{M}_2$-invariant parameters:

$$f_2 = f_2(\psi, t_1, t_2, l_1, l_2),$$

and we can reduce (7.14.8) to (7.14.7) by applying the transformation of §3.15, I. We find that

$$m(d\psi \, dl_1 \, dl_2 \, dt_1 \, dt_2) = f_2 \sin \psi \, d\psi \, dl_1 \, dl_2 \, dt_1 \, dt_2.$$

We say that $\{s_i\}$ is of *second order in hits* if for every disc $b(r, O)$ we have

$$m_2(\{(\delta_1, \delta_2) : \text{both } \delta_1 \text{ and } \delta_2 \text{ hit } b(r, O)\}) < \infty.$$

In contrast with line processes, this is a separate condition which does not follow automatically from the existence of m_2.

If the process $\{s_i\}$ is Poisson governed by the measure

$$dM \, f(l) \, dl \quad \text{with} \quad \int_0^\infty f(l) \, dl < \infty$$

then it is of second order in hits and f_2 in (7.14.8) equals

$$f_2 = f(l_1)f(l_2).$$

Example 4 Let $\{s_i\}$ be the process of edges of the mosaic on $\mathbb{R}^2$ *generated by a Poisson line process* governed by $\lambda \, dg$ $(\{s_i\} = m_2(\omega)$ of §7.13). Its first moment measure has the form

$$m(d\delta) = \exp(-2\lambda l) \, dM \, dl.$$

Its second moment measure m_2^0 is concentrated (outside the 'diagonal') on the sets which correspond to the cases shown in fig. 7.14.2.

The restrictions of m_2^0 on these sets in corresponding coordinates are written as follows:

$$m_2^0(d\delta_1 \, d\delta_2) \begin{cases} = f_A \, dg_1 \, dg_2 \, dl_1 \, dl_2 \, dt_1 \, dt_2 & \text{on } A \\ = f_B \, dM \sin \psi \, d\psi \, dl_1 \, dl_2 & \text{on } B \\ = f_C \, dM \, d\rho \, dl_1 \, dl_2 & \text{on } C \\ = f_D \, dM \, dl_1 \, dl_2 & \text{on } D. \end{cases}$$

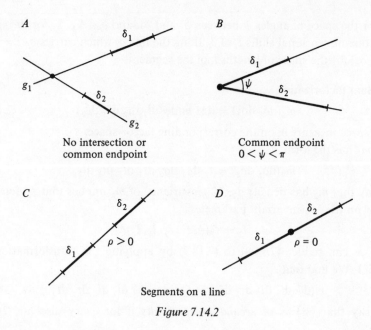

Segments on a line

Figure 7.14.2

f_A, f_B, f_C and f_D are functions depending on the corresponding $\mathbb{M}_2$-invariant parameters. The reader may easily find their exact form.

Even without exact knowledge of the expression one can see that the value of

$$m_2^0(\{\delta_1 \text{ hits } v_1, \delta_2 \text{ hits } v_2\}).$$

(where v_1 and v_2 are the vertical windows shown in fig. 6.5.1) equals $4\lambda^2 l^2$ (i.e. it coincides with (7.14.3) for the Poisson line process from which $\{s_i\}$ was derived).

Given a random mosaic, we say that the second moment measure m_2 of its process of edges *has a density* if

$$m_2(\mathrm{d}\delta_1\,\mathrm{d}\delta_2) = f(\delta_1, \delta_2)m_2^0(\mathrm{d}\delta_1\,\mathrm{d}\delta_2), \qquad (7.14.9)$$

for some f which is called the *density* of m_2.

7.15 Averaging in the space of realizations

In §6.6 and §6.8 we derived a number of integral identities for functions depending on sets m ('realizations') from certain classes. The integrals were calculated with respect to Haar measure on groups.

Suppose we have a geometrical process $m(\omega)$ which is invariant with respect to the group in question and whose realizations belong to the required class with probability 1, and in one of those identities we put

$$m = m(\omega).$$

Averaging the result with respect to P, the distribution of $m(\omega)$ sometimes leads to useful conclusions about P itself. Although the idea seems to be general enough, it has been used in a rather limited number of cases (mostly in [11] and [19] by the author).

Below we present some simple but important examples of how this method is actually applied; some other examples are scattered throughout the remaining chapters.

I Let $m(\omega)$ be a $\mathbb{T}_n$-invariant point process in $\mathbb{R}^n$; let P be its distribution; and let E_P denote the corresponding expectation (average).

Let $B \subset \mathbb{R}^n$ be a set of zero Lebesgue measure, for instance $B = \partial b(r, O)$ for some radius r. For every fixed realization $m \in \mathcal{M}_{\mathbb{R}^n}$ we have (see §6.6)

$$\int_{\bar{b}(r,O)} N(m, tB)\, dt = 0.$$

We apply E_P to both sides of this identity. By the Fubini theorem

$$E_P \int_{\bar{b}(r,O)} N(m, tB)\, dt = \int_{\bar{b}(r,O)} E_P N(m, tB)\, dt = 0.$$

By $\mathbb{T}_n$-invariance

$$E_P N(m, tB) \equiv E_P N(m, B)$$

and therefore

$$E_P N(m, B) L_n(b(r, O)) = 0.$$

We conclude that

$$L_n(B) = 0 \qquad \text{implies that} \qquad E_P N(m, B) = 0. \tag{7.15.1}$$

We note that in the finite intensity case this result follows from (7.14.1).

Because $N(m, B)$ is non-negative we also conclude that

$$L_n(B) = 0 \qquad \text{implies that} \qquad P(N(m, B) > 0) = 0. \tag{7.15.2}$$

II Let us additionally assume that $m(\omega)$ is of finite intensity λ. According to (7.15.2), (6.6.7) and (6.6.8) are satisfied with probability 1. Therefore we can average these identities with respect to P.

For the moment we suppose that the interchange of the order of the lim and E_P operations is legitimate, i.e. that

$$E_P \lim \int_{\bar{b}(r,O)} I_1(m, tD)\, dt = \lim E_P \int_{\bar{b}(r,O)} I_1(m, tD)\, dt \tag{7.15.3}$$

and

$$E_P \lim \int_{\bar{b}(r,O)} \sum_{k=2}^{\infty} k I_k(m, tD)\, dt = \lim E_P \int_{\bar{b}(r,O)} \sum_{k=2}^{\infty} k I_k(m, tD)\, dt. \tag{7.15.4}$$

Then by the Fubini theorem and the identity

$$E_P I_k(m, tD) = P\binom{D}{k}$$

we will get that as D shrinks down to O

$$\lim \frac{P\binom{D}{1}}{\|D\|} = \frac{E_P N(b(r, O), m)}{\|b(r, O)\|} = \lambda \tag{7.15.5}$$

and

$$\lim \frac{P\binom{D}{k}}{\|D\|} = 0 \qquad \text{for } k > 1. \tag{7.15.6}$$

To prove these important relations completely it remains to justify (7.15.3) and (7.15.4). For this purpose we note that for all $t \in \bar{b}(r, O)$, and whenever the diameter of D does not exceed 1, we have

$$N(m, tD) = N(m \cap b(r + 1, O), tD).$$

Therefore

$$\|D\|^{-1} \int_{\bar{b}(r, O)} N(m, tD) \, dt = \|D\|^{-1} \int_{\bar{b}(r, O)} N(m \cap b(r + 1, O), tD) \, dt$$

$$\leqslant \|D\|^{-1} \int_{\mathbb{T}_n} N(m \cap b(r + 1, O), tD) \, dt$$

$$= N(m, b(r + 1, O)),$$

where at the last step we have applied (6.6.3). Clearly

$$N(m, tD) = \sum_{k=1}^{\infty} k I_k(m, tD),$$

hence

$$I_1(m, tD) \leqslant N(m, tD)$$

and

$$\sum_{k=2}^{\infty} k I_k(m, tD) \leqslant N(m, tD).$$

Thus $N(m, b(r + 1, O))$ is an upper bound for both ratios

$$\|D\|^{-1} \int_{\bar{b}(r, O)} I_1(m, tD) \, dt$$

and

$$\|D\|^{-1} \int_{\bar{b}(r, O)} \sum_{k=2}^{\infty} k I_k(m, tD) \, dt.$$

Since by the finite intensity assumption this bound is summable, Lebesgue's dominated convergence theorem guarantees the validity of (7.15.3) and (7.15.4). Thus the relations (7.15.5) and (7.15.6) are completely proved.

III The above examples make the essence of the method very clear: the Fubini theorem in conjunction with the observation that an appropriate invariance assumption concerning $m(\omega)$ reduces the E_P of the integrand to a constant. In cases where a limiting operation is involved we have additionally to introduce conditions which guarantee interchangeability of this operation and that of taking E_P. These conditions usually take the form of a requirement that some upper bound for a sequence in question should possess a finite expectation E_P (then the Lebesgue theorem on dominated convergence provides the necessary basis). In all cases below where we follow this line of derivation it is enough

(a) to mention the identity which holds with probability 1, and
(b) to mention an appropriate upper bound,

and after that we can directly put down the result.

IV We consider the identity (6.8.8), substituting

$$m = m(\omega),$$

where $m(\omega)$ is an $\mathbb{M}_2$-invariant random segment process in $\mathbb{R}^2$, and P is its distribution. For $|\delta| < 1$ we have an inequality

$$|\delta|^{-1} \int_{\bar{b}(r,O)} I_1(\delta, m) I_\beta(\psi) \, d\delta \leqslant |\delta|^{-1} \int_{\bar{b}(r,O)} I_1(\delta, m) \, d\delta$$

$$\leqslant |\delta|^{-1} \int_{\bar{b}(r,O)} n(\delta, m) \, d\delta = |\delta|^{-1} \int_{\bar{b}(r,O)} n(\delta, m \cap b(r+1, O)) \, d\delta$$

$$\leqslant |\delta|^{-1} \int_{\Delta_2^*} n(\delta, m \cap b(r+1, O)) \, d\delta = 4\mathscr{L}(m, b(r+1, O)),$$

where $\mathscr{L}$ is the random 'length' measure introduced in §7.14; the last step was according to (6.8.1). We will have a summable upper bound, i.e.

$$E_P \mathscr{L}(m, b(r+1, O)) < \infty$$

if the segment process is of finite intensity in the sense of (7.14.2). We come to the following result.

For any $\mathbb{M}_2$-invariant segment process $m(\omega)$ in $\mathbb{R}^2$ of finite length intensity we have

$$P\left(\binom{\delta}{1} \cap \{\psi \in \beta\} \right) = \lambda \cdot |\delta| \cdot \frac{1}{4} \int_\beta |\sin \psi| \, d\psi + o(|\delta|). \qquad (7.15.7)$$

Here δ is any fixed segment in $\mathbb{R}^2$, β is any Borel set of angles, and ψ denotes the random angle between δ and the segment from $m(\omega)$ which hits δ. Here and below we use a slightly abused notation

$$\{\delta \text{ is hit by } k \text{ segments from } m(\omega)\} = \binom{\delta}{k}.$$

Lastly

$$\lambda = 2(\pi \cdot L_2(B))^{-1} E_P \mathscr{L}(m, B).$$

We remark that if $\beta = (0, 2\pi)$ then (7.15.7) reduces to (7.15.5). It follows that

the length intensity λ_1 of $m(\omega)$ and the intensity λ of the point process of intersections $\{x_i\}$ induced by $m(\omega)$ on any line g_0 are proportional, namely

$$\lambda = 2\lambda_1/\pi.$$

We stress that according to (7.15.7) the distribution law of the intersection angle ψ is *always* the same and has density $1/4|\sin \psi|$.

V We consider the identity (6.8.7) with $k \geqslant 2$, assuming that $m = m(\omega)$ is an M_2-invariant segment process in $\mathbb{R}^2$. We start with the remark that if $|\delta| < 1$

$$|\delta|^{-2} \int_{\bar{b}(r,O)} I_k(\delta, m)\, d\delta \leqslant \sum_{\substack{\delta_i, \delta_j \in m \\ \delta_i, \delta_j \text{ hit } b(r+1,O)}} |\delta|^{-2} \int I_{\delta_i \delta_j}(\delta)\, d\delta,$$

where

$$\begin{aligned} I_{\delta_i \delta_j}(\delta) &= 1 \qquad \text{if } \delta \text{ hits both } \delta_i \text{ and } \delta_j \text{ segments,} \\ &= 0 \qquad \text{otherwise.} \end{aligned}$$

If we additionally require that $m(\omega)$ should not (with probability 1) possess pairs of parallel segments, then as follows from (6.8.4) and the above,

$$|\delta|^{-2} \int_{\bar{b}(r,O)} I_k(\delta, m)\, d\delta \leqslant \sum_{\substack{\delta_i, \delta_j \in m \\ \delta_i, \delta_j \text{ hit } b(r+1,O)}} (2 + (\pi - 2\Phi_{ij}) \cot \Phi_{ij}),$$

where Φ_{ij} is the angle between δ_i and δ_j.

Thus the condition

$$E_P \sum_{\substack{\delta_i, \delta_j \in m \\ \delta_i, \delta_j \text{ hit } b(r+1,O)}} (2 + (\pi - 2\Phi_{ij}) \cot \Phi_{ij}) < \infty \tag{7.15.8}$$

will suffice for our purpose. Note that (7.15.8) can be equivalently written in terms of the second moment measure of $m(\omega)$ as a convergence condition (see §7.14, Example 3)

$$\int_{\delta_1, \delta_2 \text{ hit } b(r+1,O)} (2 + (\pi - 2\Phi_{12}) \cot \Phi_{12}) m_2(d\delta_1\, d\delta_2) < \infty.$$

The result again refers to the intersection point process $\{x_i\}$, namely to the probabilities of the events $\binom{\delta}{k}$ to have k intersections on fixed segment $\delta \subset g_0$.

Let $m(\omega)$ be an $\mathbb{M}_2$-invariant segment process in $\mathbb{R}^2$ satisfying the condition (7.15.8). For every $k \geqslant 2$ exists the limit

$$\lim_{|\delta| \to 0} |\delta|^{-2} P\binom{\delta}{k} = \frac{1}{2\pi \cdot \pi r^2} E_P \sum_{N_i \in b(r,0)} C_k(N_i). \tag{7.15.9}$$

Remark Condition (7.15.8) may be too strong. In fact it is *not* satisfied even in the cases of Poisson $\{s_i\}$ or of $\{s_i\}$ generated by Poisson lines as in §7.13 (the check is left to the reader), although in both cases the limits in question clearly exist.

A proof of (7.15.9) can be given, however, under conditions which cover both cases. It is enough to require that $\{s_i\}$ be of second order in hits and *the existence of bounded density* for the second moment measure of $\{s_i\}$ in the sense (7.14.8) or (7.14.9). The proof in the style of §§10.3 and 10.4 is left to the reader.

In general (i.e. if no additional conditions leading to (7.15.9) are imposed) we can average (6.8.7) and use the Fatou lemma. This yields the inequality

$$\liminf_{|\delta| \to 0} |\delta|^{-2} P\binom{\delta}{k} \geqslant (2\pi \cdot \pi r^2)^{-1} E_P \sum_{N_i \in b(r,O)} C_k(N_i).$$

Corollary If the intersection point process $\{x_i\}$ on g_0 is such that

$$\liminf_{|\delta| \to 0} |\delta|^{-2} P\binom{\delta}{k} = 0 \qquad \text{for all } k \geqslant 3; \tag{7.15.10}$$

then the segment process possesses with probability 1 only nodes of order two (see fig. 6.8.2).

The proof follows from the fact that a point process on $\mathbb{R}^2$ of zero intensity is with probability 1 void of points and from the remarks in §6.8, II.

VI In addition to the conditions of $\mathbb{M}_2$-invariance, absence of parallel segments and (7.15.10), we assume now that $m(\omega)$ is a *random mosaic*. This clearly excludes the nodes of angle type (see fig. 6.8.2) thus leaving the possibility of having order two nodes of only the remaining three types. The concrete random mosaics described in §7.13 fall into our class: in fact they possess nodes of cross or knot types but fail to display forks. We show now that under an additional general condition (7.15.11) *the absence of forks is a consequence of the PIAsin property* (see §7.13) of the marked intersection process $\{x_i, \psi_i\}$ induced by a random mosaic on (any) line g_0.

The proof is done by averaging the identities (6.8.9) (which under our assumptions are satisfied with probability 1). It is not difficult to show that $\pi N_1(r + 1, m)$ (where $N_1(r + 1, m)$ is the number of pairs of edges of the mosaic which hit $b(r + 1, O)$) can be an upper bound for both ratios under the limit signs in (6.8.9) (assuming that $|\delta| < 1$).

Therefore the condition

$$E_P N_1(r + 1, m) < \infty \tag{7.15.11}$$

guarantees that

$$\lim_{|\delta| \to 0} |\delta|^{-2} P \binom{\delta}{2} E_P[1 + (\pi - |\pi - \psi_1 - \psi_2|) \cot |\pi - \psi_1 - \psi_2|]^{-1}$$
$$= (4\pi \cdot \pi r^2)^{-1} \cdot (3E_P n_f + 4E_P n_c + 2E_P n_k),$$

and

$$\lim_{|\delta| \to 0} |\delta|^{-2} P \binom{\delta}{2} E_P \frac{|\pi - \psi_1 - \psi_2|}{1 + (\pi - |\pi - \psi_1 - \psi_2|) \cot |\pi - \psi_1 - \psi_2|}$$
$$= (4\pi \cdot \pi r^2)^{-1} \cdot (2\pi E_P n_f + 2\pi E_P n_c + \pi E_P n_k).$$

(In writing the left-hand sides we made use of the assumed independence of $\{x_i\}$ and $\{\psi_i\}$.)

The ratio of the two limits thus equals

$$a = \pi \frac{2E_P n_f + 2E_P n_c + E_P n_k}{3E_P n_f + 4E_P n_c + 2E_P n_k}. \tag{7.15.12}$$

On the other hand, a depends solely on the distribution of the $\{\psi_i\}$ sequence and therefore is the same for all random mosaics with the PIAsin property.

But for a Poisson line mosaic governed by dg we have

$$E_P n_f = 0, \qquad E_P n_k = 0,$$

and (7.15.12) yields

$$a = \pi/2.$$

We can see from (7.15.12) that this can be possible only if

$$E_P n_f = 0.$$

VII More results concerning M_2-invariant planar segment processes have been obtained by this method in [19], where the following problem was considered.

Let m_2 be the second moment measure of the random length measure $\mathcal{L}$ associated with a segment process $m(\omega)$ (see §7.14):

$$m_2(B_1 \times B_2) = E_P L(B_1) L(B_2),$$

and let v_2 be the second moment measure of the marked intersection process $\{(x_i, \psi_i)\}$ induced by $m(\omega)$ on any line g_0. Does v_2 determine m_2? The main result of [19] indicates certain general conditions when v_2 together with F, the distribution of the length of 'typical' segment in $m(\omega)$, actually determine m_2. If v_2 has a smooth density the problem is solved by an elegant explicit formula.

An attempt to apply this method to segment and *curve* processes in $\mathbb{R}^n$ has been made in [49]. We briefly mention the results.

The trace a curve process leaves on a hyperplane is a point process which we denote as $\{x_i\}$. Under the assumption of $\mathbb{M}_n$-invariance what can be inferred about the curve process if we know that $\{x_i\}$ is Poisson? It was shown that in three dimensions ($n = 3$) the Poisson nature of $\{x_i\}$ implies that the curve process possesses no nodes (with probability 1); if $n \geqslant 4$ the Poisson nature of $\{x_i\}$ implies that the curves of the process are of zero curvature (i.e. we have a segment process).

8

Palm distributions of point processes in $\mathbb{R}^n$

The idea of Palm distribution originated from the theory of point processes on the line [31]. For $\mathbb{R}^n$, connections of this idea with integral geometry have been identified in [32]. In this book our approach to the Palm distribution will be common for all spaces: we base it on Haar factorization. However, it is natural to present the simpler and more explored case of $\mathbb{R}^n$ in a separate chapter.

The Palm distribution of a $\mathbb{T}_n$-invariant finite intensity point process in $\mathbb{R}^n$ is often defined to be the conditional distribution of the process, given that the latter 'has a point at the origin O'. This can hardly be considered an honest definition since the conditioning event has zero probability. There are several equivalent rigorous definitions of this notion, but they can all be reduced to Lebesgue factorization. The importance of Palm distribution is rooted in the fact that, together with the value of intensity, it provides a complete probabilistic description of a point process which is alternative to that of chapter 7.

In the concluding sections we concentrate on related analytical tools, especially 'Palm formulae' and consider certain equations that relate the distribution of a point process (in the sense of chapter 7) to its Palm distribution.

8.1 Typical mark distribution

Let $\{(\mathscr{P}_i, k_i)\}$ be a $\mathbb{T}_n$-invariant marked point process in $\mathbb{R}^n$ (see §7.12) with marks k_i belonging to some space $\mathbb{K}$.

We say that $\{(\mathscr{P}_i, k_i)\}$ is of finite intensity if its $\{\mathscr{P}_i\}$ (which is a $\mathbb{T}_n$-invariant point process in $\mathbb{R}^n$) is of finite intensity λ.

For a set

$$A \subset \mathbb{R}^n \times \mathbb{K}$$

we denote

$$N(A) = \operatorname{card} \{i : (\mathscr{P}_i, k_i) \in A\}.$$

The first moment (or expectation) measure for $\{(\mathscr{P}_i, k_i)\}$ is defined to be

$$m_1(A) = EN(A) \tag{8.1.1}$$

where E is the mathematical expectation. If $\{(\mathscr{P}_i, k_i)\}$ is of finite intensity, then the values of m_1 on the product sets

$$A = B \times K, \quad B \subset \mathbb{R}^n, \quad K \subset \mathbb{K} \tag{8.1.2}$$

are finite whenever B is compact. $\mathbb{T}_n$-invariance of $\{(\mathscr{P}_i, k_i)\}$ implies $\mathbb{T}_n$-invariance of the measure m_1 (see §1.2). Hence by Lebesgue factorization

$$m_1 = \lambda L_n \times \Pi, \tag{8.1.3}$$

where Π is a *probability measure on* $\mathbb{K}$. Π is called the *distribution of a typical mark in the process* $\{(\mathscr{P}_i, k_i)\}$.

Let us calculate m_1 on the sets in (8.1.2). We have

$$N(A) = \sum_{\mathscr{P}_i \in B} I_K(k_i).$$

Applying (8.1.3) we get the following expression for Π:

$$\Pi(K) = (\lambda L_n(B))^{-1} E \sum_{\mathscr{P}_i \in B} I_K(k_i). \tag{8.1.4}$$

In particular, we see that the expression on the right-hand side does not depend on the choice of compact $B \subset \mathbb{R}^n$. We can obtain another useful interpretation for $\Pi(K)$ if we replace B in (8.1.4) by a sequence B_m of 'small' domains which converge to the origin $O \in \mathbb{R}^n$.

Then, as shown in §§7.15, II, we will have

$$P\binom{B_m}{1} = \lambda \cdot L_n(B_m) + o(L_n(B_m)) \tag{8.1.5}$$

and

$$\sum_{l=2}^{\infty} lP\binom{B_m}{l} = o(L_n(B_m))$$

so that the limiting form (as $m \to \infty$) of (8.1.4) will be

$$\Pi(K) = \lim_{m \to \infty} \frac{P(\{k \in K\})}{P\binom{B_m}{1}}, \tag{8.1.6}$$

where k is the mark of that point which lies in B_m. Let us look more closely at the ratio under the limit. Since, by the definition of k

$$\{k \in K\} \subset \binom{B_m}{1},$$

this ratio has the usual interpretation of conditional probability of the event $\{k \in K\}$ given that the event $\binom{B_m}{1}$ has occurred. Thus (8.1.6) says that $\Pi(K)$ can be defined as the *limit of the corresponding conditional probability*.

8.2 Reduction to calculation of intensities

Given $K \subset \mathbb{K}$ we perform what can be called K-thinning of $\{\mathscr{P}_i\}$ (which is the projection of $\{(\mathscr{P}_i, k_i)\}$ on $\mathbb{R}^n$). Namely, a point $\mathscr{P}_i$ is deleted if its mark k_i belongs to K^c (the complement of K) and is retained otherwise. The thinned set

$$\{\mathscr{P}_i\}_K = \{\mathscr{P}_i : k_i \in K\} \qquad (8.2.1)$$

is a point process which inherits the properties of $\mathbb{T}_n$-invariance and of finite intensity. The number of points of the process (8.2.1) in a set $B \subset \mathbb{R}^n$ can be represented as

$$\sum_{\mathscr{P}_i \in B} I_K(k_i).$$

Hence it follows from (8.1.4) that the intensity of $\{\mathscr{P}_i\}_K$ equals

$$\Lambda(K) = \lambda \Pi(K).$$

Thus $\Pi(K)$ has a very simple expression in terms of intensities:

$$\Pi(K) = \lambda^{-1} \Lambda(K). \qquad (8.2.2)$$

This expression can be used for the statistical estimation of $\Pi(K)$ in cases where both λ and $\Lambda(K)$ can be reasonably replaced by numbers of points from $\{\mathscr{P}_i\}$ and $\{\mathscr{P}_i\}_K$ within big volumes (i.e. in ergodic cases).

Remark We did not assume above that the marks are independent. In fact our aim in the following sections will be to apply the definition of Π to the cases where there is strong dependence between points and marks.

In cases where marks are independent, however, and each is distributed according to some probability law F (as for instance in the examples in §7.12), both (8.1.6) and (8.2.2) show that

$$\Pi = F.$$

8.3 The space of anchored realizations

Our approach to Palm distributions of point processes in $\mathbb{R}^n$ as outlined in the next section is based on marked point processes with marks in the space

$$\mathbb{K} = \mathscr{M}_{\mathbb{R}^n}^*,$$

which we call 'the space of anchored realizations'. By definition,

$$\mathscr{M}_{\mathbb{R}^n}^* = \{m \in \mathscr{M}_{\mathbb{R}^n} : m \text{ has a point at } O \text{ (i.e. is 'anchored')}\}.$$

It can be proved [18] that $\mathscr{M}_{\mathbb{R}^n}^*$ is a measurable subset of $\mathscr{M}_{\mathbb{R}^n}$ (i.e. $\mathscr{M}_{\mathbb{R}^n}^* \in \mathscr{A}$), thus $\mathscr{M}_{\mathbb{R}^n}^*$ inherits a measurability structure from $\mathscr{M}_{\mathbb{R}^n}$. In particular, every measurable $A \subset \mathscr{M}_{\mathbb{R}^n}$ has a measurable counterpart A^* in $\mathscr{M}_{\mathbb{R}^n}^*$, namely

$$A^* = A \cap \mathscr{M}_{\mathbb{R}^n}^*.$$

In fact any probability measure P on $\mathscr{M}_{\mathbb{R}^n}^*$ can be considered as a probability on $\mathscr{M}_{\mathbb{R}^n}$ which has the property

$$P(\mathcal{M}_{\mathbb{R}^n}^*) = 1,$$

and in this case of course

$$P \text{ (there is a point of realization at } O) = 1. \tag{8.3.1}$$

Therefore on $\mathcal{M}_{\mathbb{R}^n}^*$ no $\mathbb{T}_n$-invariant probability measures exist (see §7.15, I).

Each fixed $m \in \mathcal{M}_{\mathbb{R}^n}$ naturally generates countably many points in $\mathcal{M}_{\mathbb{R}^n}^*$: namely, if

$$m = \{\mathcal{P}_i\}$$

then, with each $\mathcal{P}_i \in m$, we associate the element

$$k_i = t_i m, \qquad \text{where} \qquad t_i = \overrightarrow{\mathcal{P}_i O} \tag{8.3.2}$$

(t_i shifts m in such a way that $\mathcal{P}_i$ goes into O). In other words, we map each $m \in \mathcal{M}_{\mathbb{R}^n}$ into a countable system of points in the product space $\mathbb{R}^n \times \mathcal{M}_{\mathbb{R}^n}^*$:

$$m \to \{(\mathcal{P}_i, k_i)\}. \tag{8.3.3}$$

The number of points from the set $\{(\mathcal{P}_i, k_i)\}$ which fall inside a set $A \subset \mathbb{R}^n \times \mathcal{M}_{\mathbb{R}^n}^*$ depends both on A and the generating m. We denote this number by $N^*(A, m)$:

$$N^*(A, m) = \text{the number of points in } A \cap \{(\mathcal{P}_i, k_i)\}.$$

(Note the essential difference between N^* and the quantity $N(B, m)$ defined in §7.1.)

For certain sets A, the function N^* can be represented as an indicator function of some sets in $\mathcal{M}_{\mathbb{R}^n}$. The simplest are the following examples.

Let D be a domain in $\mathbb{R}^n$. We put

$$B = \begin{pmatrix} D \\ 1 \end{pmatrix} = \{m : N(D, m) = 1\}$$

and

$$A = \left\{ (\mathcal{P}, k) : \mathcal{P} \in D, k \in \begin{pmatrix} tD \\ 1 \end{pmatrix} \right\},$$

where

$$t = \overrightarrow{\mathcal{P}O}.$$

Then it is an easy matter to check that

$$N^*(A, m) \equiv I_B(m).$$

In fact the indicator of every set of the form

$$B = \begin{pmatrix} D_1 & D_s \\ l_1 & \cdots & l_s \end{pmatrix},$$

where each $l_i = 1$ or 0, $\sum l_i > 0$, admits a similar representation i.e.

$$I_B(m) = N^*(A, m) \qquad \text{for some } A.$$

In particular this is true for every $B \in \mathcal{H}_{\mathbb{R}^n}$ (see §7.5) provided that $\sum l_i > 0$.

This observation has an important corollary in the theory of point processes, i.e. when we assume that

$$m = m(\omega),$$

in which case $\{(\mathscr{P}_i, k_i)\}$ becomes a *marked point process* in $\mathbb{R}^n$ with marks in $\mathscr{M}_{\mathbb{R}^n}^*$.

It follows that probabilities of all sets from the class $\mathscr{H}_{\mathbb{R}^n}$ can in principle be determined from knowledge of the first moment measure $E_P N^*$ of $\{(\mathscr{P}_i, k_i)\}$. In view of the results of §7.5 this means in turn that $E_P N^*$ *determines the distribution P of $m(\omega)$ in a unique way.*

8.4 Palm distribution

The ideas of the previous sections can be successfully applied for the purpose of defining the Palm distribution of a $\mathbb{T}_n$-invariant point process in $\mathbb{R}^n$ of finite intensity λ. Let

$$m(\omega) = \{\mathscr{P}_i\}$$

be such a process. We transform $m(\omega)$ into a marked point process,

$$m(\omega) \to \{(\mathscr{P}_i, k_i)\},$$

by applying (8.3.3) for each ω.

Definition (8.3.2) renders the marks k_i invariant under shifts. Therefore for every $t \in \mathbb{T}_n$

$$tm(\omega) = \{(t\mathscr{P}_i, k_i)\},$$

and the assumed $\mathbb{T}_n$-invariance of $m(\omega)$ implies $\mathbb{T}_n$-invariance of $\{(\mathscr{P}_i, k_i)\}$. The assumed finite intensity of $\{\mathscr{P}_i\}$ implies the existence (i.e. local-finiteness) of the first moment measure of the $\{(\mathscr{P}_i, k_i)\}$ process. Thus our two basic assumptions concerning $m(\omega)$ enable us to speak about the distribution of a typical mark in $\{(\mathscr{P}_i, k_i)\}$.

The probability measure Π on $\mathscr{M}_{\mathbb{R}^n}^*$ which gives the distribution of the typical mark in $\{(\mathscr{P}_i, k_i)\}$ where the marks k_i are defined by (8.3.2) is called the Palm distribution of $m(\omega)$.

Both interpretations of Palm distribution which correspond to (8.1.6) and (8.2.2) remain valid. Even more directly, the definition of Π can be given by means of Lebesgue factorization:

$$E_P N^* = \lambda \cdot L_n \times \Pi. \tag{8.4.1}$$

From this and the closing remarks of the previous section we conclude that λ and Π *together completely determine the distribution P of the point process $m(\omega)$.* In §8.8 we arrive at the same result from another point of view.

8.5 A continuity assumption

Let us consider formula (8.1.6) in the context of Palm distribution, i.e. taking

$$\mathbb{K} = \mathscr{M}_{\mathbb{R}^n}^*.$$

We now choose the set K to be

$$K = \begin{pmatrix} D_1 \\ k_1 \end{pmatrix}, \ldots, \begin{pmatrix} D_s \\ k_s \end{pmatrix} = \begin{pmatrix} \bar{D} \\ \bar{k} \end{pmatrix}$$

(see §7.5), where $D_i \subset \mathbb{R}^n$ are open domains whose closures do not contain the origin O. This K can be viewed both as a subset of $\mathscr{M}_{\mathbb{R}^n}$ and $\mathscr{M}_{\mathbb{R}^n}^*$. The event $\{k \in K\}$ can be now written as

$$\left\{ tm \in \begin{pmatrix} \bar{D} \\ \bar{k} \end{pmatrix} \right\},$$

where tm is the mark of that point $\mathscr{P} \in m(\omega)$ which in accordance to the condition $\begin{pmatrix} B_m \\ 1 \end{pmatrix}$ lies in the domain B_m; we put $t = \mathscr{P}O$. Since B_m converges to 0, the shift t is infinitesimal. Therefore we can expect that under some continuity condition we will have

$$P\left(\begin{pmatrix} B_m \\ 1 \end{pmatrix} \cap \left\{ tm \in \begin{pmatrix} \bar{D} \\ \bar{k} \end{pmatrix} \right\} \right) = P\left(\begin{pmatrix} B_m \\ 1 \end{pmatrix} \cap \begin{pmatrix} \bar{D} \\ \bar{k} \end{pmatrix} \right) + o(L_n(B_m)).$$

A condition for the validity of the above can be derived from the conditions which guarantee (6.6.9). It follows from (8.1.4) that if

$$\Pi \begin{pmatrix} Z \\ 0 \end{pmatrix} = 1 \tag{8.5.1}$$

for any set $Z \subset \mathbb{R}^n$ of Lebesgue measure zero which does not contain O, then both conditions (a) and (b) in §6.6 are fulfilled with probability 1. We replace in (6.6.9) the ball $b(\varepsilon, O)$ by B_m and average in the space of realizations with upper bound $N(m, b(r, O))$ (compare with §7.15, II). Then using (8.1.5) we obtain

$$\lim_{m \to \infty} \frac{P\left(\begin{pmatrix} B_m \\ 1 \end{pmatrix} \cap \begin{pmatrix} \bar{D} \\ \bar{k} \end{pmatrix} \right)}{P\begin{pmatrix} B_m \\ 1 \end{pmatrix}} = (\lambda L_n(b(r, O)))^{-1} E_P \sum_{\mathscr{P}_i \in m \cap b(r, O)} I_{\left(\frac{\bar{D}}{\bar{k}} \right)}(t_i m) \tag{8.5.2}$$

By comparison with (8.1.4) we easily recognize on the right-hand side the value of Palm probability Π of the event $\begin{pmatrix} \bar{D} \\ \bar{k} \end{pmatrix}$. Thus (8.5.2) implies that this value can be obtained as a limit of conditional probability

$$\Pi \begin{pmatrix} \bar{D} \\ \bar{k} \end{pmatrix} = \lim_{m \to \infty} P\left(\begin{matrix} \bar{D} \\ \bar{k} \end{matrix} \middle| \begin{matrix} B_m \\ 1 \end{matrix} \right) \tag{8.5.3}$$

The condition (8.5.1) can be given an equivalent form

$$E_\Pi N(m, Z) = 0 \qquad \text{whenever } L_n(Z) = 0, \qquad (8.5.4)$$

in particular, the first moment measure of Π (if it exists) should be absolutely continuous, with respect to Lebesgue measure L_n.

A sufficient condition for (8.5.4) can be given in terms of the second moment measure m_2 of P (assuming m_2 exists). According to (8.1.4) we have

$$E_\Pi N(m, Z) = \sum k \Pi \binom{Z}{k}$$

$$= (\lambda L_n(B))^{-1} E_P \sum_{\mathscr{P}_i \in B} N(t_i m, Z)$$

$$= (\lambda L_n(B))^{-1} E_P N(m^2, Z^*)$$

where m^2 is the square of the realization m (see fig. 7.14.1) and

$$Z^* = \{(x_1, x_2) \in \mathbb{R}^n \times \mathbb{R}^n : x_1 \in B, x_2 \in tZ\}$$

where t is the shift $\overline{Ox}$.

Hence (8.5.4) is equivalent to the condition that m_2 on sets of the above type be zero. A sufficient condition for this (and therefore for (8.5.1)) is that 'outside diagonal' m_2 should be absolutely continuous with respect to $L_n \times L_n$ (possess a density).

Example If the second moment measure of a $\mathbb{T}_1$-invariant point process in $\mathbb{R}$ exists then it necessarily has the form

$$m_2(dx_1 \, dx_2) = dt \, m(dl)$$

where dt is the Haar measure on $\mathbb{T}_1$, l is the distance between the points x_1, x_2, m is a measure on $(0, \infty)$.

Assume that $D_1, \ldots, D_s \subset \mathbb{R}$ are intervals. Then Z reduces to a finite set of points and (8.5.1) will follow from the condition that the measure $m(dl)$ does not charge any individual point on $(0, \infty)$. Recall that the set of such 'heavy' points can be at most countable.

If (8.5.1) is violated, we cannot expect (8.5.3) to hold for every collection of domains $\bar{D}$ as described above. For instance, (8.5.2) does not hold identically in Example 1 of the next section. We note that (8.5.1) (and therefore (8.5.3)) hold for $\mathbb{T}_n$-invariant Poisson processes in $\mathbb{R}^n$.

Condition (8.5.1) permits the modification of the right-hand side of (8.5.3), i.e. by replacing $\bar{D}$ by

$$\bar{D}^{(m)} = (D_1^{(m)}, \ldots, D_s^{(m)})$$

assuming that, as $m \to \infty$, $\bar{D}^{(m)}$ converges to $\bar{D} = (D_1, \ldots, D_s)$. Then a sufficient additional condition to have

$$\Pi\left(\frac{\bar{D}}{k}\right) = \lim_{m \to \infty} P\left(\binom{B_m}{1} \cap \binom{\bar{D}^{(m)}}{k}\right)\left[P\binom{B_m}{1}\right]^{-1}$$

can be that the interior of B_m contains O and does not intersect with the interior of any of the $D^{(m)}$'s. However, O may lie on the boundary of one of the limiting domains D_i. The proof follows from an appropriate modification of (6.6.9). We use this possibility in the derivation of the Palm formulae in §§8.7 and 8.8.

8.6 Some examples

Example 1 Let $m(\omega)$ be the point process described in §7.8. We have identically

$$t_i t m_0 = m_0,$$

where t_i has the form $\vec{\mathscr{P}_i O}$, $\mathscr{P}_i \in t m_0$. This means that marks (in the sense of §8.1) are non-random and coincide with m_0. We conclude that Π is concentrated on the lattice m_0, i.e.

$$\Pi(\{m_0\}) = 1.$$

Example 2 Let $m(\omega)$ be the point process described in §7.9. For any $t_i = \vec{\mathscr{P}_i O}$, $\mathscr{P}_i \in wt m_0$, we have

$$t_i wt m_0 = wt'_i t m_0, \qquad (8.6.1)$$

where t'_i is the solution of the equation (see §3.5)

$$t_i w = wt'_i.$$

Both sets in (8.6.1) contain O, therefore $t'_i t m_0 = m_0$. Thus for all i

$$t_i wt m_0 = wm_0,$$

i.e. Π is the distribution of the randomly rotated lattice m_0.

Example 3 Let P be the distribution of a Poisson process in $\mathbb{R}^n$ (see §7.5). To find its Π we use (8.5.3). For any system of pairwise non-intersecting domains $\bar{D} = (D_1, \ldots, D_s)$ such that $O \bar{\in} D_i$, $i = 1, \ldots, s$, we get from (7.5.2)

$$\Pi\left(\frac{\bar{D}}{k}\right) = P\left(\frac{\bar{D}}{k}\right).$$

This means that, *outside O, Π is Poisson*. Since always

$$\Pi \text{ (there is a realization point at } O) = 1$$

the complete description of Π in question will be

$$\Pi = \Delta * P, \qquad (8.6.2)$$

where Δ is the distribution of the point process consisting with probability 1 of a single point at O, and $*$ is the sign of composition (corresponding to the superposition of independent point processes).

Example 4 Let $m(\omega)$ be the Poisson cluster process discussed in §7.12, IV. If the total number of points in each cluster has finite expectation, then $m(\omega)$ is of finite intensity. The Palm distribution Π of $m(\omega)$ satisfies a relation which is a direct generalization of (8.6.2), namely

$$\Pi = \Theta * P, \tag{8.6.3}$$

where P is the distribution of $m(\omega)$, and Θ is the distribution of some point process m_1 in $\mathbb{R}^n$ with properties

$$\Theta\ (m_1 \text{ has finite number of points in } \mathbb{R}^n) = 1$$

$$\Theta\ (m_1 \text{ has a point at } O) = 1.$$

The probability distribution Θ depends on that of the clusters. For instance, if the clusters are *non-random*, i.e. each is represented by the same set of points

$$\{Q_1, \ldots, Q_l\},$$

then m_1 can be obtained as a randomly 'anchored' version of the above set, i.e.

$$m_1 = t_i\{Q_1, \ldots, Q_l\}, \tag{8.6.4}$$

where $t_i = \vec{Q_i O}$, and the point Q_i is chosen at random from $\{Q_1, \ldots, Q_l\}$ (each point can be chosen with probability $1/l$). If the clusters are random, but the total number l of points in a cluster remains non-random, then (8.6.4) still applies. We choose the index i as above and independent of the realization of $\{Q_1, \ldots, Q_l\}$.

In the general case, we obtain the *conditional* distribution of m_1 from the *conditional* distribution of a cluster using (8.6.4), when conditioning is upon l. Let $\{P_l\}$ be the probability distribution of the total number of points in a cluster. Then its counterpart for m_1 will be $\{lp_l/\sum lp_l\}$. With this, Θ is completely described. Proofs are left to the reader.

8.7 Palm formulae in one dimension

The derivation of Palm formulae in one dimension is directly based upon (8.5.3) and proceeds as follows. We consider the events happening in the intervals (see fig. 8.7.1)

$$B = \left(-\frac{\delta}{2}, \frac{\delta}{2}\right) \quad \text{and} \quad D = D_\delta = \left(\frac{\delta}{2}, x + \frac{\delta}{2}\right),$$

and we assume the brief notations

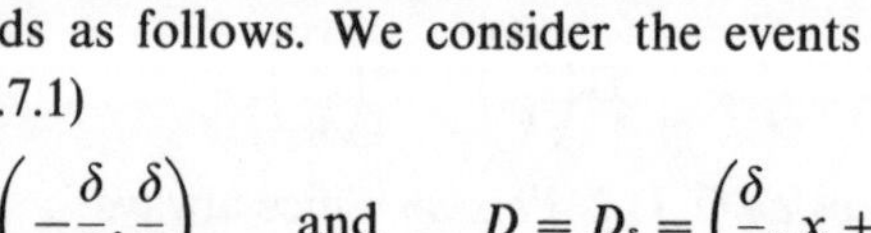

Figure 8.7.1

$$p_k(x) = P\binom{D}{k}, \qquad p_k(x + \delta) = P\binom{D \cup B}{k},$$

$$p_{l,k}(\delta, x) = P\binom{B \ \ D}{l \ , \ k}.$$

Because the probability of having a point of realization at $\delta/2$ is zero (this follows from (8.1.5)) we have

$$p_k(x) = \sum_{l=0}^{\infty} p_{l,k}(\delta, x),$$

$$p_k(x + \delta) = \sum_{l=0}^{k} p_{l,k-l}(\delta, x).$$

We are interested in summands which are $O(1)$ or $O(\delta)$ by order, as $\delta \to 0$. Because of (8.1.5) the above can be rewritten as

$$\begin{aligned} p_k(x) &= p_{0,k}(\delta, x) + p_{1,k}(\delta, x) + o(\delta), \\ p_k(x + \delta) &= p_{0,k}(\delta, x) + p_{1,k-1}(\delta, x) + o(\delta). \end{aligned} \qquad (8.7.1)$$

In the present situation, (8.5.1) can be replaced by the condition

$$\Pi(\text{there is a point of a realization at } x) = 0 \qquad (8.7.2)$$

for every point $x \in \mathbb{R}$. Let us assume that (8.7.2) holds. Then using (8.1.5), (8.5.2) and the remark at the end of §8.5 we identify the limits which follow as values of Palm distribution:

$$\lim_{\delta \to 0} p_{1,k}(\delta, x) \cdot \delta^{-1} = \lambda \cdot \Pi\binom{(0, x)}{k} \qquad (8.7.3)$$

(Π is no longer $\mathbb{T}_1$-invariant, and therefore when under Π we show the interval completely). From (8.7.1) the limit of $\delta^{-1}(p_k(x + \delta) - p_k(x))$ is thus found:

$$\frac{dp_k(x)}{dx} = \lambda \left[\Pi\binom{(0, x)}{k - 1} - \Pi\binom{(0, x)}{k} \right], \qquad k = 0, 1, 2, \ldots, \qquad (8.7.4)$$

where we have to put

$$\Pi\binom{(0, x)}{-1} \equiv 0.$$

Strictly speaking, we have shown above the existence of the one-sided derivatives $d^+ p_k(x)/dx$. However, considering the intervals

$$B = \left(-\frac{\delta}{2}, \frac{\delta}{2}\right), \qquad D = D_\delta = \left(\frac{\delta}{2}, x - \frac{\delta}{2}\right),$$

and by making use of the remark at the end of §8.5 we can similarly derive the same expression for $d^- p_k(x)/dx$. Thus (8.7.2) guarantees the existence of $dp_k(x)/dx$ and the validity of (8.7.4) for every x. Equations (8.7.4) are called Palm formulae in differential form. They permit us to find the functions $p_k(x)$

in terms of Π:

$$p_k(x) = p_k(0) + \lambda \int_0^x \left[\Pi\binom{(0, u)}{k-1} - \Pi\binom{(0, u)}{k} \right] du, \qquad (8.7.5)$$

where the initial conditions are as follows:

$$p_k(0) = 1 \quad \text{if } k = 0$$
$$= 0 \quad \text{if } k > 0.$$

Equations (8.7.5) are called Palm formulae in integral form. In contrast with (8.7.4) for their validity the assumption (8.7.2) is no longer needed. A proof of (8.7.5) for the general case can be based on the fact that (8.7.2) is satisfied for almost all values of x (if the second moment measure exists this easily follows from the remarks in the Example in §8.5). Thus (8.7.4) is valid for almost all values of x.

It remains to show that the functions $p_k(x)$ are absolutely continuous. We consider their sums,

$$F_k(x) = \sum_{i=k}^{\infty} p_i(x),$$

which are monotone increasing in x (they may be interpreted as the distribution functions of certain 'waiting times'). We have

$$\sum_{k=1}^{\infty} F_k(x) = \sum_{k=1}^{\infty} kp_k(x) = \lambda x.$$

Applying the Lebesgue decomposition theorem, we conclude that the functions $F_k(x)$ do not possess discrete or singular components. Therefore the same is true for

$$p_k(x) = F_k(x) - F_{k+1}(x).$$

8.8 Several intervals

The approach of the previous section can be applied to the probabilities

$$P\left(\frac{\bar{I}}{k}\right) = P\left(\frac{I_1}{k_1}, \ldots, \frac{I_s}{k_s}\right),$$

where we assume that the intervals $I_1, \ldots, I_s$ do not overlap. The same arguments as used in the derivation of (8.7.4) yield a similar equation:

$$\frac{dP\left(\frac{\bar{I}}{k}\right)}{dI_r} = \lambda \left[\Pi\left(\frac{t_r\bar{I}}{k - \bar{1}_r}\right) - \Pi\left(\frac{t_r\bar{I}}{k}\right) \right]. \qquad (8.8.1)$$

The derivation is taken with respect to the length of the rth interval; the shift t_r sends the left end (say) of I_r to zero and $\bar{1}_r = (0, \ldots, 0, 1, 0, \ldots, 0)$, where 1 occurs at the rth locus. The value of $P\left(\frac{\bar{I}}{k}\right)$ now can be found by integration, using the initial condition

$$P\left(\frac{\overline{I}}{k}\right)\Bigg|_{\text{at } I_r = 0} = 0 \qquad \text{if } k_r > 0$$

$$= P\left(\begin{matrix} I_1 \\ k_1 \end{matrix}, \ldots, \begin{matrix} I_{r-1} \\ k_{r-1} \end{matrix}, \begin{matrix} I_{r+1} \\ k_{r+1} \end{matrix}, \ldots, \begin{matrix} I_s \\ k_s \end{matrix}\right) \qquad \text{if } k_r = 0.$$

The event under P on the right-hand side depends on $s - 1$ intervals. Therefore, by repeating the procedure we will end up with the expression of $P\left(\dfrac{\overline{I}}{k}\right)$ in terms of Π and λ. This confirms (in one dimension) the conclusion of §8.4.

Remark The choice of the number r, as well as the endpoint of I_r in (8.8.1), was arbitrary. This leads (by means of integration of (8.8.1)) to several different representations for $P\left(\dfrac{\overline{I}}{k}\right)$. The independence of the result of integration of the right-hand side of (8.8.1) is a necessary and sufficient condition for a probability Π on $\mathcal{M}_0$ to be a Palm distribution (see [50] and [51]).

8.9 $\mathbb{T}_1$-invariant renewal processes

There is an important class of point processes on a line $\mathbb{R}$ which are called *renewal* processes. They are usually described as processes in which the *intervals between consecutive points in a realization are independent identically distributed random variables.* Yet, strictly speaking, this description is incomplete for it does not explain how, given a realization of interval lengths, we construct the corresponding set of points. In particular the situation is not clear if the process is intended to be $\mathbb{T}_1$-invariant (which is of most interest for us).

However, there is no difficulty in defining what is an *anchored renewal* point process. Let

$$\ldots, \xi_{-2}, \xi_{-1}, \xi_1, \xi_2, \ldots \tag{8.9.1}$$

be a sequence of independent positive random variables with a common distribution function $F(x)$. Starting from O, we plot the values in (8.9.1) as shown on fig. 8.9.1 and thus obtain a *random* realization $m \in \mathcal{M}_{\mathbb{R}}^*$. For this point process probabilities of all events can in principle be expressed in

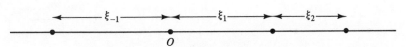

Figure 8.9.1 The solid points form a realization of the anchored renewal point process (there is a point of realization at O)

terms of F. For instance,

$$P\binom{(0, x)}{0} = 1 - F(x),$$

$$\sum_{i=0}^{k-1} P\binom{(0, x)}{i} = 1 - F^{(k)}(x),$$

where $F^{(k)}$ is the kth convolution of F with itself (the distribution function of the sum $\xi_1 + \cdots + \xi_k$).

In the case of finite intensity, a definition of a $\mathbb{T}_1$-invariant renewal process can be as follows.

Let $m(\omega)$ be a point process on $\mathbb{R}$, let it be $\mathbb{T}_1$-invariant and of finite intensity.

We call $m(\omega)$ renewal if its Palm distribution is the distribution of an anchored renewal process.

Which anchored renewal processes can appear as Palm distributions for $\mathbb{T}_1$-invariant renewal processes? The answer to this question is that the corresponding F should possess a finite mean, i.e.

$$\lambda^{-1} = \int x \, dF(x) < \infty;$$

the condition is both necessary and sufficient (the reciprocal value λ becomes the intensity of the $\mathbb{T}_1$-invariant process in question).

The following three-step stochastic construction yields the desired $\mathbb{T}_1$-invariant renewal point process.

Step (1) Sample a pair of random variables τ_{-1} and τ_1 whose joint distribution function is given by the integral

$$P(\tau_1 > x_1, \tau_1 > x_2) = 1 - \lambda \int_0^{x_1 + x_2} (1 - F(u)) \, du. \quad (8.9.2)$$

Step (2) Sample an independent realization of the sequence (8.9.1).
Step (3) Plot the obtained values as shown on fig. 8.9.2.

A proof of $\mathbb{T}_1$-invariance of this construction would take too much space. We remark only that the special form of (8.9.2) follows from Palm formulae and therefore is necessary.

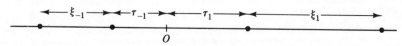

Figure 8.9.2 The solid points form a realization of the $\mathbb{T}_1$-invariant renewal point process (there is no point at O)

Figure 8.9.3 With probability 1, O coincides with the left endpoint of a black interval

There is a class of *segment processes* on $\mathbb{R}$ which we also call *renewal*. Their loose description involves the following conditions:

(a) the lengths of the segments of the process are independent identically distributed random variables;
(b) the segments of the process do not overlap with probability 1;
(c) the lengths of the gaps between consecutive segments of the process are independent identically distributed random variables.

For convenience, we will call the segments of the process *black* and the gaps *white* segments. Let two distribution functions be given as

$$F_b(x) = P\{\text{the length of a black segment is less than } x\}$$

$$F_w(x) = P\{\text{the length of a white segment is less than } x\}.$$

We sample two random sequences

... $\delta_{-1}, \delta_1, \delta_2, \ldots$: independent lengths, distribution of each is F_b;

... $v_{-1}, v_1, v_2, \ldots$: independent lengths, distribution of each is F_w.

and plot them on a line starting from O as shown on fig. 8.9.3.

We call the resulting process of black segments *anchored renewal* corresponding to the given F_b and F_w.

Now an exact definition of a $\mathbb{T}_1$-invariant renewal segment process can be given. We use the idea of the relative Palm distribution (see §9.1 below for the corresponding definition).

Given a process of (black) segments, by $\{l_i\}$ we denote the point process of the left endpoints of the black segments.

A $\mathbb{T}_1$-invariant segment process for which $\{l_i\}$ is of finite intensity is called *renewal* if its relative Palm distribution with respect to $\{l_i\}$ coincides with the distribution of some anchored renewal (black) segment process.

Example Let $\{s_i\}$ be a $\mathbb{T}_1$-invariant *Poisson* segment process on $\mathbb{R}$; it can alternatively be described (see §7.12) as a marked point process

$$\{s_i\} = \{(x_i, |s_i|)\},$$

where the process $\{x_i\}$ of segment left endpoints is Poisson governed by $\lambda \cdot L_1$, and the corresponding segment lengths $|s_i|$ are independent and identically distributed according to some distribution function $G(x)$. We assume that

$$\int x \, dG(x) < \infty. \tag{8.9.3}$$

We construct the so-called Boolean model, which by definition is the random set

$$U = \bigcup s_i.$$

If we think that each s_i is colored black, then U is the resulting black subset of $\mathbb{R}$; we call its complement, U^c, the white set. It can be shown that if (8.9.3) is violated then with probability 1 $U = \mathbb{R}$; if (8.9.3) holds then U is a union of countably many non-contacting black segments which we denote as δ_i:

$$U = \bigcup \delta_i, \qquad \delta_i \cap \delta_j = \varnothing \qquad \text{if } i \neq j,$$

and in fact $\{\delta_i\}$ is a *segment process*.

The process $\{\delta_i\}$ of black segments is $\mathbb{T}_1$-invariant renewal and we always have

$$F_w(x) = 1 - e^{-\lambda x},$$

where λ is the intensity of $\{x_i\}$.

Instead of giving a full proof of the first assertion we demonstrate a somewhat lesser fact that the relative Palm distribution of $\{\delta_i\}$ with respect to $\{l_i\}$ has the property that the parts of realization belonging to $(0, \infty)$ and $(-\infty, 0)$ are *independent*.

We split the space of segments (which is now $\mathbb{R} \times (0, \infty)$) into three non-overlapping subsets

I – the segments which lie entirely on the right of O,
II – the segments which lie entirely on the left of O,
III – the segments which cover O.

The parts of $\{s_i\}$ in these sets are independent Poisson processes governed by the restrictions of the measure $\lambda \cdot dx \cdot dG$ to the corresponding sets.

The limiting (as $\varepsilon \to 0$) conditional distribution of $\{s_i\}$, conditional upon the event

$$\{\text{there are no segments in III}\} \cap \{\text{there is a left endpoint in } (0, \varepsilon)\}$$

has the form

$$\Delta_1 * P_1. \tag{8.9.4}$$

Here Δ_1 is the distribution of a segment process which, with probability 1, consists of one segment whose left endpoint is at O with random length and is G-distributed; P_1 is the distribution of the Poisson process on I $\cup$ II governed by the restriction of $\lambda \cdot dx \cdot dG$ on this union. Because it is a Poisson process, its parts on the sets I and II are independent. The segments from the set II we represent now by means of their *right* endpoints y and lengths l. This corresponds to Cavalieri transformation

$$y = x + l$$
$$l = l$$

which maps II into a reflected copy of the set I. The image of the measure will be again $\lambda\,dx\,dG$.

We conclude that P_1 corresponds to two independent Poisson segment processes, one on $(0, \infty)$ and the other on $(-\infty, 0)$ with distributions related via reflection. The relative Palm distribution of $\{\delta_i\}$ with respect to $\{l_i\}$ corresponds to the Boolean model of the segment process (8.9.4) and inherits the desired independence property from (8.9.4).

We also deduce from the above remarks that the white length to the left of 0 has exponential distribution with parameter λ.

Renewal $\mathbb{T}_1$-invariant segment processes with exponential F_w also arise in connection with Poisson–Boolean models in many dimensions.

Let, for instance, $\{b_i\}$ be an $\mathbb{M}_n$-invariant Poisson process of balls (see §7.12) in $\mathbb{R}^n$. The union of the balls from $\{b_i\}$, i.e. the random set

$$U = \bigcup b_i, \tag{8.9.5}$$

is called the Boolean model for the ball process. It is again useful to call U the *black set* and its complement the *white set*.

The trace of U on any line g_0 is a $\mathbb{T}_1$-invariant renewal process of black segments with exponential F_w.

The proof follows from the observation that the process of chords $b_i \cap g_0$ is $\mathbb{T}_1$-invariant Poisson.

It is important to note that the above statements remain valid for Boolean models generated by rather general $\mathbb{M}_n$-invariant Poisson processes of *convex* domains in $\mathbb{R}^n$.

Are the mentioned properies of 'trace' processes characteristic for Poisson–Boolean models? In chapter 10 we obtain some partial results concerning this question in the planar case. Our analysis there will involve two-dimensional marks which we attach to the black segments of the trace set induced on a line by the planar U as in (8.9.5). Namely, we consider the marked segment processes

$$\{(\delta_i, \psi_i', \psi_i'')\},$$

where the (black) segments δ_i are the non-contacting components of the trace set $U \cap g_0$, and the mark (ψ_i', ψ_i'') consists of the angles shown on fig. 8.9.4.

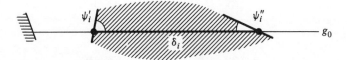

Figure 8.9.4 The angles are between g_0 and ∂U at the endpoints of the black segments

In this situation the above can be complemented by the following statement.

If a black set U is the Boolean model for an M_2-invariant Poisson disc process then the triads $(|\delta_i|, \psi_i', \psi_i'')$ for different values of i are independent.

(More exact formulation should, of course, be in terms of the anchored version of the process $\{(\delta_i, \psi_i', \psi_i'')\}$). Independence of the pairs (ψ_i', ψ_i'') with different i's follows from the independence of the elevations of the centers of the discs from $\{b_i\}$ which actually hit g_0 above this line.

8.10 Palm formulae for balls in $\mathbb{R}^n$

Let b_V be the ball of volume V centered at O. We consider the annuli

$$a_{V,h}^+ = b_{V+h} \backslash b_V \qquad \text{or} \qquad a_{V,h}^- = b_V \backslash b_{V-h}.$$

As mentioned in §6.6, fig. 6.6.2, (6.6.7) and (6.6.8) remain valid if we replace D by these annuli and let h tend to zero. By averaging in the space of realizations we conclude that (7.15.5) and (7.15.6) remain valid if D is replaced by our thin annuli.

By an easy modification of the method which led to (8.5.2) we can show the existence of the limits (see [32])

$$\lim_{h \to 0} h^{-1} P\left(\binom{a^{\pm}}{1} \cap \binom{b_V}{s} \right) = \lambda \pi_s(V).$$

The functions π above may be interpreted as the conditional probabilities of the event $\binom{b_V}{s}$, the condition being that there is a point of $m(\omega)$ on the boundary ∂b_V:

$$\pi_s(V) = |\partial b_V|^{-1} \int_{\partial b_V} \Pi\binom{tb_V}{s} \, \mathrm{d}t^*. \tag{8.10.1}$$

Here tb_V is the shift of b_V by the vector $t \in \partial b_V$ (so that always $O \in \partial(tb_V)$), $\mathrm{d}t^*$ is the area measure on the surface of b_V, and $|\partial b_V|$ is the total 'area' of ∂b_V. In other words, t is distributed uniformly over ∂b_V.

By considering events which happen in b_{V+h} and b_V and applying essentially the same argument as in derivation of (8.7.4) we come to the desired Palm formulae

$$\frac{\mathrm{d}p_s(V)}{\mathrm{d}V} = \lambda[\pi_{s-1}(V) - \pi_s(V)], \tag{8.10.2}$$

where

$$p_s(V) = P\binom{b_V}{s}, \qquad \pi_{-1}(V) = 0.$$

The derivation of formulae similar to (8.10.2) for several balls or n-dimensional intervals is left to the reader. We use (8.10.2) in the next two sections.

8.11 The equation $\Pi = \Theta * P$

In this section we consider the relation

$$\Pi = \Theta * P \qquad (8.11.1)$$

as an *equation* where Θ is a distribution of some point process in $\mathbb{R}^n$ and is assumed known. The problem is to find a distribution of a $\mathbb{T}_n$-invariant point process in $\mathbb{R}^n$ of finite intensity whose Palm distribution Π can be calculated by means of (8.11.1) (the composition sign $*$ corresponds to the superposition of *independent* point processes). We saw in §8.6 that Poisson cluster processes satisfy (8.11.1); hence, at least for some distributions Θ, (8.11.1) indeed has solutions. Let us show how the probabilities $P\begin{pmatrix} b_V \\ l \end{pmatrix}$ can be found for all balls from (8.11.1).

We introduce the following generating functions:

$$P(z, B) = \sum P\begin{pmatrix} B \\ k \end{pmatrix} z^k;$$

$$\Pi(z, B) = \sum \Pi\begin{pmatrix} B \\ k \end{pmatrix} z^k;$$

$$\Theta(z, B) = \sum \Theta\begin{pmatrix} B \\ k \end{pmatrix} z^k;$$

$$\pi(z, V) = \sum \pi_k(V) z^k;$$

(the probabilities $\pi_k(V)$ have been defined in (8.10.1)). Equation (8.11.1) implies that

$$\Pi(z, tb_V) = \Theta(z, tb_V) P(z, b_V)$$

for every shift t. Integrating this over ∂b_V and using (8.10.1) yields

$$\pi(z, V) = P(z, b_V) \tilde{\Theta}(z, b_V),$$

where

$$\tilde{\Theta}(z, b_V) = |\partial b_V|^{-1} \int \Theta(z, tb_V) \, dt^*.$$

Let us formally apply (8.10.2), which in terms of generating functions is now written as

$$\frac{dP(z, b_V)}{dV} = \lambda(z - 1) P(z, b_V) \tilde{\Theta}(z, b_V).$$

Its solution satisfying the initial condition

$$P(z, b_V)|_{V=0} = 1$$

is clearly

$$P(z, b_V) = \exp\{\lambda(z - 1) \int_0^V \tilde{\Theta}(z, b_u) \, du\}. \qquad (8.11.2)$$

Of course this is a partial result, since for a complete description of P we need to find the probabilities of the type

$$P\begin{pmatrix} I_1 & & I_s \\ k_1 & , \cdots, & k_s \end{pmatrix},$$

where the I_i's are n-dimensional intervals.

We see from (8.11.2) that the solutions of (8.11.1) cannot be unique (if they exist); in fact we have to deal with families of solutions depending on the parameter λ (the intensity).

Example Let us consider the equation

$$\Pi = \Delta * P, \tag{8.11.3}$$

where Δ is the distribution of a point process which has only one point in the whole of $\mathbb{R}^n$ placed at O. In this case

$$\Theta(z, tb_u) \equiv 1 \quad \text{for } t \in \partial b_u$$
$$\tilde{\Theta}(z, b_u) \equiv 1,$$

and from (8.11.2) we find that

$$P(z, b_V) = \exp\{\lambda(z-1)V\}. \tag{8.11.4}$$

By applying n-dimensional versions of (8.8.1) we can show that (8.11.3) *has only Poisson solutions.*

8.12 Asymptotic Poisson distribution

In this section we consider a rather special but important feature of a point process – the distribution of the number $N(m, b_V)$ of points within a ball b_V. In fact to derive the Poisson property of this variable (i.e. (8.11.4)) much less than (8.11.3) is needed as (8.11.4) follows merely from the equation

$$p_s(V) = \pi_s(V). \tag{8.12.1}$$

We repeat that $\{\pi_s(V)\}$ may be interpreted as the conditional distribution of the number of points in b_V, given that there is a point of the process on ∂b_V, see (8.10.1).

Can we make any conclusions concerning the distributions in (8.12.1) under less stringent assumptions then their coincidence? We show now that the Palm formulae (8.10.2) make this possible.

We rewrite (8.10.2) in the form

$$\frac{dp_s(V)}{dV} = \lambda[p_{s-1}(V) - p_s(V)] + \lambda[\delta_{s-1}(V) - \delta_s(V)], \tag{8.12.2}$$

where

$$\delta_s(V) = \pi_s(V) - p_s(V).$$

In terms of the generating functions

$$p(z, V) = \sum p_s(V)z^s$$
$$\delta(z, V) = \sum \delta_s(V)z^s,$$

(8.12.2) becomes

$$\frac{dp(z, V)}{dV} = \lambda(z - 1)p(z, V) + \lambda(z - 1)\delta(z, V) \qquad (8.12.3)$$

or, after solving with respect to $p(z, V)$,

$$p(z, V) = e^{\lambda(z-1)V} + \lambda(z - 1) \int_0^V e^{\lambda(z-1)(V-u)}\delta(z, u)\, du. \qquad (8.12.4)$$

The characteristic function of the random variable

$$\eta_V = \frac{N(\mathscr{m}, b_V) - \lambda V}{\sqrt{(\lambda V)}} \qquad (8.12.5)$$

equals

$$f(\varphi) = e^{-i\varphi\sqrt{(\lambda V)}}p(e^{i\varphi/\sqrt{(\lambda V)}}, V)$$

and therefore can be obtained from the right-hand side of (8.12.4) by substituting

$$z = e^{i\varphi/\sqrt{(\lambda V)}}$$

and multiplying the result by $e^{i\varphi/\sqrt{(\lambda V)}}$. If we let V tend to ∞, then the term corresponding to $e^{\lambda(z-1)V}$ will tend to $e^{-\varphi^2/2}$, the characteristic function of the standard normal distribution (recall the well known result that Poisson distributions are asymptotically normal).

Therefore it is natural to look for conditions under which the term corresponding to the integral in (8.12.4) vanishes. Since our substitution renders $e^{\lambda(z-1)(V-u)}$ bounded (we obtain the characteristic function of a Poisson distribution), the problem reduces to the evaluation of

$$\lambda|z - 1| \int_0^V |\delta(z, u)|\, du \leqslant \lambda|z - 1| \int_0^V \sum |\pi_s(u) - p_s(u)|\, du$$

$$= \lambda|z - 1| \int_0^V \rho(u)\, du.$$

The function

$$\rho(u) = \sum |\pi_s(u) - p_s(u)|$$

is the *variational distance* between the distributions $\{\pi_s(u)\}$ and $\{p_s(u)\}$. Since as $V \to \infty$ we have

$$z \approx 1 + \frac{i\varphi}{\sqrt{(\lambda V)}}$$

the required condition will be

$$\int_0^V \rho(u)\, du = o(\sqrt{V}) \quad \text{as } V \to \infty. \qquad (8.12.6)$$

We have found that (8.12.6) is a sufficient condition for η_V as defined by (8.12.5) to be asymptotically normal distributed with zero mean and unit variance. Due to the specific form of (8.12.5) (normalization by the square root of expectation) this property can be called the *asymptotic Poisson* property of the distribution of $N(m, b_V)$.

In fact (8.12.6) imposes a restriction on the rate of the decrease of the distance ρ at infinity (it must be not too slow). Loosely speaking, if the condition of having a point on the boundary of b_V does not affect the distribution of the number of points in b_V for larger values of V too much, then $N(m, b_V)$ is *asymptotically Poisson* distributed [52].

8.13 Equations with Palm distribution

The success of (8.11.3) suggests that we should try to apply similar equations in more complicated stochastic situations. Below we introduce two different equations of this style. In the next section where we describe their solutions the interrelation between the two problems will become apparent.

I A natural idea is to replace P in (8.11.3) by the result P^* of some operation acting on P. Here we will define P^* as obtained from P by means of location dependent but statistically independent thinning.

More properly let us assume that a function $h(\mathscr{P})$, $0 \leqslant h(\mathscr{P}) \leqslant 1$ is defined in $\mathbb{R}^n$. Given a realization m we define m^* to be a random *subset* of m: a point $\mathscr{P}_i \in m$ belongs to m^* with probability $h(\mathscr{P}_i)$; for different values of i the events $\{\mathscr{P}_i \in m^*\}$ are assumed independent.

Let P be the distribution of a point process $m(\omega)$. Then by P^* we denote the distribution of the thinned point process $m^*(\omega)$. The desired version of the equation (8.11.3) will be

$$\Pi = \Delta * P^*. \qquad (8.13.1)$$

Here the problem is to describe conditions on λ and on the function $h(\mathscr{P})$ which ensure the existence of a $\mathbb{T}_n$-invariant probability P of finite intensity for which (8.13.1) holds as well as to clarify the uniqueness questions.

II The second equation provides a description of processes of non-intersecting balls of unit radius which otherwise 'do not interact'. These two seemingly contradictory requirements – non-intersection and non-interaction – can be reconciled as follows.

We assume that *the centers of the balls constitute a $\mathbb{T}_n$-invariant process* $m(\omega)$. The centers cannot be closer than distance 2 apart, thus $m(\omega)$ is necessarily of finite intensity, and we can speak about the Palm distribution Π of $m(\omega)$.

Since always

$$\Pi(A_1) = 1,$$

where

$$A_1 = \{\text{there is a point of } m(\omega) \text{ at } O\},$$

it follows that the condition

$$P\{\text{the unit balls with centers in } m(\omega) \text{ do not intersect}\} = 1$$

implies

$$\Pi(A \cap A_1) = 1,$$

where

$$A = \begin{pmatrix} B(2) \backslash O \\ 0 \end{pmatrix},$$

and $B(2)$ is the ball of radius 2 centered at O. We will write $P(\cdot/A)$ for the distribution of $m(\omega)$ *conditional upon* A.

We have

$$\Pi \equiv P(\cdot/A_1) = P(\cdot/A_1 \cap A). \tag{8.13.2}$$

We postulate that the mathematical expression of the 'no interaction' property is that 'outside O' we have

$$P(\cdot/A_1 \cap A) = \Delta * P(\cdot/A). \tag{8.13.3}$$

Intuitively speaking, this equation means that if there is enough space somewhere in the realization to put a ball there (condition A), the actual placing of a new ball (condition $A \cap A_1$) does not affect other balls of the realization.

Equations (8.13.2) and (8.13.3) can be put together to give

$$\Pi = \Delta * P(\cdot/A), \tag{8.13.4}$$

which is another direct generalization of (8.11.3).

8.14 Solution by means of density functions

I First we consider the equation (8.13.1). We are looking for the solution P in the class of point processes in $\mathbb{R}^m$ generated by continuous densities $\varphi(x_1, \ldots, x_n)$, see §7.4. We will use the interpretation (7.4.7) with dx_i now denoting infinitesimal volumes in $\mathbb{R}^m$. Note that because of $\mathbb{T}_m$-invariance and by (7.15.5) we have

$$\varphi(x_1) \equiv \lambda, \qquad \text{the intensity of } P.$$

How are the density functions of the distributions Π and P^* expressed in terms of the functions $\varphi(x_1, \ldots, x_n)$?

It follows from the construction of P^* by means of thinning that

$$P^*\begin{pmatrix} dx_1 & dx_n \\ 1 & ,\ldots, & 1 \end{pmatrix} = P\begin{pmatrix} dx_1 & dx_n \\ 1 & ,\ldots, & 1 \end{pmatrix} \prod_{i=1}^{n} h(x_i)$$

and we conclude that the density functions for P^* are necessarily of the form

$$\varphi^*(x_1, \ldots, x_n) = \varphi(x_1, \ldots, x_n) \prod_{i=1}^{n} h(x_i).$$

Denote by $\varphi_\Pi(x_1, \ldots, x_n)$ the density functions of Π. We base on a heuristic relation

$$\Pi\begin{pmatrix} dx_1 \\ 1 \end{pmatrix}, \ldots, \begin{pmatrix} dx_n \\ 1 \end{pmatrix} = (\lambda \, dO)^{-1} P\begin{pmatrix} dO & dx_1 \\ 1 & 1 \end{pmatrix}, \ldots, \begin{pmatrix} dx_n \\ 1 \end{pmatrix}$$

which is a version of (8.5.2) in which the domains D_i are taken infinitesimal and which can be easily justified under the assumption (7.4.9) concerning densities φ. This implies

$$\varphi_\Pi(x_1, \ldots, x_n) = \lambda^{-1}\varphi(0, x_1, \ldots, x_n). \tag{8.14.1}$$

By $\mathbb{T}_m$-invariance of P

$$\varphi(x_1, x_2, \ldots, x_{n+1}) = \varphi(0, x_2 - x_1, \ldots, x_{n+1} - x_1)$$
$$= \lambda\varphi_\Pi(x_2 - x_1, \ldots, x_{n+1} - x_1).$$

The equation (8.13.1) implies the coincidence of densities φ^* and φ_Π i.e.

$$\varphi(x_1, \ldots, x_{n+1}) = \lambda\varphi(x_2 - x_1, \ldots, x_{n+1} - x_1) \prod_{i=2}^{n+1} h(x_i - x_1).$$

We easily derive from this that

$$\varphi(x_1, \ldots, x_n) = \lambda^n \prod_{i<j} h(x_i - x_j). \tag{8.14.2}$$

In order to guarantee that $\varphi(x_1, \ldots, x_n)$ be symmetrical functions it is enough to require (and we do) that

$$h(-x) = h(x).$$

Also the condition

$$\varphi(x_1, \ldots, x_n) \leqslant \lambda^n$$

is satisfied because $0 \leqslant h \leqslant 1$. Thus it remains to find out when condition (2) of §7.4 is satisfied. For then the corresponding P will become a solution of (8.13.1) and unique within the class we consider. The following proposition whose complete proof can be found in [64] provides the answer.

Assume

$$C = \int_{\mathbb{R}^m} (1 - h(u)) \, du < \infty.$$

Then for the values of λ in the interval $0 < \lambda < (eC)^{-1}$ (e is the base of natural logarithms) (8.14.1) are density functions of a $\mathbb{T}_m$-invariant point process of finite intensity λ. Its distribution P satisfies (8.13.1).

In one dimension the case

$$h(x) = \begin{cases} 0 & \text{if } |x| < 1 \\ 1 & \text{otherwise} \end{cases}$$

can be treated by means of Palm formulae (8.7.4). We have

$$\Pi\begin{pmatrix}(0, x)\\0\end{pmatrix} = \begin{cases}1 & \text{if } x < 1\\p_0(x-1) & \text{if } x > 1.\end{cases}$$

Substituting this in (8.7.4) for $k = 0$ we get

$$\frac{dp_0(x)}{dx} = -\lambda p_0(x-1) \qquad \text{if } x > 1$$

$$\tag{8.14.3}$$

$$p_0(x) = 1 - \lambda x \qquad \text{for } 0 < x < 1.$$

The above statement implies that the solution of (8.14.3) remains non-negative whenever $\lambda < e^{-1}$. This particular result was obtained in [65] with no probabilistic interpretation.

II Instead of (8.13.4) we prefer to consider a somewhat more general equation

$$\Pi^*(dm) = c\Theta(m)P(dm) \tag{8.14.4}$$

which expresses the idea that Π has a density $c\Theta(m)$ with respect to P. Here again P is a $\mathbb{T}_m$-invariant probability on $\mathcal{M}_{\mathbb{R}^m}$ of finite intensity λ, Π^* is obtained from the Palm distribution of P by discarding the point at O.

We obtain (8.13.4) from (8.14.4) when we take

$$\Theta(m) = 1 \qquad \text{if } m \in \begin{pmatrix}B(2)\\0\end{pmatrix}, \text{ and } 0 \text{ otherwise}$$

and

$$c = \left(P\begin{pmatrix}B(2)\\0\end{pmatrix}\right)^{-1}.$$

For a given function $\Theta(m)$ we can expect a family of solutions for (8.14.4) (if any exist at all) and the constant c depends on the particular solution we choose since it has to satisfy the norming condition

$$c^{-1} = \int \Theta(m)P(dm). \tag{8.14.5}$$

We try to find the solutions of (8.14.4) in the class of probabilities P generated by continuous densities $\varphi(x_1, \ldots, x_n)$. From (8.14.4) it follows that

$$\varphi_{\Pi^*}(x_1)\,dx_1 = \Pi^*\begin{pmatrix}dx_1\\1\end{pmatrix} = c\int_{\begin{pmatrix}dx_1\\1\end{pmatrix}}\Theta(m)P(dm). \tag{8.14.6}$$

On the set $\begin{pmatrix}dx_1\\1\end{pmatrix}$ the variable m can be represented as

$$m = x_1 \cup m'$$

and on the same set $P(dm)$ can be written as

$$P(dm) = \lambda\,dx_1\,\Pi^*_{x_1}(dm')$$

which is a heuristic analog of (8.5.3). Here Π^*_x is the image of Π^* under the shift map

$$m \to t_x m \qquad \text{where} \qquad t_x = \vec{Ox}$$

and satisfies the equation

$$\Pi_x^*(dm) = \Theta_x(m)P(dm) \tag{8.14.7}$$

where

$$\Theta_x(m) = \Theta(t_{-x}m).$$

Using (8.14.1) we find further

$$\lambda^{-1}\varphi(0, x_1) = \varphi_{\Pi^*}(x_1) = c\lambda \int \Theta(m \cup x_1)\Pi_{x_1}^*(dm)$$

$$= c^2\lambda \int \Theta_0(m \cup x_1)\Theta_{x_1}(m)P(dm)$$

where of course $\Theta_0 \equiv \Theta$. Replacing here 0 by x_1 and x_1 by x_2 we obtain

$$\varphi(x_1, x_2) = (c\lambda)^2 \int \Theta_{x_1}(m \cup x_2)\Theta_{x_2}(m)P(dm).$$

Acting similarly we find a general expression

$$\varphi(x_1, \ldots, x_n)$$

$$= (c\lambda)^n \int \Theta_{x_1}(m \cup \{x_2, \ldots, x_n\})\Theta_{x_2}(m \cup \{x_3, \ldots, x_n\})\Theta_{x_n}(m)P(dm). \tag{8.14.8}$$

Because of (8.14.5) this yields in the case $n = 1$

$$\varphi(x_1) \equiv \lambda.$$

The relation (8.14.8) remains unexplored for general Θ. We will concentrate on the case (which includes (8.13.4))

$$\Theta_x(m) = \prod_{\mathscr{P}_i \in m} h(x - \mathscr{P}_i)$$

where h is an even function defined in $\mathbb{R}^m$. Clearly in this case for $j = 1, \ldots, n$

$$\Theta_{x_j}(m \cup \{x_{j+1}, \ldots, x_n\}) = \prod_{k=j+1}^{n} h(x_j - x_k) \cdot \prod_{\mathscr{P}_i \in m} h(x_j - \mathscr{P}_i).$$

Therefore (8.14.8) reduces to

$$\varphi(x_1, \ldots, x_n) = \int a(x_1, \ldots, x_n)b_m(x_1, \ldots, x_n)P(dm) \tag{8.14.9}$$

where

$$a(x_1, \ldots, x_n) = \lambda_1^n \prod_{j<k} h(x_j - x_k), \qquad \lambda_1 = c\lambda,$$

$$b_m(x_1, \ldots, x_n) = \prod_{j=1}^{n} \prod_{\mathscr{P}_i \in m} h(x_j - \mathscr{P}_i).$$

If λ_1 and h satisfy the conditions of the proposition in part I of this section then $a(x_1, \ldots, x_n)$ happen to be density functions of a probability P_0 on $\mathcal{M}_{\mathbb{R}^m}$. Because now $0 \leqslant h(x) \leqslant 1$ the products

$$a(x_1, \ldots, x_n) b_m(x_1, \ldots, x_n)$$

receive interpretation of density functions of a point process which is obtained from P_0 by independent thinning. The thinning procedure is as follows.

Let $m_0(\omega)$ be a point process whose distribution is P_0. A point $Q_1 \in m_0(\omega)$ survives with probability $\prod_{\mathcal{P}_j \in m} h(Q_i - \mathcal{P}_j)$ (this probability depends on m) and for different points $Q_i, Q_s \in m_0$ the events of survival are independent.

The distribution of the thinned process (i.e. of the points which survive) we denote by $P_0^{(m)}$. Then (8.14.9) can be replaced by an equivalent relation

$$P = \int P_0^{(m)} P(\mathrm{d}m). \tag{8.14.10}$$

This equation has the advantage that it can be solved by successive approximations. Namely we start with some P_1 (it can be $\mathbb{T}_m$-invariant Poisson), and define

$$P_{n+1} = \int P_0^{(m)} P_n(\mathrm{d}m), \qquad n = 1, 2, \ldots.$$

The problem is to show that the sequences of probabilities P_n converges to a limit P which is a solution of both (8.14.10) and (8.14.4) and which does not depend on the choice of P_1. Conditions for such a convergence were found in [67].

We now complement the above by a brief outline of an explicit result concerning (8.13.4) in one dimension.

We consider the functions

$$p_0(x) = P\begin{pmatrix} (0, x) \\ 0 \end{pmatrix} \quad \text{and} \quad \pi_0(x) = \Pi\begin{pmatrix} (0, x) \\ 0 \end{pmatrix}.$$

We assume that $\pi_0(x)$ is continuous. Then the Palm formulae yield

$$\frac{\mathrm{d}p_0(x)}{\mathrm{d}x} = -\lambda \pi_0(x). \tag{8.14.11}$$

The basic relation (8.13.4) implies

$$\pi_0(x) = \begin{cases} 1 & \text{if } x < 2 \\ \dfrac{p_0(x + 2)}{p_0(4)} & \text{if } x > 2. \end{cases}$$

Therefore for $x > 2$

$$\pi_0(x + 2) = -\frac{p_0(4)}{\lambda} \frac{\mathrm{d}}{\mathrm{d}x} \pi_0(x). \tag{8.14.12}$$

This relation enables us to determine $\pi_0(x)$ for all values $x > 4$ if $\pi_0(x)$ is in some way given on the interval $(2, 4)$. But $\pi_0(x)$ cannot be arbitrary on $(2, 4)$. For every $x \in (2, \infty)$ and every $k > 0$

$$\pi_0(x + 2k) = \left(-\frac{p_0(4)}{\lambda}\right)^k \frac{d^k}{(dx)^k} \pi_0(x).$$

The function $\pi_0(x)$ must remain non-negative. It follows that $\pi_0(x)$ should be *absolutely monotone* on $(2, \infty)$; in other words a measure μ on $(0, \infty)$ exists such that

$$\pi_0(x) = \int e^{-\alpha(x-2)}\mu(d\alpha), \qquad x > 2$$

The values on a half-axis of Laplace transformation of a measure defines the latter uniquely. Therefore we conclude from (8.14.12) that for every interval $I \subset (0, \infty)$

$$\int_I \alpha e^{2\alpha}\mu(d\alpha) = \frac{\lambda}{p_0(4)} \int_I \mu(d\alpha). \qquad (8.14.13)$$

The equation

$$\alpha e^{2\alpha} = \text{const}$$

has only one solution. Considering (8.14.13) for small intervals I we conclude, that μ is concentrated on a single α which is the solution of the equation

$$\alpha e^{2\alpha} = \frac{\lambda}{p_0(4)}.$$

This means that

$$\pi_0(x) = \begin{cases} 1 & \text{if } x < 2 \\ \exp(-\alpha(x-2)) & \text{if } x \geqslant 2. \end{cases} \qquad (8.14.14)$$

In fact we can choose $\alpha > 0$ arbitrarily. Then the parameters λ and $p_0(4)$ should be determined from the equations

$$\lambda^{-1} = \int_0^\infty \pi_0(u) \, du = 2 + \alpha^{-1}$$

$$p_0(x) = 1 - \lambda \int_0^x \pi_0(u) \, du \qquad (8.14.15)$$

A deeper analysis (see [66]) shows that the point process in question happens to be a *renewal* point process with the distribution of the space between the consecutive points given by (8.14.14). It corresponds to a segment process which is renewal with constant segment length 2 and exponential distribution of intersegment spaces (see §8.9).

9

Poisson-generated geometrical processes

One application of Haar factorization leads to the concept of *relative* Palm distribution of a group-invariant geometrical process. By associating different point processes on corresponding groups, we obtain different Palm-type distributions for the same geometrical process.

When the geometrical process is Poisson, its relative Palm distributions can be calculated, and the result is usually reminiscent of (8.11.1). Much of this chapter will be devoted to different corollaries of such results. The purpose is to derive size–shape distributions (or at least to give the corresponding stochastic constructions) of typical configurations generated by Poisson processes.

9.1 Relative Palm distribution

Let $\mathbb{X}$ be a group of transformations of a space $\mathbb{Y}$. For instance, $\mathbb{X}$ can be $\mathbb{T}_n$, $\mathbb{M}_n$ or $\mathbb{A}_n$, $n = 2, 3 \ldots$ and $\mathbb{Y}$ can be the corresponding $\mathbb{R}^n$, or $\mathbb{G}$, Γ, $\mathbb{E}$ etc. We write $x \in \mathbb{X}$, $y \in \mathbb{Y}$. Suppose we have two *jointly distributed* point processes

$$m_1(\omega) \subset \mathbb{X} \qquad \text{and} \qquad m_2(\omega) \subset \mathbb{Y}. \qquad (9.1.1)$$

We can speak of simultaneous transformations of *both* $m_1(\omega)$ and $m_2(\omega)$ by any $x \in \mathbb{X}$:

$$m_1(\omega), m_2(\omega) \to x m_1(\omega), x m_2(\omega).$$

We assume below that the *joint distribution of the processes* (9.1.1) *is invariant*, that is *for any* $x \in \mathbb{X}$ the pairs

$$m_1(\omega), m_2(\omega) \qquad \text{and} \qquad x m_1(\omega), x m_2(\omega).$$

possess the same joint distribution.

Another basic assumption is that $m_1(\omega)$ *is of finite intensity*. This means that the mean number of points from $m_1(\omega)$ in a domain $D \subset \mathbb{X}$ is $\lambda H(D)$, where H is the Haar measure on $\mathbb{X}$ (in all cases we consider, H is actually bi-invariant). If both assumptions are fulfilled, we can construct the relative Palm distribution of $m_2(\omega)$ with respect to $m_1(\omega)$. We base this construction upon Haar factorization.

We consider a marked point process on $\mathbb{X}$:

$$\{(x_i, x_i^{-1}\{y_i\})\} \qquad \text{where} \qquad \{x_i\} = m_1(\omega), \{y_i\} = m_2(\omega) \qquad (9.1.2)$$

with marks belonging to the space of realizations $\mathcal{M}_\mathbb{Y}$.

Let us consider the counting function

$N(B_1, B_2, \omega) = $ the number of points $x_i \in m_1(\omega)$ which lie in $B_1 \subset \mathbb{X}$ and whose marks lie within $B_2 \subset \mathcal{M}_\mathbb{Y}$.

The marked point process (9.1.2) is invariant with respect to $\mathbb{X}$. This property is inherited by the expectation of N, therefore we have the Haar factorization (see §1.3)

$$EN(B_1, B_2, \omega) = \lambda H(B_1)\Pi_{m_1}(B_2), \qquad (9.1.3)$$

We choose the value of λ in (9.1.3) from the condition that Π_{m_1} be a probability distribution, in which case λ coincides with the intensity of $\{x_i\}$.

We call the probability distribution Π_{m_1} on $\mathcal{M}_\mathbb{Y}$ defined by (9.1.3) the *relative Palm distribution* of $\{y_i\}$ with respect to $\{x_i\}$. We rewrite (9.1.3) as

$$\lambda \int\int I_{B_1 \times B_2}(x, m_2)H(\mathrm{d}x)\Pi_{m_1}(\mathrm{d}m_2) = E \sum_{x_i \in m_1} I_{B_1 \times B_2}(x_i, x_i^{-1}m_2),$$

where we denote by m_2 the elements of $\mathcal{M}_\mathbb{Y}$. Using linearity, we derive from this the following:

$$\lambda \int\int V(x, m_2)H(\mathrm{d}x)\Pi_{m_1}(\mathrm{d}m_2) = E \sum_{x_i \in m_1} V(x_i, x_i^{-1}m_2), \qquad (9.1.4)$$

first for the functions V which are linear combinations of indicator functions or $I_{B_1 \times B_2}$ type. Then by standard measure-theoretic argument we establish that (9.1.4) is *valid for any measurable function V which maps $\mathbb{X} \times \mathcal{M}_\mathbb{Y}$ onto the real axis* (compare with (7.14.5)).

We get the simplest example of a Palm distribution as defined by (9.1.3) when we assume that $\mathbb{Y} = \mathbb{X}$ and that $m_2(\omega) = m_1(\omega)$. In this case $\Pi_{m_1} = \Pi$ defines what can be called the Palm distribution of a left invariant point process $\{x_i\}$ on a group $\mathbb{X}$ (assuming that $\{x_i\}$ is of finite intensity).

The general topic of point processes on groups (especially non-Abelian ones) remains largely unexplored. In the sections that follow, processes on groups appear merely as specific tools for the study of particular geometrical processes. We start each time with a geometrical process $m_2(\omega)$ (which we wish to study). Then we construct the auxilliary point process $m_1(\omega)$ by applying an appropriate realization-to-realization mapping

$$m_2 \to m_1, \qquad \text{where} \qquad m_2 \in \mathcal{M}_\mathbb{Y}, m_1 \in \mathcal{M}_\mathbb{X}.$$

Whenever m_2 happens to depend on ω (is random) a mapping of this kind automatically induces dependence of m_1 on ω. Hence the joint distribution of the processes $m_2(\omega)$ and $m_1(\omega)$ will be well defined each time and we will be able to apply the above concepts.

9.2 Extracting point processes on groups

The following examples will make clear the general idea of how a point process on the group $\mathbb{M}_2$ can be associated with a given geometrical process in $\mathbb{R}^2$. A case where we 'extract' a point process on $\mathbb{T}_2$ is considered in §§ 9.3 and 9.4.

Example 1 Let $\{g_i\}$ be a line process on $\mathbb{R}^2$ (a point process in $\overline{\mathbb{G}}$). On each line g_i of the process we plant an independent Poisson point process each governed by L_1 (the one-dimensional Lebesgue measure). Each of the points thus obtained together with the direction of the carrying line g_i determines a motion $M_s \in \mathbb{M}_2$ (see §2.13). These motions constitute a point process on $\mathbb{M}_2$ which we represent as $\{M_s\}$. The pair of dependent processes will be

$$m_1(\omega) = \{M_s\}, \qquad m_2(\omega) = \{g_i\}.$$

The property of $\mathbb{M}_2$-invariance of $\{g_i\}$ directly implies invariance of the pair $m_1(\omega)$ and $m_2(\omega)$ in the sense of the previous section, and $\{M_s\}$ happens to be of finite intensity whenever $\{g_i\}$ is of finite intensity (see §7.14).

In the case $\{g_i\}$ is a process on $\mathbb{G}$ (nondirected lines) we can apply the same construction by converting g_i into a process on $\overline{\mathbb{G}}$. This can be done by random equiprobable assignment of one of the two possible directions to each g_i; the choice of the directions for different lines can be independent.

Example 2 Again let the geometrical process be

$$m_2(\omega) = \{g_i\} \subset \overline{\mathbb{G}}.$$

Let us assume that with probability 1 there are no pairs of parallel lines in $\{g_i\}$. Then each pair of lines g_i and g_j has a single intersection point

$$Q_{ij} = g_i \cap g_j.$$

For each Q_{ij} we construct four unit length directed segments as shown in fig. 9.2.1. We denote these segments by M_s. Again each M_s can be identified with an element of the group $\mathbb{M}_2$. The obtained random set of motions $\{M_s\}'$

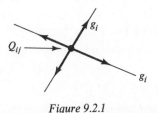

Figure 9.2.1

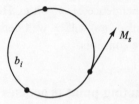

Figure 9.2.2 The segment M_s emerges from a point on a ∂b_i in the direction tangent to ∂b_i. The disc b_i remains in the left halfplane with respect to continuation of M_s

is the desired process on $\mathbb{M}_2$;

$$m_1(\omega) = \{M_s\}'.$$

The conditions of §9.1 will be satisfied whenever $\{g_i\}$ is $\mathbb{M}_2$-invariant and of second order.

In the above examples the marks $M_i^{-1}m_2$ are different in nature. In both cases they contain the x-axis. In the second example, $M_i^{-1}m_2$ necessarily contains yet another line through O whose direction is not specified and Π_{m_1} can be considered as a distribution in the space

$$\mathbb{S}_1 \times \mathcal{M}_{\overline{\mathbb{G}}}$$

where the circle $\mathbb{S}_1$ is appropriated for the direction of this line.

Example 3 Let $\{b_i\}$ be a disc process in $\mathbb{R}^2$. On the boundary of each b_i we plant an independent Poisson process governed by the arc length measure. Each of these points determines a directed segment (equivalently, a motion) $M_s \in \mathbb{M}_2$ in the manner shown in fig. 9.2.2. In this way we obtain a point process $\{M_s\}''$ on $\mathbb{M}_2$ and we put

$$m_1(\omega) = \{M_s\}'', \qquad m_2(\omega) = \{b_i\}.$$

We have the situation described in §9.1 if $\{b_i\}$ is $\mathbb{M}_2$-invariant, the process of centers of b_i's is of finite intensity and the typical radius distribution (see §8.1) has a finite mean. Each $M_s^{-1}m_2$ contains a disc tangential to the x axis at O which lies in the upper halfplane.

Example 4 Let $\{s_i\}$ be a segment process on $\mathbb{R}^2$. With each s_i from a realization we associate an $M_i \in \mathbb{M}_2$ (we again think of M_i as a directed segment): the source of M_i lies at the midpoint of s_i; the direction of M_i coincides (mod π) with the direction of s_i. We choose at random, with probability $1/2$, each of the remaining two possibilities to fix the arrow on M_i, according to the outcome of an independent experiment of tossing a coin.

We put

$$m_1(\omega) = \{M_i\}, \qquad m_2(\omega) = \{s_i\}.$$

If $\{s_i\}$ is $\mathbb{M}_2$-invariant and the process of the midpoints is of finite intensity then the conditions of §9.1 will be satisfied. Each $M_i^{-1}m_2$ will contain a segment on the x axis with the midpoint at O. We note that Π_{m_1} can be considered as a distribution in the space

$$(0, \infty) \times \mathcal{M}_{\Delta_2},$$

where the factor space $(0, \infty)$ is appropriated for the length of the mentioned segment.

We leave it to the reader to prove that *the projection of* Π_{m_1} *on* $(0, \infty)$ *coincides with the distribution of the length of the typical segment* in $\{s_i\}$ (see §8.1).

9.3 Equally weighted typical polygon in a Poisson line mosaic

Let $\{g_i\}$ be a Poisson line process on $\mathbb{R}^2$ governed by an M_2-invariant measure on G (i.e. by $\lambda \, dg$). The lines of $\{g_i\}$ split the plane in non-overlapping convex polygons π_i; thus the collection $\{\pi_i\}$ is a *planar random mosaic* (see fig. 9.3.1).

Problem (Loose formulation): find the 'distribution' of the polygon randomly chosen from the collection $\{\pi_i\}$. The choice should give 'equal weight' to 'each' polygon of the mosaic.

We called the above formulation 'loose' because in the present situation the notion of random choice 'with equal weights' is ambiguous: the number of polygons π_i is infinite with probability 1. Also, because of the absence of uniform distribution in the whole of $\mathbb{R}^2$, we can at most hope to obtain a probability distribution of the T_n-invariant characteristics of the randomly chosen polygon.

Therefore it is natural to look for a solution in the space of 'anchored' polygons.

Below we call a polygon π *anchored* if

(1) the origin $O \in \mathbb{R}^2$ is a vertex of π;
(2) π has no other vertices on the $y = 0$ axis;
(3) the interior of π lies on the right of the $y = 0$ axis (see fig. 9.3.2).

We put

$$\mathbb{K} = \text{the space of anchored polygons.} \qquad (9.3.1)$$

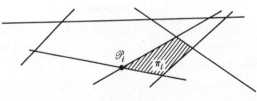

Figure 9.3.1

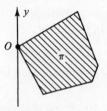

Figure 9.3.2

Using the ideas of §8.1 it is possible to *reformulate* the above problem so as to give it an absolutely precise meaning.

We consider the point process $\{\mathscr{P}_i\}$ of nodes generated by $\{g_i\}$: each $\mathscr{P}_i$ (a node) is the point of intersection of two lines from $\{g_i\}$. We associate with each $\mathscr{P}_i$ a mark

$$k_i = t_i \pi_i,$$

where $t_i = \vec{\mathscr{P}_i O}$, π_i is the polygon of the mosaic which has $\mathscr{P}_i$ for its *utmost left vertex* (such a vertex is unique with probability 1, see fig. 9.3.1).

Clearly each k_i is an element of $\mathbb{K}$ defined by (9.3.1). We obtain a marked point process

$$\{(\mathscr{P}_i, k_i)\}$$

which is $\mathbb{T}_2$-invariant (follows from $\mathbb{M}_2$-invariant of $\{g_i\}$) and of finite intensity (follows from the observation that the number of nodes in a circle never exceeds the number of pairs of lines from $\{g_i\}$ which hit the circle; the latter number has finite expectation).

Exact formulation

Find the probability distribution **p** of the typical mark in $\{(\mathscr{P}_i, k_i)\}$. This formulation is a plausible interpretation of the earlier loose one because in our construction each polygon of the mosaic is represented in $\{\mathscr{P}_i\}$ by exactly one point (with probability 1).

In this sense all polygons receive equal weights. Cases of non-equal weights will be considered in §9.6.

9.4 Solution

To solve the problem of the previous section in its exact formulation we use the idea of relative Palm distribution. We have a point process on the group $\mathbb{T}_2$, namely the process of nodes

$$\{\mathscr{P}_i\} = \{t_i\}, \qquad t_i = \vec{O\mathscr{P}_i}$$

and a point process on the space $\mathbb{G}$, namely Poisson $\{g_i\}$ (the two processes are strongly dependent). We mentioned in §9.3 that $\{t_i\}$ is invariant and of finite intensity, therefore the relative Palm distribution of $\{g_i\}$ with respect to $\{t_i\}$ is well defined. We denote it as Π_1.

The Palm distribution is concentrated on the set of realizations

$$\mathcal{M}_{\mathbb{G}}^0 = \{m \in \mathcal{M}_{\mathbb{G}} : m \text{ has a node at } O\}.$$

We stress that a complete description of a realization from this set includes specification of the directions of two lines through O. In fact we can assume that

$$\mathcal{M}_{\mathbb{G}}^0 = [(0, \pi) \times (0, \pi)]/2 \times \mathcal{M}_{\mathbb{G}},$$

where $[(0, \pi) \times (0, \pi)]/2$ denotes the space of unordered pairs of directions. In the mosaic generated by any $m \in \mathcal{M}_{\mathbb{G}}^0$ there is only one anchored polygon (see §9.3). The map

$$\text{realization} \rightarrow \text{its anchored polygon} \tag{9.4.1}$$

is a kind of projection of $\mathcal{M}_{\mathbb{G}}^0$ on $\mathbb{K}$ as defined by (9.3.1). In fact the distribution $\mathbf{p}$ of the equally weighted typical polygon as described in §9.3 *is the image of Π_1 under the map* (9.4.1), i.e.

$$\mathbf{p} = \text{projection of } \Pi_1 \text{ on } \mathbb{K}. \tag{9.4.2}$$

Proof By coincidence of the sets

$$\{\mathcal{P}_i : k_i \in A\} = \{\mathcal{P}_i : t_i m \in A_1\},$$

where

$$A \subset \mathbb{K},$$

$A_1 = \{m \in \mathcal{M}_{\mathbb{G}}^0 : \text{the image of } m \text{ under (9.4.1) belongs to } A\}$. Thus, (9.4.2) reduces the problem to that of the description of Π_1.

In the next section we show that

$$\Pi_1 = \Delta_1 * P \tag{9.4.3}$$

Here Δ_1 is the distribution of a line process which can be constructed as follows. We put two lines through O, say g' and g''. The direction of g' has uniform distribution on $(0, \pi)$, and the angle ψ between g' and g'' is random independent and has probability density $1/2 \sin \psi$ on $(0, \pi)$. P is the distribution of the line process $\{g_i\}$ with which we started, i.e. it is Poisson governed by $\lambda \, dg$, and the $*$ corresponds to the superposition of independent processes. Hence we come to the following stochastic construction suggested by Miles [33]: we construct the random lines g' and g'' and superpose them on an independent realization of (Poisson) $\{g_i\}$. In the resulting random mosaic we take its (unique) anchored polygon (doubly shaded in fig. 9.6.2). The distribution of the latter coincides with the distribution of the typical equally weighted polygon in the mosaic generated by $\{g_i\}$, i.e. it is $\mathbf{p}$.

9.5 Derivation of the basic relation

Let us give a derivation of (9.4.3). The problem is in the factorization of the first moment measure m_1 of the marked point process $\{(\mathscr{P}_i, t_1^{-1}m)\}$, namely in the separation of a Lebesgue measure in $\mathbb{R}^2$ and in the description of the factor measure in the space of marks (normalized to become a probability measure).

We choose two infinitesimal domains $\widehat{dg_1}$ and $\widehat{dg_2}$ in $\mathbb{G}$ whose invariant measures we denote as dg_1 and dg_2 and an event

$$A = \begin{pmatrix} D_1 \\ k_1 \end{pmatrix}, \dots, \begin{pmatrix} D_s \\ k_s \end{pmatrix}.$$

We assume that the union of the domains $D_1, \dots, D_s$ covers neither $\widehat{dg_1}$ nor $\widehat{dg_2}$. According to (7.5.2)

$$P\left(\begin{pmatrix} \widehat{dg_1}, & \widehat{dg_2} \\ 1 & 1 \end{pmatrix} \cap A\right) = \lambda^2\, dg_1\, dg_2\, P(A). \tag{9.5.1}$$

Let $d\widehat{\mathscr{P}} \times \widehat{d\varphi_1} \times \widehat{d\varphi_2}$ be the infinitesimal domain in the space $\mathbb{R}^2 \times (0, \pi) \times (0, \pi)$ corresponding to $\widehat{dg_1} \times \widehat{dg_2}$ via the map

$$(g_1, g_2) \to (\mathscr{P}, \varphi_1, \varphi_2),$$

where $\mathscr{P} = g_1 \cap g_2$, and the angle φ_i is the direction of g_i, see §2.1.

The event under P on the left-hand side of (9.5.1) can be expressed in terms of the marked point process $\{(\mathscr{P}_i, t_i^{-1}m)\}$:

$$\begin{pmatrix} \widehat{dg_1}, & \widehat{dg_2} \\ 1 & 1 \end{pmatrix} \cap A = \begin{pmatrix} d\widehat{\mathscr{P}} \\ 1 \end{pmatrix} \times \{t_0^{-1}m \in B\},$$

where $t_0^{-1}m$ is the mark of the vertex which lies in $d\widehat{\mathscr{P}}$ and

$$B = \widehat{d\varphi_1} \times \widehat{d\varphi_2} \times (t_0^{-1}A) \subset \mathscr{M}_{\mathbb{G}}^0$$

We have (see (8.1.1) and §3.15, I)

$$m_1(d\widehat{\mathscr{P}} \times B) = EN(d\widehat{\mathscr{P}} \times B)$$

$$= P\left(\begin{pmatrix} d\widehat{\mathscr{P}} \\ 1 \end{pmatrix} \times \{t_0^{-1}m \in B\}\right) + o(d\widehat{\mathscr{P}})$$

$$= \lambda^2\, dg_1\, dg_2\, P(A) + o(d\mathscr{P})$$

$$= \lambda^2 d\mathscr{P} \sin\psi\, d\psi\, d\varphi\, P(A) \tag{9.5.2}$$

(actually $o(d\mathscr{P}) \equiv 0$ because the above expectation is strictly proportional to $d\mathscr{P}$). We have

$$\frac{1}{2}\int_0^\pi \int_0^\pi \sin\psi\, d\psi\, d\varphi = \pi$$

and the desired factorization receives its final form

$$m_1(d\mathscr{P} \times B) = \lambda_1 d\mathscr{P}\pi^{-1} \sin\psi\, d\psi\, d\varphi\, P(A)$$

with

$$\lambda_1 = \pi \lambda^2.$$

Hence the result (9.4.3). The factor $\pi^{-1} \sin \psi \, d\psi \, d\varphi$ (which is a probability distribution on $[(0, \pi) \times (0, \pi)]/2$) corresponds to Δ_1. We mention that λ_1 is the intensity of the point process $\{\mathscr{P}_i\}$ of vertices (nodes) generated by Poisson line process governed by $\lambda \, dg$. The same value appears in another context in §9.11.

9.6 Further weightings

In this section we briefly describe a development of the method of the previous two sections which we then use to answer some further questions concerning the random mosaic generated by Poisson line processes governed by $\lambda \, dg$. But first we explain the main idea of weighting in terms of a general $\mathbb{M}_2$-invariant random mosaic $\{\pi_i\}$.

Let ξ be some $\mathbb{M}_2$-invariant parameter of a polygon (such as area, perimeter length, etc). We define the distribution of a ξ-weighted polygon in the mosaic as follows. With each polygon π_i of the mosaic we associate a cluster of points (see §7.12, IV) $c_i(\omega)$ on the group $\mathbb{M}_2$. An essential condition

$$EN(c_i(\omega), \mathbb{M}_2) = \xi \text{ for every } i \tag{9.6.1}$$

should be satisfied (the expected number of points in the whole space equals ξ).
   We put

$$\{M_s\} = \bigcup c_i(\omega).$$

By construction, each point of this point process is related to its 'parent' polygon π_k:

$$f : M_s \to \pi_k. \tag{9.6.2}$$

We will refer to f as a *parent map*.
   The second assumption is that the marked point process on $\mathbb{M}_2$:

$$\{M_s, M_s^{-1}\pi_k\} \tag{9.6.3}$$

(where $\pi_k = f(M_s)$) should be *invariant and of finite intensity*.
   If these assumptions are satisfied, then the distribution $\mathbf{p}_1$ of the typical mark in (9.6.3) is well defined: we call this $\mathbf{p}_1$ the distribution of ξ-weighted typical polygons of the mosaic. Our definition of $\mathbf{p}_1$ does not depend on the details of construction of the processes $c_i(\omega)$ (beyond those mentioned above). We leave the proof of this logically necessary proposition to the reader.
   In concrete problems it is natural to try to use the freedom in the construction of the $c_i(\omega)$'s in an 'optimal' way. In the solution of the problems that follow these constructions are devised so as to obtain simpler relative Palm distributions of $\{g_i\}$ with respect to $\{M_i\}$. The expressions (9.6.4)–(9.6.6) can be derived by procedures similar to one used in §9.4, and they strongly rely on the assumption of Poisson distribution of $\{g_i\}$.

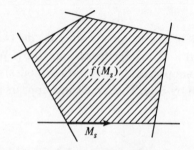

Figure 9.6.1 The segment representing M_s begins at a vertex and follows a side of the polygon $f(M_s)$. The latter lies in the left halfplane with respect to continuation of M_s

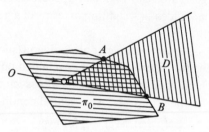

Figure 9.6.2 The polygon in question is shown by double shading (refers to Problem 1 below and to §9.4)

The step from these expressions to stochastic constructions of random polygons in question is by means of projections of probability measures similar to (9.4.2).

Problem 1 Find the distribution of the polygon randomly chosen from the Poisson line mosaic if the weight given to each polygon is

$$\xi = \text{the number of vertices of the polygon.}$$

The $\{M_i\}$ process: we construct this as described in §9.2, Example 2.

The parent map (9.6.2) is explained by fig. 9.6.1. Clearly the number of 'points' M_i which have the same parent polygons *equals* ξ, the number of vertices of the polygon.

The relative Palm distributions Π_2 is given by

$$\Pi_2 = \Delta_2 * P, \tag{9.6.4}$$

where Δ_2 is the distribution of the line process consisting of two lines through O: one is fixed to be the x axis, the direction ψ of the other is random; ψ has probability density $1/2 \sin \psi \, d\psi$; and P is the distribution of the original (Poisson) line process $\{g_i\}$.

The stochastic construction: We take an angular domain D of random opening ψ. In an independent realization of $\{g_i\}$ we take the polygon π_0 which covers O. The intersection of the two will be the polygon in question (see fig. 9.6.2).

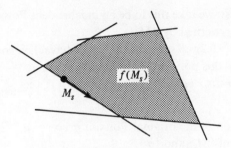

Figure 9.6.3 The segment representing M_s has its source on and follows a side of $f(M_s)$. The latter polygon lies in the left halfplane with respect to continuation of M_s

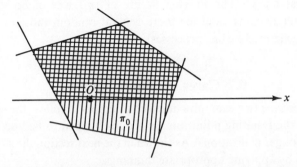

Figure 9.6.4 The polygon in question is shown by double shading

Problem 2 The formulation is the same as Problem 1 but with

$$\xi = \text{perimeter length of the polygon.}$$

The $\{M_i\}$ process: we construct this as described in §9.2, Example 1.

The parent map (9.6.2) is explained by fig. 9.6.3. Clearly the number of 'points' M_i which have the same parent polygon is random, distributed according to Poisson law with mean ξ.

The relative Palm distribution Π_3 is given by

$$\Pi_3 = \Delta_3 * P, \tag{9.6.5}$$

where Δ_3 is the distribution of the line process consisting of a single line, namely the x axis, and P is the distribution of the original (Poisson) line process $\{g_i\}$.

The stochastic construction: in a realization of the Poisson line mosaic we take the polygon π_0 which covers O. The part of π_0 which lies above the x axis will be the polygon in question (see fig. 9.6.4).

Problem 3 The formulation is the same as Problem 1 but with

$$\xi = \text{area of polygon.}$$

The $\{M_i\}$ process: we take this to be an independent Poisson point process on the group $\mathbb{M}_2$ governed by dM.

The parent map (9.6.2): all segments M_i which have their source in a polygon of the mosaic have this polygon for their parent.

The relative Palm distribution Π_4 coincides with P, i.e.

$$\Pi_4 = P \qquad (9.6.6)$$

P is the distribution of the original Poisson process $\{g_i\}$. This follows from the independence of $\{M_i\}$ and $\{g_i\}$.

The stochastic construction: in a realization of the Poisson line mosaic we take the polygon π_0 which covers O. It coincides (in distribution) with the polygon in question.

Remark. In the latter construction we could replace $\{M_i\}$ by a Poisson point process on $\mathbb{T}_2$. The Poisson nature of $\{g_i\}$ was of no significance, and the result remains valid for more general random mosaics (those not necessarily generated by line processes).

9.7 Cases of infinite intensity

Below we describe two examples in which attempts to follow the style of §9.2 fail because the resulting point sets on groups happen to be 'too dense' (i.e. they are no longer realizations). As shown in the next section the situation can be remedied by applying appropriate thinnings.

I Let

$$m(\omega) = \{g_i\}$$

be an $\mathbb{M}_2$-invariant Poisson line process on $\mathbb{R}^2$ governed by $\lambda \, dg$. We consider the triads

$$\pi_s = \{g_{i_1}, g_{i_2}, g_{i_3}\}$$

of lines from $\{g_i\}$. Since with probability 1 there are no pairs of parallel lines in $\{g_i\}$, each π_s can be interpreted as a triangle. Modifying the description of the triangles given in §3.16 we write

$$\pi_s = (M_s, h_s, \sigma_s),$$

where $M_s \in \mathbb{M}_2, h \in (0, \infty)$ is the perimeter length and $\sigma \in \Sigma_2/6$ is the non-label shape of the triangle. We also put

$$m_s = m \setminus \{\text{the three lines from } m \text{ which determine } \pi_s\}.$$

The random set

$$\{(M_s, h_s, \sigma_s, M_s^{-1} m_s)\} \qquad (9.7.1)$$

represents a point process in the space

$$\mathbb{M}_2 \times (0, \infty) \times \Sigma_2/6 \times \mathcal{M}_G.$$

By an argument similar to that of §9.5 we can show that the first moment measure of this process equals

$$\lambda^3 \, dM \, h \, dh \, m_h(d\sigma)P(d\mathcal{m}). \tag{9.7.2}$$

Here P denotes the distribution of a Poisson line process governed by $\lambda \, dg$ (i.e. P coincides with the distribution of $\{g_i\}$ itself).

The projection of the measure (9.7.2) on $\mathbb{M}_2$ is not totally finite (because $\int h \, dh = \infty$). Therefore in this context the perimeter-shape distribution of a typical triangle remains undefined. The measure $h \, dh \, m_h(d\sigma)P(d\mathcal{m})$ in (9.7.2) is an example of what we call a *relative Palm measure* (if the latter is totally finite it is proportional to relative Palm distribution).

II Let

$$\mathcal{m}(\omega) = \{\mathscr{P}_i\}$$

be a $\mathbb{T}_n$-invariant Poisson point process in $\mathbb{R}^n$ of intensity λ. We consider m-subsets from $\{\mathscr{P}_i\}$ $m > n + 1$. Each such m-subset $\{\mathscr{P}_{i_1}, \ldots, \mathscr{P}_{i_m}\}$ we convert into a random sequence which we denote as $(\{\mathscr{P}_{i_1}, \ldots, \mathscr{P}_{i_m}\})$, by applying random ordering of the points so that each of $m!$ different ordering can be chosen with probability $1/m!$. We also assume that the choices of orderings for different m-subsets are independent.

Using the ideas of §4.15 we represent $(\{\mathscr{P}_{i_1}, \ldots, \mathscr{P}_{i_m}\})$ in the form

$$(\{\mathscr{P}_{i_1}, \ldots, \mathscr{P}_{i_m}\}) = (A_s, v_s, \tau_s) \tag{9.7.3}$$

where A_s corresponds to the sequence of the first $n + 1$ points in $(\{\mathscr{P}_{i_1}, \ldots, \mathscr{P}_{i_m}\})$, v_s is the Lebesgue measure of the minimal convex hull of the set $\{\mathscr{P}_{i_1}, \ldots, \mathscr{P}_{i_m}\}$, τ_s is the affine shape of $(\{\mathscr{P}_{i_1}, \ldots, \mathscr{P}_{i_m}\})$. We stress that for fixed $\{\mathscr{P}_{i_1}, \ldots, \mathscr{P}_{i_m}\}$ the quantities A_s and τ_s in (9.7.3) are random (they depend on random ordering of the m-set). Put

$$\mathcal{m}_s = \mathcal{m} \backslash \{\text{the } m \text{ points from } \mathcal{m} \text{ which correspond to } (A_s, v_s, \tau_s)\}.$$

The set

$$\{(A_s, v_s, \tau_s, A_s^{-1} \mathcal{m}_s)\} \tag{9.7.4}$$

is a point process in the space

$$\mathbb{A}_n \times (0, \infty) \times \tau_{n,m} \times \mathcal{M}_{\mathbb{R}^n}.$$

Acting as in §9.5 we find that the first moment measure of this process is equal to

$$\frac{1}{m!} \lambda^m c_{n,m} \cdot dA \, v^{n+m-2} \, dv \, P_{n,m}(d\tau)P(d\mathcal{m}). \tag{9.7.5}$$

where the constants $c_{n,m}$ generalize c_m in (4.15.2) for the case where the points are in $\mathbb{R}^n$.

Here P is the distribution of the Poisson point process $\{\mathscr{P}_i\}$ with which we started, and the probabilities $P_{n,m}$ have been considered in §§4.8, 4.14 and 4.15.

The presence of a totally-infinite measure $v^{n+m-2}\,dv$ shows that the relative Palm measure in this case is totally infinite. Thus the projection of the process (9.7.4) on $\mathbb{A}_n$ is 'too dense', and it is impossible to speak about the distribution of typical v and τ values in this situation.

9.8 Thinnings yield probability distributions

By applying various thinning procedures to the processes of the previous section, invariant marked processes of finite intensity on groups can be derived.

One possibility is to impose restrictions on the domain of the size parameters in question.

For instance, instead of $\{\pi_s\}$ (a point process in the space $\mathbb{M}_2 \times (0, \infty) \times \Sigma_2/6$, see I in the previous section) we can consider its thinned version

$$\{(M_s, h_s, \delta_s)\}_B = \{\pi_s\}_B, \qquad B \subset (0, \infty), \tag{9.8.1}$$

where the thinning depends on a set B and is defined as follows:

$$\{\pi_s\}_B = \{\pi_s : \text{the perimeter length } h_s \text{ falls in } B\}.$$

since

$$\{\pi_s\}_B = \{\pi_s\} \cap (\mathbb{M}_2 \times B \times \Sigma_2/6),$$

the first moment measure of $\{\pi_s\}_B$ coincides with the restriction of $\lambda^3\,dg_1\,dg_2\,dg_3$ (the first moment measure of $\{\pi_s\}$) to the set $\mathbb{M}_2 \times B \times \Sigma_2/6$, i.e. it is given by

$$\lambda^3 I_B(h)\,dg_1\,dg_2\,dg_3 = \lambda^3 I_B(h)\,dM\,h\,dh\,m_h(d\sigma),$$

where I_B is the indicator of the set B.

Since $m_h(\Sigma_2) < \infty$ (see §3.17) it follows that if

$$\int_B h\,dh = c(B) < \infty$$

then the projection of the process (9.8.1) on $\mathbb{M}_2$ becomes an invariant point process of finite intensity $\lambda^3 c(B)m_h(\Sigma_2/6)$, see (3.17.1). The perimeter–shape probability distribution for the typical triangle in (9.8.1) is then well defined and is proportional to

$$I_B(h)h\,dh\,m_h(d\sigma).$$

In words:

Let the triangle process (9.8.1) be derived from an $\mathbb{M}_2$-invariant Poisson line process $\{g_i\}$ by a choice of B with $c(B) < \infty$. Then for a typical triangle the variables h and σ are independent and the distribution of σ is proportional to m_h.

We have a similar situation with the process $\{(A_s, v_s, \tau_s)\}$ (see (9.7.3)). Taking its thinned version

$$\{(A_s, v_s, \tau_s)\}_B, \tag{9.8.2}$$

which we define by the condition $v_s \in B$, we come to the following result.

Let the m-tuple process (9.8.2) be derived from a $\mathbb{T}_n$-invariant Poisson $\{\mathscr{P}_i\}$ by means of a thinning set B for which

$$\int_B V^{n+m-2}\, dV < \infty.$$

Then the projection of the process (9.8.2) on $\mathbb{A}_n$ is of finite intensity; the affine shape τ and v of the typical m-tuple are independent; the affine shape is always distributed according to $P_{n,m}(d\tau)$.

Now let us consider other types of thinnings which do not depend on arbitrary truncations of the parameter space.

From the process $\{\pi_s\}$ we delete all triangles which fail to satisfy the condition

the interior of π_s is intersected by l lines from $\{g_i\}$.

What remains after this thinning operation we represent by

$$\{\pi_s\}_l = \{(M_s, h_s, \sigma_s)\}_l. \tag{9.8.3}$$

(If $l = 0$ this is the collection of all triangular tiles of the mosaic formed by $\{g_i\}$).

Our process $\{\pi_s\}_l$ can be obtained from the marked point process (9.7.1) by two operations:

(a) restricting the process (9.7.1) to the set

$$a_l = \{(\pi, m) : l \text{ lines from } m \text{ hit the triangle } \pi\};$$

and

(b) projecting the resulting process on the space of triangles.

In terms of the first moment measures, operation (a) corresponds to a passage from (9.7.2) to

$$I_{a_l}(\pi, m)\lambda^3\, dM\, h\, dh\, m_h(d\sigma)P(dm).$$

Operation (b) corresponds to the integration of the above by $P(dm)$. We get

$$\lambda^3\, dM\, h\, dh\, m_h(d\sigma) \int I_{a_l}(\pi, m)P(dm)$$

$$= \lambda^3\, dM\, h\, dh\, m_h(d\sigma)(\lambda h)^l(l!)^{-1} \exp(-\lambda h),$$

which is the first moment measure of $\{\pi_s\}_l$. The expression

$$(m_h(\Sigma_2/6))^{-1}(\lambda h)^{l+1}\frac{1}{(l+1)!}\exp(-\lambda h)\lambda\, dh\, m_h(d\sigma) \tag{9.8.4}$$

(proportional to the previous expression) is a probability measure in the space $(0, \infty) \times \Sigma_2/6$. This means that (9.8.4) presents the perimeter–shape distribution of the typical triangle in $\{\pi_s\}_l$. We stress that perimeter and shape are

independent and the shape distribution does not depend on l. Lastly, the
intensity of the projection $\{M_i\}$ of (9.8.3) on $\mathbb{M}_2$ equals (see (3.17.1))

$$\frac{1}{6}(l+1)\lambda \cdot m_h(\Sigma_2) = (l+1)\lambda\frac{12-\pi^2}{12}.$$

Let us consider a similar thinning of the process (9.7.4). In the space $\mathbb{A}_n \times (0, \infty) \times \tau_{n,m} \times \mathcal{M}_{\mathbb{R}^n}$ we take the set

$b_l = \{(A, v, \tau, m): \text{there are } l \text{ points from } m \text{ in the domain whose volume}$
    we denoted by $v\}$.

We define the thinned process to be

$$\{(A_s, v_s, \tau_s, A_s^{-1}m_s)\}_l = b_l \cap \{(A_s, v_s, \tau_s, A_s^{-1}m_s)\}. \qquad (9.8.5)$$

The first moment measure of this process is the restriction of the measure
(9.7.5) on b_l. The projection of this restriction on the space $\mathbb{A}_n \times (0, \infty) \times \tau_{n,m}$
is

$$\frac{\lambda^m}{m!}c_{n,m}\,dA(\lambda v)^l(l!)^{-1}\exp(-\lambda v)v^{n+m-2}\,dv\,P_{n,m}(d\tau). \qquad (9.8.6)$$

Because projections of first moment measures correspond to projections of
point processes, we conclude that (9.8.6) coincides with the first moment
measure of the process

$\{(A_s, v_s, \tau_s)\}_l = \{$those m-tuples from $\{\mathscr{P}_i\}$ whose minimal convex hull
                contains, apart from the points of the m-tuple itself, l
                points from $\mathscr{P}_i\}$.

In this way we come to the following conclusion. Given a $\mathbb{T}_n$-invariant Poisson
point process $\{\mathscr{P}_i\}$ in $\mathbb{R}^n$, we construct the thinned process of m-tuples

$$\{(\mathscr{P}_{i_1}, \ldots, \mathscr{P}_{i_m})\}_l = \{(A_s, v_s, \tau_s)\}_l. \qquad (9.8.7)$$

The distribution of typical mark (v, τ) is well defined for this: v and τ are
independent, v is gamma-distributed, the distribution of τ does not depend on l
and is always given by $P_{n,m}$. The first moment measure of the projection of the
point process (9.8.7) on the space (group) $\mathbb{A}_n$ equals

$$c_{n,m}\lambda^{1-n}\frac{(l+m+n-2)!}{l!m!}\,dA$$

In particular (see (4.8.2) and (4.14.2)) this equals

$$\frac{5(l+1)(l+2)(l+3)(l+4)}{6\lambda}\,dA \qquad \text{in the case } n=2, m=4$$

and

$$\frac{189}{\lambda^2}\prod_{i=1}^{6}(l+i)\,dA \qquad \text{in the case } n=3, m=5.$$

Projections of these measures on $\mathbb{R}^n$ are not locally-finite. In order to obtain
locally-finite projection measures, thinnings in the space of 'Euclidean' shapes
are necessary. An example of such thinning can be found in the next section.

9.9 Simplices in the Poisson point processes in $\mathbb{R}^n$

Here again $\{\mathscr{P}_i\}$ denotes a Poisson point process in $\mathbb{R}^n$ governed by the Lebesgue measure $\lambda \, d\mathscr{P}$, and P denotes its distribution.

Results concerning the shape of the simplices generated by $\{\mathscr{P}_i\}$ can be obtained by reducing the problem to marked point processes on the Euclidean group $\mathbb{M}_n$.

We describe a simplex θ (a set of $n + 1$ points) as

$$\theta = (M, v, \sigma),$$

where $M \in \mathbb{M}_n$ determines the position of θ, $v > 0$ is the volume of θ, and σ is its shape with no label. In the space $\mathbb{M}_2 \times (0, \infty) \times \Sigma_n/(n + 1)! \times \mathscr{M}_{\mathbb{R}^n}$ we consider the point process.

$$\{(M_s, v_s, \sigma_s, M_s^{-1} m_s)\}, \tag{9.9.1}$$

where each (M_s, v_s, σ_s) corresponds to a simplex θ_s with vertices from $\{\mathscr{P}_i\}$ and

$$m_s = \{\mathscr{P}_i\} \setminus \text{the vertices of the simplex } \theta_s.$$

Applying the reasoning of §9.5, the first moment measure of (9.9.1) is found to be

$$\lambda^{n+1} \, dM \, v^{n-1} \, dv \, m_v(d\sigma) P(dm). \tag{9.9.2}$$

Here $m_v(d\sigma)$ is a measure in the space $\Sigma_n/(n + 1)!$ of simplex shapes in $\mathbb{R}^n$ determined by the decomposition

$$d\mathscr{P}_1 \ldots d\mathscr{P}_{n+1} = dM \, v^{n-1} \, dv \, m_v(d\sigma),$$

where $d\mathscr{P}_i$ are Lebesgue measure elements. Explicit expressions for m_v have been obtained for the cases $n = 2, 3$ in §4.6 and §4.12. Clearly (9.9.2) can be considered as another instance of calculation of the relative Palm measure (which is now totally infinite). *The value $m_v(\Sigma_n)$ of the whole space of shapes is infinite* (consult §4.6 and §4.12). We come to the conclusion that, in order to obtain a marked point process on $\mathbb{M}_n$ of *finite intensity*, thinning of (9.9.1) by imposing the conditions

(a) v_s belongs to a fixed bounded set *or*
(a') there are exactly l points from $\{\mathscr{P}_i\}$ within θ_s.

is not enough. In order to obtain finite intensity, it is sufficient to supplement (a) (or (a')) by an additional condition of the type

(b) σ_s belongs to a fixed subset of the space of shapes, whose m_v-measure is finite.

In the resulting thinned processes

$$\{M_s, v_s, \sigma_s\}_{a,b} \quad \text{or} \quad \{M_s, v_s, \sigma_s\}_{a',b}$$

the shape distribution of a typical simplex will be *proportional to the restriction of the measure m_v to the set mentioned in* (b). In this sense (which is usual in classical geometrical probability) the distribution in question is governed by m_V.

Perhaps more convenient is another type of thinning which we describe for $n = 2$.

Assume that our Poisson $\{\mathscr{P}_i\}$ is superposed by an *independent line process* $\{g_i\}$ which is also Poisson and governed by $\lambda_1 \, dg$. We consider all triangles $\theta_s = \mathscr{P}_{i_1}, \mathscr{P}_{i_2}, \mathscr{P}_{i_3}$ which are hit by exactly l lines from $\{g_i\}$. We denote this set of triangles by $\{\theta_s\}_l$. Note that $\{\theta_s\}_{l=0}$ is just the set of triangles θ_s, each of which lies within a tile of the mosaic generated by $\{g_i\}$. Now we describe a triangle by the parameters

$$\theta = (M, h, \sigma),$$

where M and σ are the same as before, and h is the *perimeter length* of θ. In the space $\mathbb{M}_2 \times (0, \infty) \times \Sigma_2/6$ we consider the point process

$$\{(M_s, h_s, \theta_s)\} = \{\sigma_s\}_l. \tag{9.9.3}$$

Its intensity measure can be easily seen to be (see §3.16, III)

$$\lambda^3 \, dM \, h^3 \, dh \, v_h(d\sigma)(\lambda_1 h)^l (l!)^{-1} \exp(-\lambda_1 h). \tag{9.9.4}$$

By integration we find that the projection of the process (9.9.3) on $\mathbb{M}_2$ has the first moment measure (see §6.11)

$$\frac{\pi}{126} \lambda^3 \lambda_1^{-4} (l + 1)(l + 2)(l + 3) \, dM.$$

From (9.9.4) we conclude that *the perimeter length h and the shape σ of a typical triangle in $\{\theta_s\}_l$ are independent*; h is gamma-distributed, and the law of $\sigma \in \Sigma_2/6$ for every l is $(126/\pi) \cdot v_h$.

9.10 Voronoi mosaics

Similar ideas can be applied in the study of the so-called *Voronoi mosaics* generated by $\mathbb{T}_2$-invariant Poisson processes on $\mathbb{R}^2$. First we recall their definition.

Given a realization set m of points in $\mathbb{R}^2$, we associate a convex polygon π_i with each point $\mathscr{P}_i \in m$: by definition π_i consists of all points of $\mathbb{R}^2$ which lie closer to $\mathscr{P}_i$ than to any other point of m. An equivalent construction of π_i is the following.

Let us denote by g_{ij} the line which is perpendicular to the $\mathscr{P}_i\mathscr{P}_j$ segment in its midpoint (the midperpendicular), and let

$$\{g_{ij}\}_i = \{g_{ij} : i \text{ is fixed}, \mathscr{P}_j \in m\}.$$

If h_{ij} denotes the halfplane bounded by g_{ij} which contains the point $\mathscr{P}_i$ then

$$\pi_i = \bigcap_j h_{ij}. \tag{9.10.1}$$

Now let $m = m(\omega) = \{\mathscr{P}_i\}$ be random, namely a $\mathbb{T}_2$-invariant point process of finite intensity. Then the above construction yields a polygon process $\{\pi_i\}$ which in fact is a random mosaic (the Voronoi mosaic for $m(\omega)$).

It is natural to consider the marked point process

$$\{(\mathscr{P}_i, t_i\{g_{ij}\}_i)\}, \qquad \text{where} \qquad t_i = \vec{\mathscr{P}_i O} \qquad (9.10.2)$$

(marks are line processes). It is $\mathbb{T}_2$-invariant and of finite intensity. The distribution of its typical mark will be that of a stochastic line process; we denote it by $\mathbf{p}$.

A line process with distribution $\mathbf{p}$ can be constructed in two steps:

(a) take a point process $\{Q_s\}$ whose distribution coincides with the Palm distribution of $\{\mathscr{P}_i\}$;

(b) take the set $\{g_{Os}\}$ of midperpendicular lines for all pairs O, Q_s.

The proof that $\{g_{Os}\}$ is the desired line process follows directly from the definition in §8.1.

The line process $\{g_{Os}\}$ generates a random mosaic, and in it the polygon which covers O is well defined. *The latter polygon coincides in distribution with the typical equally weighted polygon in the Voronoi mosaic $\{\pi_i\}$ generated by $\{\mathscr{P}_i\}$.*

The distribution $\mathbf{p}$ can be further specified if we assume that $\{\mathscr{P}_i\}$ is Poisson governed by $\lambda \cdot L_2$. In this case outside O, the process $\{Q_i\}$ will also be Poisson governed by λL_2. Representing the points Q_i in the usual polar coordinates we write

$$\{Q_i\} = \{(r_i, \varphi_i)\}.$$

Clearly $\{(r_i, \varphi_i)\}$ will be a Poisson point process in the space $(0, \infty) \times \mathbb{S}_1$ governed by the measure $\lambda r \, dr \, d\varphi$. Corresponding to that, the line process $\{g_{Os}\}$ will be Poisson governed by $4\lambda p \, dp \, d\varphi$ (see §2.1). We conclude that

If $\{\mathscr{P}_i\}$ is Poisson governed by $\lambda \cdot L_2$ then the typical equally weighted polygon in the Voronoi mosaic $\{\pi_i\}$ coincides in distribution with the polygon which covers O in the mosaic generated by Poisson line process governed by the measure $4\lambda p \, dp \, d\varphi = 4\lambda p \, dg$.

If $\{\mathscr{P}_i\}$ is Poisson, we can also solve the problem of the probabilistic description of the shape of a typical vertex in the corresponding Voronoi mosaic.

There is an elementary geometrical characterization of the collection of vertices in terms of the set $\{\mathscr{P}_i\}$. Namely, a point $v \in \mathbb{R}^2$ is a vertex of $\{\pi_i\}$ if and only if in $\{\mathscr{P}_i\}$ there is a triad $\mathscr{P}_i, \mathscr{P}_j, \mathscr{P}_s$ for which

(a) v is the center of the circle through $\mathscr{P}_i, \mathscr{P}_j, \mathscr{P}_s$;

(b) the interior of this circle does not contain any points from $\{\mathscr{P}_i\}$.

We denote by $\{v_r\}$ the set of vertices of Voronoi $\{\pi_i\}$. If $\{\mathscr{P}_i\}$ is Poisson (and this we now assume), there are no quadruples of points in $\{\mathscr{P}_i\}$ which lie on a

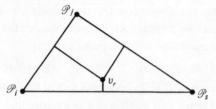

Figure 9.10.1

circle with probability 1; it follows from (a) and (b) that with probability 1 each v_r is of fork type (see fig. 9.10.1). The shape of v_r is determined by two angles between edges emanating from v_r and is completely determined by the shape of the triangle $\mathscr{P}_i$, $\mathscr{P}_j$, $\mathscr{P}_s$ for which v_r is the circumcircle (see fig. 9.10.1). Therefore it is enough to describe the shape of the typical triangle in the triangle process

$$\{\mathscr{P}_i, \mathscr{P}_j, \mathscr{P}_s\}_{(b)}, \tag{9.10.3}$$

(which is the set of triads from $\{\mathscr{P}_i\}$ which satisfy the above condition (b).

The problem can be solved by applying the machinery of the previous sections in the following steps:

(1) we consider triads $\{\mathscr{P}_i, \mathscr{P}_j, \mathscr{P}_s\}$ of points from $\{\mathscr{P}_i\}$. We represent the rth triad by means of parameters $M_r \in \mathbb{M}_2$ which determine the position of the triad; $A_r =$ the area of the circle through the points of the rth triad and the triangular shape $\sigma_r \in \Sigma_2/6$; $\Sigma_2/6$ is the space of triangular shapes without labels.

(2) calculation of the first moment measure of the point process $\{(M_r, A_r, \sigma_r, M_r^{-1}m_r\}$ where $m_r = \{\mathscr{P}_i\} \setminus$ the points of the rth triad. This measure happens to be

$$\lambda^3 \, dM \, A \, dA \, v_A(d\sigma)P(dm)$$

where v_A is given by (3.16.9), P is Poisson governed by λL_2 (i.e. P coincides with the distribution of $\{\mathscr{P}_i\}$);

(3) restriction of this moment measure to the set

$$\{m \in \mathscr{M}_{\mathbb{R}^2}: \text{no point from } m \text{ lies within area } A\}$$

and projection of the result on the space $\mathbb{M}_2 \times (0, \infty) \times \Sigma_2/6$ (where $(0, \infty)$ is the range of A) with the result

$$\lambda^3 \, dM \, e^{-\lambda A}A \, dA \, v_A(d\sigma) = \lambda^3 \, d\mathscr{P} \, d\varphi \, e^{-\lambda A}A \, dA \, v_A(d\sigma). \tag{9.10.4}$$

Using (3.17.2) we find the projection of this measure on $\mathbb{R}^2$ to be

$$2\lambda \, d\mathscr{P}.$$

This is the first moment measure of the process of vertices of our Voronoi mosaic. We also gather from (9.10.4) that the shape of the typical vertex is distributed according to the density

$$\pi \sin \alpha_1 \sin \alpha_2 \sin \alpha_3 \, d\xi_1 \, d\xi_2,$$

in the space $\Sigma_2/6$.

The value 2λ for the intensity of the vertex process agrees with the value 6 for the mean number of vertices in a typical polygon in Voronoi mosaic obtained by a different method in the next section.

9.11 Mean values for random polygons

The stochastic constructions of random polygons of §§9.4, 9.6 and 9.10 can be applied in Monte-Carlo calculations. They also permit exact evaluation of the expectations of various random variables depending on these polygons. We show this in the examples of

A – the area of the polygon;

H – its perimeter length;

n – the number of its vertices.

Let us start with the derivation of certain integral representations for expectations EA, EH and En for

π_0 – random polygon which covers the origin O in a Poisson line process on $\mathbb{R}^2$ governed by a measure $f(g) \, dg$.

We will assume that the density $f(g)$ is continuous; in our stochastic constructions we met two cases: $f \equiv \lambda$ and $f = 4\lambda p$.

We consider random measures

$A(B) = $ area of $\pi_0 \cap B$, $B \subset \mathbb{R}^2$ is a Borel set;

$H(B) = |\partial\pi_0 \cap B|$, the total length of $\partial\pi_0 \cap B$;

$n(B) = $ the number of vertices of π_0 within B;

$\mathscr{L}(B) = \sum |g_i \cap B|$, where summation is over all lines g_i from the realization of the line process;

$N(B) = $ number of $g_i \cap g_j$ points in B.

Clearly

$$A = A(\mathbb{R}^2) = \int A(dQ),$$

$$H = H(\mathbb{R}^2) = \int H(dQ),$$

$$n = n(\mathbb{R}^2) = \int n(dQ),$$

The application of averaging and the Fubini theorem yields

$$EA = \int EA(dQ),$$

$$EH = \int EH(dQ), \qquad (9.11.1)$$

$$En = \int En(dQ),$$

where $EA(dQ)$, $EH(dQ)$ and $En(dQ)$ are the corresponding averaged measures. In our case these have densities with respect to dQ. The purpose now is to calculate these densities. We note that for areas dQ which shrink down to a point Q we have (up to summands which are $o(dQ)$ in the average)

$$A(dQ) = I_0(OQ) \cdot dQ,$$

$$H(dQ) = I_0(OQ)\mathscr{L}(dQ), \qquad (9.11.2)$$

$$n(dQ) = I_0(OQ) \cdot N(dQ),$$

where $I_0(OQ) = 1$ if the segment OQ is not intersected by the lines of the process which avoid hitting dQ, and zero otherwise. The random variable $I_0(OQ)$ is independent of either dQ, $\mathscr{L}(dQ)$ or $N(dQ)$ (this follows from the independence properties of the Poisson line process, see (7.5.2)). Taking expectations of (9.11.2) we find

$$EA(dQ) = P\binom{(OQ)}{0} \cdot dQ,$$

$$EH(dQ) = P\binom{(OQ)}{0} E\mathscr{L}(dQ), \qquad (9.11.3)$$

$$En(dQ) = P\binom{(OQ)}{0} EN(dQ).$$

Because the process is Poisson

$$P\binom{(OQ)}{0} = \exp\left(-\int_{[OQ]} f(g)\, dg\right), \qquad (9.11.4)$$

where as usual $[OQ]$ is the set of lines which hit OQ. The main term of $E\mathscr{L}(dQ)$ is contributed by the event in which only one line hits dQ, i.e.

$$E\mathscr{L}(dQ) = \int \chi f(g)\, dg + o(dQ),$$

where χ is the length of the chord $g \cap dQ$. Using considerations suggested by fig. 6.3.1 we find

$$E\mathscr{L}(dQ) = \left(\int_{[Q]} f(g)\, d\varphi\right) \cdot dQ, \qquad (9.11.5)$$

where $[Q]$ denotes the bundle of lines through Q.

Lastly, using the notation of §3.15, I,

$$EN(dQ) = \frac{1}{2}\left(\iint_{g_1 \cap g_2 = Q} f(g_1)f(g_2) \sin \psi \, d\psi \, d\varphi\right) dQ \qquad (9.11.6)$$

(the factor $\frac{1}{2}$ appears here because the two lines that hit Q should be considered as unordered).

Formulae (9.11.1) and (9.11.3)–(9.11.6) reduce the problem to performing integrations over a plane.

The simplest is the case in which $f = \lambda$. In this case,

$$P\binom{(OQ)}{0} = \exp(-2\lambda r)$$

(we use polar coordinates $Q = (r, \theta)$), and

$$E\mathcal{L}(dQ) = \pi \cdot \lambda \cdot dQ,$$
$$EN(dQ) = \pi \cdot \lambda^2 \, dQ.$$

Therefore

$$EA_3 = \iint \exp(-2\lambda r)r \, dr \, d\theta = \pi \cdot (2\lambda^2)^{-1},$$

$$EH_3 = \pi\lambda \iint \exp(-2\lambda r) \, r \, dr \, d\theta = \pi^2(2\lambda)^{-1}, \qquad (9.11.7)$$

$$En_3 = \pi\lambda^2 \iint \exp(-2\lambda r)r \, dr \, d\theta = \frac{1}{2}\pi^2.$$

These mean values correspond to the random polygon of Problem 3 in §9.6. The same values can also be found in [1], section 6.3.

The integrations in the case $f(g) = 4\lambda p$ are somewhat more complicated but still tractable. Below we present the results for this case.

For a Poisson line process governed by $4\lambda p \, dg$, (9.11.4)–(9.11.6) give

$$P\binom{(OQ)}{0} = \exp(-\lambda\pi r^2)$$

since

$$\int_{[OQ]} f(g) \, dg = 4\lambda \int_0^r dx \int_0^\pi x \sin^2 \psi \, d\psi = \lambda\pi \cdot r^2;$$

$$E\mathcal{L}(dQ) = 4\lambda \cdot dQ \int_0^\pi r \sin \varphi \, d\varphi = 8\lambda r \, dQ;$$

$$EN(dQ) = dQ \, 8\lambda^2 r^2 \int_0^\pi \sin \varphi \, d\varphi \int_0^\pi |\sin(\psi + \varphi)| \sin \psi \, d\psi$$

$$= dQ \, 8\lambda^2 r^2 \int_0^\pi \sin \varphi \, d\varphi \cdot \left(\left(\frac{\pi}{2} - \varphi\right) \cos \varphi + \sin \varphi\right)$$

$$= 6\pi\lambda^2 r^2 \, dQ.$$

Formulae (9.11.3) and (9.11.1) now yield

$$EA\hat{V} = 2\pi \int_0^\infty r \exp(-\lambda \pi r^2)\,dr = \lambda^{-1};$$

$$EH\hat{V} = 8\lambda \cdot 2\pi \int_0^\infty r^2 \exp(-\lambda \pi r^2)\,dr = 4/\sqrt{\lambda};$$

$$En\hat{V} = 6\pi\lambda^2 \cdot 2\pi \int_0^\infty r^3 \exp(-\lambda \pi r^2)\,dr = 6.$$

According to the results of the previous section these mean values refer to the typical polygon in the Voronoi mosaic generated by Poisson point process on $\mathbb{R}^2$ governed by $\lambda\,d\mathscr{P}$.

The values (9.11.7) can be used for direct calculation of similar expectations for the random polygon of Problem 2, §9.6 (we call them EA_2, EH_2 and En_2):

$$EA_2 = \frac{EA}{2} = \frac{\pi}{4\lambda^2};$$

$$EH_2 = \frac{EH}{2} \text{ plus the average length of the side on the } Ox \text{ line}$$

$$= \frac{\pi^2}{4\lambda} + \frac{1}{\lambda};$$

$$En_2 = \frac{En}{2} + 2 = 2 + \frac{\pi^2}{4}.$$

By slightly modifying the derivation which led to (9.11.7) we can obtain the values of EA, EH and En, for both the random polygons of §9.4 and of Problem 1, §9.6. The counterpart of (9.11.2) now has the form (see fig. 9.6.2)

$$A(dQ) = I_D(Q)I_0(OQ)\,dQ,$$

$$H(dQ) = I_D(Q)I_0(OQ)\mathscr{L}(dQ),$$

$$n(dQ) = I_D(Q)I_0(OQ)N(dQ).$$

Except for A, the total values of these random measures do not represent the desired random quantities. We have in fact

$$A = \int A(dQ),$$

$$H = \int H(dQ) + |OA| + |OB|,$$

$$n = \int n(dQ) + 3.$$

Let us consider the representation for A. Because of the independence of I_D, the result of averaging will be

$$EA = \int_0^\infty \int_{-\pi/2}^{\pi/2} p(\theta) \exp(-2\lambda r) r \, dr \, d\theta$$

$$= (4\lambda^2)^{-1} \int_{-\pi/2}^{\pi/2} p(\theta) \, d\theta,$$

where $p(\theta)$ is the probability that the point $Q = (r, \theta)$ will fall in the random angular domain D. To calculate the last integral we note that

$$\int_{-\pi/2}^{\pi/2} p(\theta) \, d\theta = \int_{-\pi/2}^{\pi/2} d\theta \, \mathrm{E} I_D(\theta) = \mathrm{E} \int_{-\pi/2}^{\pi/2} I_D(\theta) \, d\theta = \mathrm{E}|D|,$$

where $|D|$ denotes the (random) opening of the angular domain D. The reader may easily verify that in the case of §9.4 $|D|$ has a probability density function

$$\pi^{-1}(\pi - \alpha) \sin \alpha \, d\alpha, \qquad 0 < \alpha < \pi;$$

hence for equally weighted typical polygon in Poisson $\{g_i\}$

$$\mathrm{E}|D| = \pi^{-1} \int_0^\pi \alpha(\pi - \alpha) \sin \alpha \, d\alpha = \frac{4}{\pi},$$

and

$$EA = (4\lambda^2)^{-1} \cdot 4\pi^{-1} = \pi^{-1}\lambda^{-2}.$$

With all the above calculations we find easily

$$EH = \pi\lambda(4\lambda^2)^{-1} \cdot 4\pi^{-1} + \lambda^{-1} = 2\lambda^{-1},$$

$$En = \pi\lambda^2(4\lambda^2)^{-1} \cdot 4\pi^{-1} + 3 = 4.$$

These values can be found in [2], p. 57.

In the case of Problem 1 in §9.6, $|D|$ has a probability density function

$$\frac{1}{2} \sin \alpha \, d\alpha, \qquad 0 < \alpha < 1.$$

This yields the following values (the index 1 stands for Problem 1):

$$EA_1 = \frac{\pi}{8\lambda^2},$$

$$EH_1 = \lambda^{-1}\left(\frac{\pi^2}{8} + 1\right),$$

$$En_1 = \frac{\pi^2}{8} + 3.$$

10

Sections through planar geometrical processes

We now consider M_2-invariant geometrical processes of a general nature (we do not assume they are Poisson or Poisson generated), namely line processes, random mosaics and Boolean models for disc processes. Our interest lies in intersection processes induced on a 'test line' (the x axis, say). The essential feature of the analysis is that it includes the angles under which the intersections occur. In other words, we study the induced one-dimensional *marked* point or segment processes with angular marks.

The first three sections are partly preparational; here we develop the ideas of chapters 8 and 9. The main results of this chapter are obtained in the latter five sections where a principally new tool is introduced: the averaging of combinatorial decompositions of chapter 5. The basic relations which we derive (such as (10.4.11)) are of stereological significance. They permit the conclusion that M_2-invariance implies strong restrictions on the possible distributions of intersection processes. In particular they imply that additional renewal-type properties alone may determine the functional form (exponentiality) of distribution functions of certain length variables.

10.1 Palm distribution of line processes on $\mathbb{R}^2$

Let $\{g_i\}$ be a line process on $\mathbb{R}^2$ (a point process in G; see §7.5, Poisson nature is no longer assumed), and let P be its distribution. We assume $\{g_i\}$ to be M_2-invariant and of finite intensity. The latter condition means that for every compact $B \subset G$

$$E_P N(B, m) < \infty, \qquad m = \{g_i\}.$$

Because this expectation is a measure in G and inherits the property of M_2-invariance from P, we conclude that

$$E_P N(B, m) = \lambda \cdot \mu(B), \qquad (10.1.1)$$

where μ is the M_2-invariant measure on G described in §3.6. We stress that

(10.1.1) follows from the uniqueness of such μ. $\lambda < \infty$ is called the *intensity* of $\{g_i\}$. First using $\{g_i\}$ we construct a point process $\{M_s\}$ on the group $\mathbb{M}_2$, namely that described in §9.2, Example 1.

By construction, the processes

$$m_1(\omega) = \{M_s\}, \qquad m_2(\omega) = \{g_i\}$$

are jointly $\mathbb{M}_2$-invariant. Hence the relative Palm distribution $\Pi_{\{M_s\}}$ of $\{g_i\}$ is well defined. We call it the *Palm distribution of* $\{g_i\}$ and denote it simply by Π.

Let us mention from the beginning that

$$\Pi\{\text{the } x \text{ axis belongs to a realization}\} = 1,$$

and it can be convenient to think of Π as the conditional distribution of $\{g_i\}$ given that one line from $\{g_i\}$ coincides with the x axis.

In (9.1.3) as applied to our case we put

$$B_1 = \{M \in \mathbb{M}_2 : MO \in b(r, O)\},$$

$$A = B_2 = \{k_i \text{ lines from } \{g_i\} \text{ hit } \delta_i, i = 1, \dots, s\},$$

where $b(r, O)$ is the disc centered at O of radius r, and $\delta_1, \dots, \delta_s$ are some segments fixed on the plane. We obtain

$$\Pi(A) = (\pi^2 r^2 \lambda)^{-1} \cdot E \sum_{M_s \in B_1} I_A(M_s^{-1} m). \tag{10.1.2}$$

(the intensity of $\{M_s\}$ equals $\lambda/2$). To calculate the expectation in (10.1.2) we may first take the lines $g_i \in m$ as fixed and average with respect to the positions of the segments M_s on these lines. After that the result should be averaged with respect to P, the distribution of $\{g_i\}$.

Applying the first step, we obtain an expression proportional to the right-hand side of (6.8.10), i.e.

$$\Pi(A) = (2\pi^2 r^2 \lambda)^{-1} E_P \sum_{g_i \text{ hits } b(r, O)} \int_{\chi(g_i)} (I_A(M_{i,+,t}^{-1} m) + I_A(M_{i,-,t}^{-1} m)) \, dt. \tag{10.1.3}$$

On the other hand, we can obtain the same expression by applying the expectation E_P to both sides of (6.8.10) and acting as we did in §7.15. We find in this way that

$$\Pi(A) = \lim_{l \to 0} (\lambda l^2)^{-1} d \cdot P(A \cap B), \tag{10.1.4}$$

where (see fig. 10.1.1)

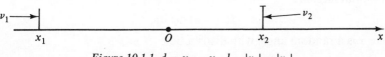

Figure 10.1.1 $d = x_2 - x_1, l = |v_1| = |v_2|$

$$B = \{\text{a line from } \{g_i\} \text{ hits both } v_1 \text{ and } v_2\}. \qquad (10.1.5)$$

In particular,

$$P(B) = \lambda d^{-1} l^2 + o(l^2). \qquad (10.1.6)$$

This permits the reinterpretation of $\Pi(A)$ as a limit of conditional probability:

$$\Pi(A) = \lim_{l \to 0} \frac{P(A \cap B)}{P(B)}. \qquad (10.1.7)$$

The right-hand side of (6.8.10) did not depend on the locations x_1 and x_2 along the x axis of the segments v_1 and v_2. Consequently, for any fixed A the limit in (10.1.4) does not depend on x_1 and x_2. This implies that

the Palm distribution Π is always invariant with respect to shifts parallel to the x axis.

Proof Using the same notation as before,

$$P(A \cap B) = P(tA \cap tB) = P(tA \cap B)$$
$$tB = \{tm : m \in B\}$$

for every shift t parallel to the x axis. Dividing both sides by $\lambda d^{-1} l^2$ and taking the limit, we obtain

$$\Pi(A) = \Pi(tA). \qquad (10.1.8)$$

Conditioning by the event (10.1.5) is not the only possible method. For instance in (10.1.7) we could take

$B = \{$a line from $\{g_i\}$ hits an interval $(x, x + dx)$ on the x axis under the angle within $(\psi, \psi + d\psi)\}$.

The corresponding limit will exist for any $\psi \neq 0$ or π and will produce a rotated version of Π, namely

$$\Pi_{x,\psi}(A) = \Pi(M_{x,\psi} A), \qquad (10.1.9)$$

where $M_{x,\psi}$ denotes the rotation around the point x by the angle ψ.

The Palm distributions for doubly stochastic Poisson line processes can be easily described. We consider two examples (proofs can be obtained by (10.1.7) and are left to the reader).

Randomly rotated $\mathbb{T}_2$-invariant Poisson line processes (These processes have already been described in §7.12, II) If the process $\{g_i\}$ is obtained by random rotation from a Poisson line process governed by the measure

$$d\mu = f(\varphi)\, d\varphi\, dp$$

then the Palm distribution of $\{g_i\}$ will be doubly stochastic Poisson governed by random measure

$$f(\varphi - \alpha)\, d\varphi\, dp,$$

where α is a random angle with distribution

$$(\pi\lambda)^{-1} f(\varphi) \, d\varphi.$$

Here λ is the intensity of $\{g_i\}$, the fact that $(\pi\lambda)^{-1}$ is a norming constant follows from (2.10.1).

Mixtures of $\mathbb{M}_2$-*invariant Poisson line processes* Let $\{g_i\}$ be a doubly stochastic Poisson line process whose governing measure with probability 1 is proportional to dg but with random proportionality coefficient λ_ω (here ω denotes an elementary event rather than a direction). Let $f(x) \, dx$ be the probability density of λ_ω.

The Palm distribution of $\{g_i\}$ will be doubly stochastic Poisson governed by $\lambda'_\omega \, dg$, where λ'_ω is a random variable with probability density

$$x f(x) \left(\int x f(x) \, dx \right)^{-1}.$$

10.2 Palm formulae for line processes

The distribution P of any $\mathbb{M}_2$-invariant line process $\{g_i\}$ of finite intensity λ is uniquely determined by λ and the corresponding Palm distribution Π. A proof suggested in [42] is by reconstruction of the probabilities

$$P\begin{pmatrix} [\delta_1] \\ k_1 \end{pmatrix}, \ldots, \begin{pmatrix} [\delta_s] \\ k_s \end{pmatrix} \tag{10.2.1}$$

in terms of λ and Π. Above $\delta_1, \ldots, \delta_s$ are intervals not necessarily on a line; $[\delta]$ is the set of lines hitting δ. We assume that the line which contains δ does not belong to Buffon set $[\delta]$. This convention will be important when we consider Palm probabilities of the events as in (10.2.1). This approach is based on the fact that the probabilities of (10.2.1) type determine the distribution of a line process uniquely, i.e. the sets $[\delta]$ can in this respect replace the 'shields' described in §7.5. This of course needs a separate proof, which we leave to the reader.

For simplicity we consider the probability

$$p_k(t) = P\begin{pmatrix} [\delta] \\ k \end{pmatrix}, \qquad t = |\delta|$$

assuming that the interval δ lies on the x axis.

The point process $\{x_i\}$ of intersections of the lines $\{g_i\}$ with the x axis will be of intensity 2λ. Applying the theory of §8.7 we find that

$$\frac{d}{dt} p_k(t) = 2\lambda [\pi_{k-1}(t) - \pi_k(t)], \tag{10.2.2}$$

where $\pi_k(t)$ is the conditional probability to have k intersections on δ given

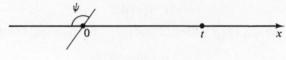

Figure 10.2.1 $\delta = (O, t)$

that an intersection has occurred at an end of δ (see fig. 10.2.1), in the sense of Palm distribution of a point process on a line. How can the probabilities $\pi_k(t)$ be expressed via the Palm distribution Π of the line process $\{g_i\}$?

The conditional probability of the event $\begin{pmatrix} [\delta] \\ k \end{pmatrix}$ given that there is an intersection at x and the intersection angle at this point equals ψ is $\Pi_{x,\psi}\begin{pmatrix} [\delta] \\ k \end{pmatrix}$, see (10.1.9); the random angle ψ has $\frac{1}{2} \sin \psi \, d\psi$ density (see §7.15, IV). Therefore

$$\pi_k(t) = \frac{1}{2} \int_0^\pi \Pi_{0,\psi}\begin{pmatrix} [\delta] \\ k \end{pmatrix} \sin \psi \, d\psi. \tag{10.2.3}$$

The two preceding equations, together with (10.1.9), solve the problem of reconstruction of $P\begin{pmatrix} [\delta] \\ k \end{pmatrix}$.

The probabilities (10.2.1) can be recovered by the method used in §8.8 (no additional difficulty stems from the possibly non-collinear position of the intervals δ_i).

One can prove similarly that the only solutions of the equation

$$\Pi = \Delta * P$$

are provided by Poisson line processes. Here

$$\Delta\{m \text{ consists of a single line, the } x \text{ axis}\} = 1.$$

10.3 Second order line processes

Second order line processes offer a number of analytical advantages, and in fact the first attempt to develop a theory of general (non-Poisson) line processes by Davidson [43] was restricted to that case. We too will assume the second order property throughout this and the following sections.

I Let P be the distribution of an $\mathbb{M}_2$-invariant line process $\{g_i\}$ of finite intensity and let Π be its Palm distribution.

If

$$E_P N(B_1, m)N(B_2, m) < \infty \quad \text{for compact} \quad B_1, B_2 \subset \mathbb{G}$$

i.e. if $\{g_i\}$ happens to be of second order then

$$E_\Pi N(B, m) < \infty \quad \text{for compact} \quad B \subset \mathbb{G}.$$

Proof Since

$$E_\Pi N(B, m) = \sum l\Pi \binom{B}{l},$$

using linear properties of (10.1.3) we find that this quantity is proportional to

$$E_P \sum \int_{\chi_i} [N(B, M_{i,+,t}^{-1}m) + N(B, M_{i,-,t}^{-1}m)] \, dt,$$

where χ_i are the chords of $b(r, O)$ generated by $\{g_i\}$ ($b(r, O)$ is the circle of radius r centered at O). We can take $B = [b(r, O)]$. Then the expression under the sign of E_P does not exceed

$$4rN(b(r, O), m)N(b(2r, O), m).$$

By assumption, E_P of this product is finite.

II Under second order assumption it is easy to prove different continuity properties of the Palm distribution. For instance, let $I_1, \ldots, I_s$ be a sequence of bounded intervals *on the x axis*, and let $k_1, \ldots, k_s$ be a sequence of non-negative intergers. Then for every $x \notin \bigcup I_i$

$$\lim_{\psi \to 0} \Pi_{x,\psi} \binom{[I_1]}{k_1}, \ldots, \binom{[I_s]}{k_s} = \Pi \binom{[I_1]}{k_1}, \ldots, \binom{[I_s]}{k_s}.$$

Clearly the above is equivalent to

$$\lim_{\psi \to 0} \Pi \binom{[\tilde{I}_1]}{k_1}, \ldots, \binom{[\tilde{I}_s]}{k_s} = \Pi \binom{[I_1]}{k_1}, \ldots, \binom{[I_s]}{k_s}. \quad (10.3.1)$$

where the intervals $\tilde{I}_1, \ldots, \tilde{I}_s$ lie on the line x, ψ and are shown in fig. 10.3.1.

Let us prove (10.3.1) assuming we have only one interval I_1. We have the inclusion

$$\{m : N([I_1], m) \neq N([\tilde{I}_1], m)\}$$
$$\subset \{m : N([b(|v_1|, A)], m) > 0\} \bigcup \{m : N([b(|v_2|, B)], m) > 0\}$$

(the segments v_1 and v_2 are shown in fig. 10.3.1). Each of the latter two sets is monotone decreasing with ψ. Therefore it is enough to show that

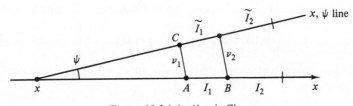

Figure 10.3.1 $|xA| = |xC| \ldots$

$\Pi\{$there is a line through A or B in a non-zero direction$\} = 0$.

This probability equals zero since, by the result of I above,

$$E_\Pi N^*([I_1], \textit{m}) = \lambda'|I_1|,$$

where $0 < \lambda' < \infty$, and N^* denotes the number of lines having non-zero direction. The case of general s in (10.3.1) yields a similar analysis.

III We consider the one-dimensional marked point process description of line processes

$$\{g_i\} = \{(x_i, \psi_i)\},$$

where x_i is the point of intersection of g_i with the x axis and ψ_i is the corresponding intersection angle (see §7.12, II).

We will be interested in the second-order $\mathbb{M}_2$-invariant line processes which satisfy the condition

$(\mathscr{I})$ *the intersection point process* $\{x_i\}$ *and the sequence of angles* $\{\psi_i\}$ *are independent*

(independence of the angles within $\{\psi_i\}$ is *not* required). This class of line processes is not empty for it contains mixtures of $\mathbb{M}_2$-invariant Poisson line processes, see §10.1.

We show now that $(\mathscr{I})$ implies strong restrictions on the structure of the process $\{x_i\}$. For the events

$$A = \begin{pmatrix} [I_1] \\ k_1 \end{pmatrix}, \dots, \begin{pmatrix} [I_s] \\ k_s \end{pmatrix}, \qquad I_i \text{ lie on the } x \text{ axis}, \qquad (10.3.2)$$

condition $(\mathscr{I})$, together with the continuity property (10.3.1), imply

$$\Pi_{x,\psi}(A) \equiv \Pi(A) \qquad (10.3.3)$$

identically for every ψ and for every x outside the union of $I_1, \dots, I_s$.

One of the consequences of (10.3.3) is that the quantity

$$\Pi_x(A) = \frac{1}{2} \int_0^\pi \Pi_{x,\psi}(A) \sin\psi \, d\psi$$

(which has the interpretation of the conditional probability of A under the condition that x is a point in $\{x_i\}$) *does not depend on* x:

$$\Pi_x(A) \equiv \Pi(A) \quad \text{for all } x \notin \bigcup I_i.$$

In particular, if we take $s = 1$ in (10.3.2), (10.2.2) and (10.3.3) may be written as

$$\frac{d}{dt} p_k(t) = 2\lambda \left[\Pi\begin{pmatrix} [I] \\ k-1 \end{pmatrix} - \Pi\begin{pmatrix} [I] \\ k \end{pmatrix} \right]. \qquad (10.3.4)$$

We say that a $\mathbb{T}_1$-invariant finite intensity process $\{x_i\} \subset \mathbb{R}$ has the *Palm-mixing property*

$$\lim_{x \to \infty} \Pi_x(A) = P(A) \qquad \text{for every } A \text{ as in (10.3.2)}$$

If we now assume that our process $\{x_i\}$ is Palm-mixing then under our other assumption it follows that

$$\Pi\binom{[I]}{k} = p_k(t).$$

Substituting this value into (10.3.4) and solving the resulting equations system (see §8.11) we conclude that

$$p_k(t) = \frac{(2\lambda t)^k}{k!}e^{-2\lambda t}.$$

Similar reasoning in conjunction with the integration method of §8.8 yields

$$P\binom{[I_1]}{k_1}, \ldots, \binom{[I_s]}{k_s}) = \prod_i \frac{(2\lambda|I_i|)^{k_i}}{k_i!}e^{-2\lambda|I_i|};$$

i.e.

if $\{x_i\}$ is Palm-mixing and $(\mathscr{I})$ holds, then $\{x_i\}$ is Poisson governed by $2\lambda L_1$.

We stress that there is no difficulty in constructing examples of second order $\mathbb{T}_1$-invariant line processes

$$\{g_i\} = \{(x_i, \psi_i)\}$$

for which both properties $(\mathscr{I})$ and of Palm-mixing of $\{x_i\}$ exist. For instance, if $\{x_i\}$ is a $\mathbb{T}_1$-invariant finite intensity renewal (see §8.9) with non-lattice-type distribution F possessing second moment, then $\{x_i\}$ is Palm-mixing (this is in fact one of the main results of the renewal theory [54]). To satisfy $(\mathscr{I})$ for $\{\psi_i\}$ we can take *any* stationary sequence of angles which is independent of $\{x_i\}$. It follows that within this class, say, $\mathbb{M}_2$-invariance of corresponding $\{g_i\}$ will necessarily imply *exponentiality* of F.

IV A family of Palm-type distributions For second order $\mathbb{M}_2$-invariant line processes $\{g_i\}$, which with probability 1 do not possess pairs of parallel lines, a limiting procedure in a sense dual to (10.1.7) leads to a family of Palm-type distributions of $\{g_i\}$ which we denote as $\tilde{\Pi}_{x_1 x_2}$.

With fig. 10.1.1 we now associate the event

$$\tilde{B} = \{m \in \mathcal{M}_G : m \text{ contains exactly one pair of lines which belongs to } B^*\},$$
$$(10.3.5)$$

where

$$B^* = \{(g_1, g_2) : g_1 \neq g_2, g_1 \text{ hits } v_1, g_2 \text{ hits } v_2\}.$$

We denote by θ_i the random angle between v_i and the line hitting v_i. $i = 1, 2$ (these angles are well defined if $\tilde{B}$ takes place). We choose an event $A \subset \mathcal{M}_G$ as in §10.1 and two arcs $J_1 \subset (0, \pi)$ and $J_2 \subset (0, \pi)$. We denote by $[J_i]$ the event $\Theta_i \in J_i$.

Let P be the distribution of a line process of the above-mentioned class, and let m_2 be its second moment measure. Then

$$P(\tilde{B}) = m_2(B^*) + o(l^2) = cl^2 + o(l^2). \qquad (10.3.6)$$

Here $l = |v_1| = |v_2|$ tends to zero, and the constant c is given by the integral (7.14.4).

The limit of the conditional probability

$$\tilde{\Pi}_{x_1 x_2}(J_1, J_2, A) = \lim_{l \to 0} \frac{P([J_1] \cap [J_2] \cap A \cap \tilde{B})}{P(\tilde{B})} \qquad (10.3.7)$$

exists and for each x_1 and x_2 determines a probability distribution in the space

$$(0, \pi) \times (0, \pi) \times \mathscr{M}_G.$$

Let us first give a proof for (10.3.6). We denote by $N_2(B^*, m)$ the number of pairs of lines from m which belong to B^* so that (see §7.14)

$$m_2(B^*) = E_P N_2(B^*, m).$$

It is not difficult to establish both the existence and coincidence of the limits

$$\lim_{l \to 0} l^{-2} \int_{b(r, 0) \times (0, 2\pi)} I_{\tilde{B}}(Mm) \, dM$$

$$= \lim_{l \to 0} l^{-2} \int_{b(r, 0) \times (0, 2\pi)} N_2(B^*, Mm) \, dM = a(m) \qquad (10.3.8)$$

for the set of realizations $m \in \mathscr{M}_G$ of probability $P = 1$. For $l < 1$ both ratios under the limit signs do not exceed

$$\sum_{g_i, g_j \text{ hit } b(r+d+1, 0)} \left[\left(\frac{\pi}{2} - \alpha_{ij} \right) \cot \alpha_{ij} + 1 \right]$$

(where α_{ij} is the angle between $g_i, g_j \in m$, see §6.5). But as we already mentioned in §7.15, V) this upper bound may have infinite expectation E_P. Therefore an attempt to directly use Lebesgue's bounded convergence theorem fails.

To remedy the situation, instead of B^* and $\tilde{B}$ we consider the sets

$B_\varepsilon^* = \{(g_1, g_2) \in B^* : \text{the angle } \alpha_{12} \text{ between } g_1 \text{ and } g_2 \text{ belongs to } (\varepsilon, \pi - \varepsilon)\},$
    $\varepsilon > 0$

and

$$\tilde{B}_\varepsilon = \{m \in \mathscr{M}_G : N_2(B_\varepsilon^*, m) = 1\}.$$

A limit similar to (10.3.8) exists with probability 1:

$$\lim_{l \to 0} l^{-2} \int_{b(r, 0) \times (0, 2\pi)} N_2(B_\varepsilon^*, Mm) \, dM = a_\varepsilon(m),$$

and the upper bound

$$\sum_{\substack{g_i, g_j \text{ hit } b(r+d+1, 0) \\ \varepsilon < \alpha_{ij} < \pi - \varepsilon}} \left[\left(\frac{\pi}{2} - \alpha_{ij} \right) \cot \alpha_{ij} + 1 \right]$$

of the expression under the limit sign has a finite expectation

$$\int_{\substack{g_1, g_2 \text{ hit } b(r+d+1,0) \\ \varepsilon < \alpha_{12} < \pi - \varepsilon}} \left[\left(\frac{\pi}{2} - \alpha_{12} \right) \cot \alpha_{12} + 1 \right] m_2(dg_1 \, dg_2)$$

(the integrand is bounded on the integration set, the latter set is bounded). By Lebesgue's theorem we conclude that

$$\lim_{l \to 0} l^{-2} m_2(B_\varepsilon^*) = (\pi r^2 \cdot 2\pi)^{-1} E_P a_\varepsilon(m).$$

Observing that $a_\varepsilon(m)$ is monotone increasing as $\varepsilon \downarrow 0$ we conclude the existence of the double limit:

$$\lim_{\varepsilon \downarrow 0} \lim_{l \to 0} l^{-2} m_2(B_\varepsilon^*) = (\pi r^2 \cdot 2\pi)^{-1} E_P a(m).$$

From the results of §6.5 we conclude that

$$m_2(B_\varepsilon^*) = l^2 \int_{\pi - \varepsilon > \alpha_{12} > \varepsilon} \left[\left(\frac{\pi}{2} - \alpha_{12} \right) \cot \alpha_{12} + 1 \right] m(d\alpha_{12})$$

and the previous double limit is directly calculated to be

$$c = l^{-2} m_2(B^*) = l^{-2} E_P N_2(B^*) = (\pi r^2 \cdot 2\pi)^{-1} E_P a(m).$$

Now we apply E_P to both sides of (10.3.8). By the Fatou lemma

$$\limsup l^{-2} P(\tilde{B}) \geqslant \liminf l^{-2} P(\tilde{B}) \geqslant (\pi r^2 \cdot 2\pi)^{-1} E_P a(m) = c. \qquad (10.3.9)$$

On the other hand, since

$$I_{\tilde{B}}(Mm) \leqslant N_2(B^*, Mm)$$

we have

$$\liminf l^{-2} P(\tilde{B}) \leqslant \limsup l^{-2} P(\tilde{B}) \leqslant \lim l^{-2} E_P N_2(B^*, m) = c. \qquad (10.3.10)$$

The inequalities (10.3.9) and (10.3.10) together imply (10.3.6). The above argument applies with minimal changes if we replace the event B by $[J_1] \cap [J_2] \cap A \cap \tilde{B}$, and thus the existence of the limit

$$\lim_{l \to 0} l^{-2} P([J_1] \cap [J_2] \cap A \cap \tilde{B})$$

can be established. From this and (10.3.6) follows (10.3.7) and the complete statement.

Remark In view of the results on pairs of segments and pairs of disc mentioned in §6.5, a natural question arises: do propositions similar to the above exist for M_2-invariant second order random segment processes (especially for random mosaics) or random disc processes? The answer is that under additional assumptions of the existence of the densities as interpreted in the examples of §7.14 the natural reformulations of (10.3.6) and (10.3.7) can be proved without substantial difficulties. At present it is not clear whether the above-mentioned additional assumptions can be removed; therefore we adopt them in our treatment of random mosaics and disc processes in §§10.6 and

§10.7. To stress the common nature of the conditioning event (each of the vertical windows is hit and the hits are produced by distinct elements of the realization) we preserve in these sections the notation $\tilde{\Pi}_{x_1 x_2}$ for the corresponding limiting $(l \to 0)$ conditional distributions.

10.4 Averaging a combinatorial decomposition

In this section we again consider general line processes $\{g_i\}$ which we assume to be $\mathbb{M}_2$-invariant, of second order and possessing with probability 1 no pairs of parallel lines. The aim is to average a combinatorial decomposition in the style of (5.4.1) and (5.4.2) where the inherent set of 'needles' depends on $\{g_i\}$, i.e. is random. The resulting quantities will depend on a small parameter l and we will be putting down their main terms which are of order l^2.

I The combinatorial decomposition We consider the segments v_1 and v_2 shown in fig. 10.4.1 as the sides of a rectangle R, the vertices of which we denote by Q_i, $i = 1, 2, 3, 4$. Let

$$\chi_i = g_i \cap R \tag{10.4.1}$$

be the random chords generated by the lines of $\{g_i\}$; we denote by $\mathscr{P}_i$ the endpoints of these chords. With probability 1 the set $\{\mathscr{P}_i\}$ is finite. The algorithm (5.1.3)–(5.1.2) can be easily applied to find the combinatorial decomposition for the set $C \cap B \in r(\{Q_i\} \cup \{P_i\})$, where

$$\begin{aligned}
C &= \{g \in \mathbb{G} : g \text{ hits both } v_1 \text{ and } v_2\} = [v_1] \cap [v_2], \\
B &= \{g \in \mathbb{G} : g \text{ does not hit any of the chords } \chi_i\}.
\end{aligned} \tag{10.4.2}$$

There is a slight complication in that with positive probability there can be collinear triads of endpoints $\mathscr{P}_i$. We overcome this by replacing the sides of R by outward circular arcs, writing the combinatorial decomposition for naturally redefined $C \cap B$ and taking the limit whilst letting the curvature of the arcs tend to zero. The final result is as follows:

$$\mu(B \cap C) = a_1 + a_2 + a_3 + a_4 + a_5, \tag{10.4.3}$$

where

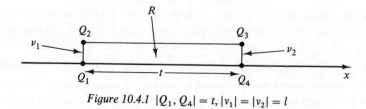

Figure 10.4.1 $|Q_1, Q_4| = t$, $|v_1| = |v_2| = l$

$$a_1 = -I_0(Q_1, Q_4)|Q_1, Q_4| - I_0(Q_2, Q_3)|Q_2, Q_3|,$$

$$a_2 = I_0(Q_1, Q_3)|Q_1, Q_3| + I_0(Q_2, Q_4)|Q_2, Q_4|,$$

$$a_3 = -2 \sum_{\chi_i \in C} |\chi_i| I_0(\chi_i),$$

$$a_4 = \sum_{\mathscr{P}_i \mathscr{P}_j \in C} |\mathscr{P}_i \mathscr{P}_j|(I_d - I_s) I_0(\mathscr{P}_i, \mathscr{P}_j),$$

$$a_5 = -\sum_{Q_i \mathscr{P}_j \in C} |Q_i \mathscr{P}_j|(I_d - I_s) I_0(Q_i, \mathscr{P}_j).$$

In these equations:

μ is the $\mathbb{M}_2$-invariant measure on $\mathbb{G}$; $|\mathscr{P}_1, \mathscr{P}_2|$ denotes the Euclidean distance between the points $\mathscr{P}_1, \mathscr{P}_2$; writing $\chi_i \in C$, $\mathscr{P}_i \mathscr{P}_j \in C$ or $Q_i \mathscr{P}_j \in C$ means that the corresponding line should belong to C; in the expression for a_4 the sum is extended over pairs $\mathscr{P}_i \mathscr{P}_j$ which fail to be endpoints to a χ_s;

$I_0(\delta) = 1$ if no closed chord χ_j hits the interior of the segment δ (if $\delta = \chi_i$ then we assume that $j \neq i$);

     0   otherwise

Lastly, the events d and s correspond to the classification introduced in §5.4 and are shown graphically in fig. 10.4.2.

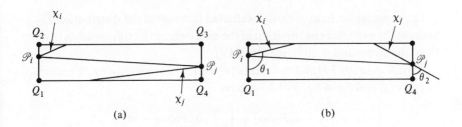

(a)   (b)

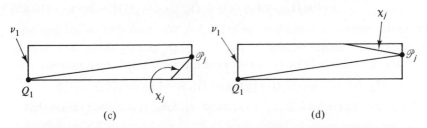

(c)   (d)

Figure 10.4.2 (a) $\mathscr{P}_i \mathscr{P}_j$ is of d-type (χ_i and χ_j lie in different halfplanes with respect to the $\mathscr{P}_i \mathscr{P}_j$ line). (b) $\mathscr{P}_i \mathscr{P}_j$ is of s-type (χ_i and χ_j lie in one halfplane with respect to the $\mathscr{P}_i \mathscr{P}_j$ line). (c) $Q_1 \mathscr{P}_j$ is of d-type (v_1 and χ_j lie in different halfplanes with respect to $Q_1 \mathscr{P}_j$). (d) $Q_1 \mathscr{P}_j$ is of s-type (v_1 and χ_j lie in one halfplane with respect to $Q_1 \mathscr{P}_j$).

II Averaging in the case of infinitesimal l The aim now is to calculate $E_P \sum a_i$, see (10.4.3). The length l of both v_1 and v_2 is assumed to be infinitesimal.

We have (see the notation of §10.2)

$$E_P \mu(B \cap C) = E_P \int_C I_B(g)\, dg = \int_C P(B)\, dg$$

$$= \int_C p_0(|R \cap g|)\, dg = t^{-1} p_0(t) l^2 + o(l^2); \qquad (10.4.4)$$

in the last step we used (3.7.4). Then

$$E_P(a_1 + a_2) = 2(t^2 + l^2)^{1/2} p_0[(t^2 + l^2)^{1/2}] - 2t p_0(t)$$

$$= (t^{-1} p_0(t) + \frac{d}{dt} p_0(t)) l^2 + o(l^2). \qquad (10.4.5)$$

To evaluate $E_P a_3$ we note that, according to (10.1.6), the contribution of the cases where the sum consists of more than one summand is $o(l^2)$. Therefore according to (10.1.7)

$$E_P a_3 = -2tP\left(A \cap \binom{C}{1}\right) + o(l^2) = -2\lambda l^2 \pi_0(t) + o(l^2), \qquad (10.4.6)$$

where

$$A = \binom{[(Q_1, Q_4)]}{0}, \qquad \pi_0(t) = \Pi(A).$$

The expectation $E_P a_4$ is easily evaluated in terms of the distribution $\tilde{\Pi}_{Q_1 Q_4}$ because the expected contribution of the cases where the sum consists of more than one summand is $o(l^2)$ (see §10.3, IV).

With $\tilde{B}$ as in (10.3.5) and the events $\{d\}$ and $\{s\}$ corresponding, respectively, to parts (a) and (b) of fig. 10.4.2, we have

$$E_P a_4 = \int_{\tilde{B} \cap \{d\}} a_4 P(dm) + \int_{\tilde{B} \cap \{s\}} a_4 P(dm) + o(l^2)$$

$$= c l^2 t [\tilde{\Pi}_{Q_1 Q_4}(A \cap \{d\}) - \tilde{\Pi}_{Q_1 Q_4}(A \cap \{s\})] + o(l^2), \qquad (10.4.7)$$

where the constant c is the same as in (10.3.6). We stress that the probabilities in the brackets do not depend on the shifts of Q_1 and Q_2 along the x axis. In the final result, (10.4.11), we will use the more emphatic notations

$$\tilde{\Pi}_{Ot} \text{ (no hits on } (O, t) \cap \{d\}) \text{ and } \tilde{\Pi}_{Ot} \text{ (no hits on } (O, t) \cap \{s\}).$$

It remains to evaluate $E_P a_5$. Here most welcome is the observation that

$$\int_{C_1} a_5 P(dm) = o(l^2), \qquad (10.4.8)$$

where

$$C_1 = \binom{[v_1]\, [v_2]}{0 \quad 1} \cup \binom{[v_1]\, [v_2]}{1 \quad 0}.$$

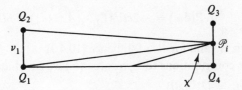

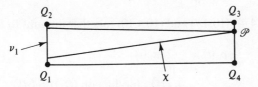

Figure 10.4.3

Figure 10.4.4 Each of the segments joining an endpoint of v_1 or v_2 with an endpoint of χ is of s type

The explanation is that if a realization

$$m \in \begin{pmatrix} [v_1] & [v_2] \\ 0 & 1 \end{pmatrix},$$

say, then we have the situation shown in fig. 10.4.3.

Because now v_1 is not hit by lines of $\{g_i\}$, the numbers of hits of the segments $Q_1\mathscr{P}_i$ and $Q_2\mathscr{P}_i$ coincide. Therefore on this set of realizations

$$a_5 = I_0(Q_1, \mathscr{P}_i)(|Q_1, \mathscr{P}_i| - |Q_2, \mathscr{P}_i|) = o(l^2)$$

as a result of subtraction of the distances. The same is true for $\begin{pmatrix} [v_1] & [v_2] \\ 1 & 0 \end{pmatrix}$, and since

$$P(C_1) = o(l),$$

(10.4.8) follows.

From fig. 10.4.4 and (10.1.6) we also conclude that

$$\int_{\binom{C}{1}} a_5 P(dm) = 4\lambda \cdot l^2 \pi_0(t) + o(l^2).$$

Lastly (compare with (10.4.7))

$$\int_{\tilde{B}} a_5 P(dm) = 2ctl^2 [\tilde{\Pi}_{Q_1Q_4}(A \cap \{s\}) - \tilde{\Pi}_{Q_1Q_4}(A \cap \{d\})] + o(l^2) \quad (10.4.9)$$

Explanation We split $\tilde{B}$ into subsets

$$\tilde{B} = (\tilde{B} \cap \{s\}) \cup (\tilde{B} \cap \{d\}).$$

As can be seen from fig. 10.4.2(a), on $\tilde{B} \cap \{d\}$ the segments $Q_2\mathscr{P}_j$ and $Q_4\mathscr{P}_i$ are hit by lines from $\{g_i\}$ with probability 1 and do not contribute to a_5, while the segments $Q_3\mathscr{P}_i$ and $Q_1\mathscr{P}_j$ are of d type. Together with (10.3.7) this yields

$$\int_{\tilde{B} \cap \{d\}} a_5 P(\mathrm{d}m) = -2ctl^2 \tilde{\Pi}_{Q_1 Q_4}(A \cap \{d\}) + o(l^2).$$

A similar analysis is valid for $\tilde{B} \cap \{s\}$, hence (10.4.9).

Since all other contributions to $E_P a_5$ are $o(l^2)$ we conclude that

$$\begin{aligned}
E_P a_5 = {}& 4\lambda l^2 \pi_0(t) \\
& - 2ctl^2 (\tilde{\Pi}_{Q_1 Q_4}(A \cap \{d\}) - \tilde{\Pi}_{Q_1 Q_4}(A \cap \{s\})) \\
& + o(l^2).
\end{aligned} \tag{10.4.10}$$

Equations (10.4.4)–(10.4.7) and (10.4.10) together amount to a basic relation

$$\begin{aligned}
\frac{\mathrm{d}}{\mathrm{d}t} p_0(t) = {}& -2\lambda \pi_0(t) \\
& + ct\tilde{\Pi}_{Ot}(\text{no hits on } (O, t) \cap \{d\}) \\
& - ct\tilde{\Pi}_{Ot}(\text{no hits on } (O, t) \cap \{s\}).
\end{aligned} \tag{10.4.11}$$

We stress that the events $\{d\}$ and $\{s\}$ can now be expressed in terms of the random angles (θ_1, θ_2) whose joint distribution is determined (marginally) by $\tilde{\Pi}_{Ot}$ of (10.3.7). If θ_1 and θ_2 are defined as shown in fig. 10.5.2 then

$$\begin{aligned}
\{d\} &= \left[\left(0, \frac{\pi}{2}\right) \times \left(0, \frac{\pi}{2}\right) \right] \cup \left[\left(\frac{\pi}{2}, \pi\right) \times \left(\frac{\pi}{2}, \pi\right) \right], \\
\{s\} &= \left[\left(0, \frac{\pi}{2}\right) \times \left(\frac{\pi}{2}, \pi\right) \right] \cup \left[\left(\frac{\pi}{2}, \pi\right) \times \left(0, \frac{\pi}{2}\right) \right].
\end{aligned} \tag{10.4.12}$$

10.5 Further remarks on line processes

I The result (10.4.11) can be generalized [44] to include probabilities of having k (rather than zero) intersections of the lines from $\{g_i\}$ with a segment of length t. This generalization can be derived by the same method starting from a combinatorial decomposition similar to (10.4.3) but written for $\mu(B_k \cap C)$, where

$$B_k = \{g \in \mathbb{G} : k \text{ chords } \chi_i \text{ hit } g\}.$$

We leave it to the reader to check (using the algorithm (5.1.3)–(5.1.2)) that the desired decomposition can be obtained by making the following changes in the quantities a_i in (10.4.3):

in a_1, a_2, replace I_0 by I_k;
in a_3, a_5, replace I_0 by $-I_{k-1} + I_k = \Delta I_{k-1}$;
in a_4, replace I_0 by $I_{k-2} - 2I_{k-1} + I_k = \Delta^2 I_{k-2}$,

where Δ and Δ^2 are the first and the second differences, and

$\quad I_k = 1 \quad$ if the line segment in question is hit by exactly k chords χ_i,

$\quad\ \ = 0 \quad$ otherwise (in each case the interpretation of a 'hit' is the same as in the description of I_0).

The procedure of averaging repeats that of the previous section with minimal changes. The result is as follows:

$$\frac{d}{dt}p_k(t) = 2\lambda(\pi_{k-1}(t) - \pi_k(t))$$

$$+ ct\Delta^2\tilde{\Pi}_{0t}(k \text{ hits on } (0, t) \cap \{d\})$$
$$- ct\Delta^2\tilde{\Pi}_{0t}(k \text{ hits on } (0, t) \cap \{s\}), \tag{10.5.1}$$

where

$$\pi_k(t) = \Pi \quad (\text{there are } k \text{ hits on } (0, t))$$

(under both $\tilde{\Pi}$ and Π the interval $(0, t)$ is taken on the x axis).

II To obtain a check for (10.4.11) let us consider the case in which $\{g_i\}$ is randomly rotated $\mathbb{T}_2$-invariant Poisson. In the notation of §7.12, II (see also §10.1) we have

$$p_0(t) = \frac{1}{\pi}\int_0^\pi \exp(-\lambda(\varphi)\cdot t)\,d\varphi,$$

$$\pi_0(t) = \frac{1}{\pi\lambda}\int_0^\pi \exp(-\lambda(\varphi)\cdot t)f(\varphi)\,d\varphi,$$

$$ct\tilde{\Pi}_{0t} (\text{no hits on } (0, t)\cap\{d\}) - ct\tilde{\Pi}_{0t} (\text{no hits on } (0, t)\cap\{s\}) \tag{10.5.2}$$

$$= \frac{t}{\pi}\int_0^\pi \exp(-\lambda(\varphi)\cdot t)(p(\varphi) - q(\varphi))^2\left(\lambda(\varphi + \frac{\pi}{2})\right)^2 d\varphi.$$

Here $p(\varphi)$ and $q(\varphi)$ are *conditional* probabilities of the events (see (10.4.12))

$$0 < \theta_1 < \frac{\pi}{2} \quad \text{and} \quad \frac{\pi}{2} < \theta_1 < \pi,$$

the condition is upon the angle of rotation. In our example the identically distributed angles θ_1 and θ_2 are conditionally independent and independent of the number of hits on the $(0, t)$ interval of the x axis.

On the other hand, as illustrated by fig. 10.5.1, we have up to $o(d\varphi)$

$$\lambda(\varphi + d\varphi) - \lambda(\varphi) = \mu([AC]) - \mu([AB])$$
$$= \mu([CB]\cap[AC]) - \mu([CB]\cap[AB])$$
$$= \lambda(\varphi + \frac{\pi}{2})(q(\varphi) - p(\varphi))\,d\varphi.$$

Hence,

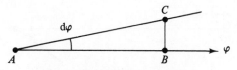

Figure 10.5.1 The length $|AB| = 1$

$$\left(\lambda(\varphi + \frac{\pi}{2})\right)^2 (q(\varphi) - p(\varphi))^2 = \left[\frac{\mathrm{d}}{\mathrm{d}\varphi}\lambda(\varphi)\right]^2. \tag{10.5.3}$$

We multiply (2.11.2) by $e^{-\lambda(\varphi)t}$ and integrate over $\mathrm{d}\varphi$. After integration by parts we obtain

$$\int \exp(-\lambda(\varphi)t)\lambda(\varphi)\,\mathrm{d}\varphi - t \int (\lambda'(\varphi))^2 \exp(-\lambda(\varphi)t)\,\mathrm{d}\varphi$$

$$= 2 \int \exp(-\lambda(\varphi)t)f(\varphi)\,\mathrm{d}\varphi.$$

In view of (10.5.2)–(10.5.3) this coincides with (10.4.11). We could also obtain (10.5.1) if we used $\exp(\lambda(\varphi)(z - 1)t)$ instead of $\exp(-\lambda(\varphi)t)$.

III As mentioned in §7.12 the marked intersection process $\{(x_i, \psi_i)\}$ induced by $\{g_i\}$ on the x axis provides an adequate description of $\{g_i\}$. In particular the sum of $\tilde{\Pi}_{0t}$ terms in (10.4.11) or (10.5.1) (and with it the probabilities $\pi_k(t)$) can be expressed in terms of the distribution of $\{(x_i, \psi_i)\}$. We now outline the corresponding derivation.

Let us consider a pair of horizontal windows

$$\delta_1 = (-l, 0) \qquad \text{and} \qquad \delta_2 = (t, t + l)$$

situated on the x axis. The event

$F = \{m \in \mathcal{M}_G$: there is exactly one pair of lines $g_1, g_2 \in m$ such that g_1 hits δ_1, g_2 hits $\delta_2\}$

is a horizontal window counterpart of the event $\tilde{B}$ in §10.3. We also consider two angles ψ_1, ψ_2 which are well defined whenever $m \in F$ (see fig. 10.5.2). The same argument which we used to demonstrate (10.3.6) applies in the proof that

$$P(F) = cl^2 + o(l^2)$$

with the same constant c as in (10.3.6). (The coincidence of the constants follows from the remark in §6.5 concerning the horizontal windows.)

We define the horizontal window counterpart of the distribution $\tilde{\Pi}_{0t}$ as

$$\Pi^*_{0t}(J_1, J_2, A) = \lim_{l \to 0} \frac{P([J_1] \cap [J_2] \cap A)}{P(F)}$$

where J_i is a set in the space of values of ψ_i, $[J_i]$ is the corresponding event in $\mathcal{M}_G, i = 1, 2$ and $A \subset \mathcal{M}_G$.

Figure 10.5.2

The desired relation between $\tilde{\Pi}_{Ot}$ and Π_{Ot}^* follows from the observations

(a) we have obvious set relations in the space $\mathbb{G}$

$\{g : g \text{ hits } v_i \text{ under angle } \theta_i\}$

$= \{g : g \text{ hits a horizontal window of length } |\cot \psi_i| \cdot l \text{ under angle } \psi_i\}$

(b) we have an elementary identity

$$|\cot \psi_1 \cot \psi_2| \cdot (I_d - I_s) = \cot \psi_1 \cot \psi_2.$$

Hence (see (10.4.12))

$$\tilde{\Pi}_{Ot}(A \cap \{d\}) - \tilde{\Pi}_{Ot}(A \cap \{s\}) = E^* \cot \psi_1 \cot \psi_2 \, I_A(m) \qquad (10.5.4)$$

where E^* is the expectation corresponding to Π_{Ot}^*. This is a solution to our problem. We note that similar relations are valid also for the probabilities $\tilde{\Pi}_{Ot}$ which appear in the problems of §10.6 and §10.7.

Among other things (10.5.4) implies the existence of $E^* \cot \psi_1 \cot \psi_2$. In the case in which the angles ψ_1 and ψ_2 are independent this expectation reduces to zero. (We have

$$E^* \cot \psi_i = E \cot \psi_i = 0, \qquad i = 1, 2$$

because each ψ_i has $\tfrac{1}{2} \sin \psi \, d\psi$ density on $(0, \pi)$. see (7.15.7).)

10.6 Extension to random mosaics

A remarkable feature of the calculations of §10.4 is that *they remain valid for a broad class of random mosaics* not necessarily generated by line processes. Apart from the $\mathbb{M}_2$-invariance property this class is determined by the further conditions (a), (b) and (c):

(a) With probability 1 a mosaic should possess *no vertices* of 'multiple T type' (see fig. 10.6.1).
(b) The process of edge midpoints of a mosaic should be of *finite intensity* and the distribution of the length of typical edge should possess finite mean.
(c) The process of edges of a mosaic should not possess parallel pairs and should be of second order. Moreover, its second moment measure should possess a density in the sense of §7.14, Example 4.

We shall denote the process of edges of the mosaic (a segment process) by $\{s_i\}$.

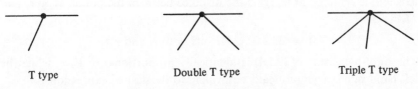

T type   Double T type   Triple T type

Figure 10.6.1 Multiplicities 1, 2, 3

We briefly outline the necessary reinterpretations and changes from §10.4.

Our first aim is to write the combinatorial decomposition for the set $C \cap B$, where C is as in (10.4.2) and

$$B = \{g \in \mathbb{G}: \text{within } R, g \text{ does not hit any edge } s_i\}.$$

By virtue of (a) and the absence of edges on a line, *this decomposition coincides with* (10.4.3).

Proof Let us denote by $\mathscr{P}_i$ the points of intersection of edges s_i with ∂R, by $\{v_i\}$ the set of vertices of the mosaic which lie within R, and let Q_i, $i = 1, \ldots,$ 4 be as in fig. 10.4.1. Clearly, using the notation of §5.1,

$$B \cap C \in r(\{\mathscr{P}_i\} \cup \{v_i\} \cup \{Q_i\}).$$

The four-indicator formula (5.1.2) yields zero whenever at least one of the points in a pair is of v_i type (this could be not the case if we had vertices of multiple T type). Thus the decomposition will actually depend only on the pairs of points from $\{\mathscr{P}_i\} \cup \{Q_i\}$, and the coefficients of the decomposition turn out to be the same as in (10.4.3). The explanations given there for the terms a_1, a_2, a_4 and a_5 carry forward if we now put

$$\chi_i = R \cap s_i$$

(compare with (10.4.1)). For a_3, we obtain another expression:

$$a_3 = -2 \sum_{\substack{s_i \text{ hits both} \\ v_1 \text{ and } v_2}} |\chi_i|, \tag{10.6.1}$$

which coincides with the old one if we consider the line process as a mosaic.

Condition (b) allows us to conclude that

$$E_P a_3 = -2\lambda_0 l^2 E_{\mathbf{p}}(|s| - t)^+ + o(l^2), \tag{10.6.2}$$

where λ_0 is $(2\pi)^{-1}$ times the intensity of edge centers, and $\mathbf{p}$ denotes the distribution of the typical edge length of the mosaic,

$$x^+ = x \quad \text{if } x > 0$$
$$\quad = 0 \quad \text{otherwise.}$$

Proof We apply the construction of Example 4 in §9.2 to the edge process $\{s_i\} = m_2(\omega)$ and obtain its relative Palm distribution Π_{m_1}.

For it we write (9.1.4) with

$$V(M, m) = I_A(M) \cdot |\chi| \cdot I_{\{|s| > |\chi|\}}(|s|),$$

where s denotes that segment of the realization $M^{-1}\{s_i\}$ which lies on the x axis and is centered at O; $|\chi|$ is the distance between the points $M_s \cap v_1$ and $M_s \cap v_2$;

$$A = \{M \in \mathbb{M}_2: Ms \text{ hits both } v_1 \text{ and } v_2\}.$$

The right-hand side of (9.1.4) is identically proportional to $E_P a_3$ while the left-hand side is asymptotically equivalent to the desired expression.

Expressions of the terms $E_P(a_1 + a_2)$ and $E_P a_4$ retain the forms (10.4.5) and (10.4.7) with interpretations

$p_0(t) = P$(no intersections on (O, t) with $\{s_i\}$), the segment (O, t) lies on the x axis,

$\tilde{\Pi}_{Ot}$ is the limiting $(l \to 0)$ distribution of $\{s_i\}$ conditional upon the event {there is one hit on each v_i, $i = 1, 2$ produced by different edges} whose asymptotical probability is $c_1 l^2$, see Remark to §10.3, IV.

The events $\{d\}$ and $\{s\}$ are defined in terms of intersection angles θ_1, θ_2 as in (10.4.12).

For $E_P a_5$ it is essential to check, that the considerations that led to (10.4.8) and are used in further analysis remain valid. For instance (10.4.8) now follows from the observation that if on fig. 10.4.3 v_1 is not hit by the edges of a mosaic then

$$I_0(Q_1, \mathscr{P}_j) = I_0(Q_2, \mathscr{P}_j).$$

(If an edge of a mosaic hits a side of a triangle then at least one of the two other sides of the triangle is also hit by the edges of the mosaic). Therefore we again have

$$E_P a_5 = 2(E_P a_3 - E_P a_4) + o(l^2).$$

In this way we come to a *relation valid for random* M_2-*invariant mosaics satisfying conditions* (a), (b), (c):

$$\frac{d}{dt} p_0(t) = 2\lambda_0 E_\mathbf{p}(|s| - t)^+$$

$$+ c_1 t \tilde{\Pi}_{Ot}(\text{no hits on } (O, t) \cap \{d\})$$
$$- c_1 t \tilde{\Pi}_{Ot}(\text{no hits on } (O, t) \cap \{s\}). \tag{10.6.3}$$

By virtue of a general relation

$$\frac{d}{dt} E_\mathbf{p}(|s| - t)^+ = \mathbf{p}(|s| > t), \tag{10.6.4}$$

(10.6.3) allows the calculation the distribution of the typical edge length s of the random mosaic in terms of the marked point process $\{(x_i, \psi_i)\}$ induced on the x axis (or any other test line).

In particular, if in the $\{x_i, \psi_i\}$ process we have

$$p_0(t) = \exp(-ht), \qquad h > 0$$

and the angles ψ_i are independent, then the $\tilde{\Pi}$ terms in (10.6.3) sum to zero (compare with (10.5.4)) and (10.6.4) yields

$$h^2 \exp(-ht) = 2\lambda_0 \mathbf{p}(|s| > t).$$

From this we conclude that

$$\mathbf{p} \text{ (typical edge length exceeds } t) = \exp(-ht),$$

$$\text{edge center intensity} = \pi h^2.$$

A result equivalent to (10.6.3) has been obtained in a less rigorous setting in [3] based upon the Pleijel identity (6.9.5). Some corollaries for typical polygons of the mosaic have also been considered in [3].

10.7 Boolean models for disc processes

The method of averaging the combinatorial decompositions seems to be far from exhausted on consideration of the preceding examples. A natural domain for expansion constitutes line and plane processes in $\mathbb{R}^3$ but no work has been done in this area yet. However another possibility is to apply the technique to domain processes as suggested by the existence of combinatorial decompositions for members of Sylvester rings (see §5.4). Some work in this direction has already been started in chapter 10 of [3]. Below we consider essentially the same problem but we use a more rigorous approach in which we analyze the events that occur in a narrow rectangle R (see fig. 10.4.1).

Let $\{b_i\}$ be an $\mathbb{M}_2$-invariant process of unit radius discs, and let P be its distribution. For simplicity *we exclude multiple tangencies*, i.e. we assume that with probability 1 in a realization no three discs have a common tangent line.

We consider the Boolean model, i.e. the set $\bigcup b_i$. Boolean models for Poisson disc processes have been considered in §8.9. Now our $\{b_i\}$ is *not* assumed to be Poisson; however, we retain the white and black sets terminology, i.e. we call $\bigcup b_i$ the black set and its complement white. The first step is to write the combinatorial decomposition for the set $B \cap C$, where C is as in (10.4.2) and

$$B = \{g \in G : g \cap R \text{ is completely white}\}.$$

Clearly $B \cap C$ belongs to the Sylvester ring generated by the segments v_1 and v_2 and the traces $b_i \cap R$ of the discs on R.

The corresponding combinatorial decomposition (a version of (5.4.3)) will be of the form

$$\mu(B \cap C) = \sum_{i=1}^{5} a_i + \sum_{i=1}^{3} \alpha_i, \qquad (10.7.1)$$

where $a_1, \ldots, a_5$ are versions of the terms in (10.4.3) while the terms α_1, α_2 and α_3 are principally new. In the present situation the expressions for a_1, a_2, a_4 and a_5 given in §10.4 undergo only minor reinterpretations which reduce to a convention that now the points $\{\mathscr{P}_i\}$ are understood to be the intersection points of the circumferences of the discs with v_1 or v_2 and

$$\chi_i = b_i \cap R.$$

The new expression for a_3 is as follows (compare with (5.4.3)):

$$a_3 = -\sum_i \int_{\partial b_i \cap R} I_w(t(\varphi)) I_C(t(\varphi)) \, d\varphi, \qquad (10.7.2)$$

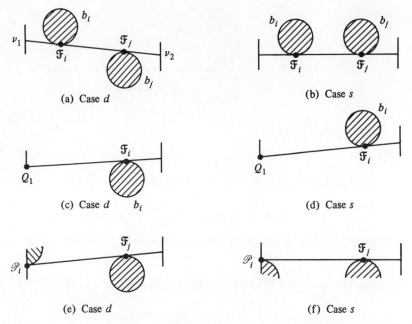

Figure 10.7.1 We refer to the configurations (a)–(b), (c)–(d) or (e)–(f) by mentioning the pairs $\mathcal{F}_i$, $\mathcal{F}_j$; Q_i, $\mathcal{F}_j$; or $\mathcal{P}_i$, $\mathcal{F}_j$, respectively

where $t(\varphi)$ is the line tangent to ∂b_i at the point with angular coordinate φ, and $I_w = 1$ if $t(\varphi) \cap R$ is completely white, zero otherwise ($w = \partial B$ if the discs are assumed open).

We describe the terms α_1, α_2 and α_3 using fig. 10.7.1.

We have

$$\alpha_1 = \sum |\mathcal{F}_i, \mathcal{F}_j| I_w(\mathcal{F}_i, \mathcal{F}_j) I_C(\mathcal{F}_i, \mathcal{F}_j)(I_d - I_s),$$
$$\alpha_2 = -\sum |Q_i, \mathcal{F}_j| I_w(Q_i, \mathcal{F}_j) I_C(Q_i, \mathcal{F}_j)(I_d - I_s),$$
$$\alpha_3 = \sum |\mathcal{P}_i, \mathcal{F}_j| I_w(\mathcal{P}_i, \mathcal{F}_j) I_C(\mathcal{P}_i, \mathcal{F}_j)(I_d - I_s). \tag{10.7.3}$$

Here $I_w = 1$ if the continuation within R of the segment in the argument is completely white, zero otherwise; $I_C = 1$ if the continuation of the same segment hits both (closures of) v_1 and v_2.

Under the assumption that the process of the disc centers is of second order and its second moment measure possesses a bounded density (in the sense of §7.14, Example 1), (10.7.1) can be averaged and the terms which are l^2 by order can be separated. This procedure resembles similar operations in §10.4. The result depends on several Palm-type distributions of our disc process $\{b_i\}$ which can be obtained as limiting conditional distributions with conditioning events corresponding to the above diagrams.

We denote by $\lambda < \infty$ the intensity of black disc centers and start to consider separately the mean values of the summands in (10.7.1).

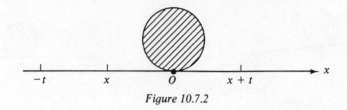

Figure 10.7.2

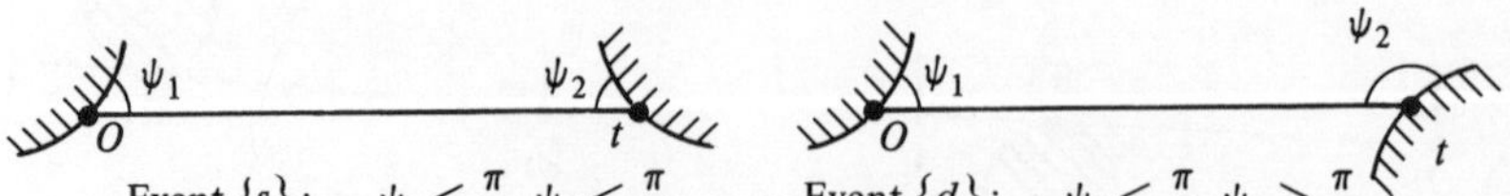

Event $\{s\}$: $\psi_1 < \dfrac{\pi}{2}, \psi_2 < \dfrac{\pi}{2}$

or $\psi_1 > \dfrac{\pi}{2}, \psi_2 > \dfrac{\pi}{2}$

Event $\{d\}$: $\psi_1 < \dfrac{\pi}{2}, \psi_2 > \dfrac{\pi}{2}$

or $\psi_1 > \dfrac{\pi}{2}, \psi_2 < \dfrac{\pi}{2}$

Figure 10.7.3 If (O, t) is white then the events $\{s\}$ and $\{d\}$ are well defined

The case of $E(a_1 + a_2)$ can be treated as in §10.4. Then

$$Ea_3 = -\frac{2\lambda}{t} l^2 \int_{-t}^{0} \Pi((x, x + t) \text{ is white}) \, dx + o(l^2)$$

where Π is the relative Palm distribution of $\{b_i\}$ with respect to the process $\{M_i\}$ on the group $\mathbb{M}_2$ constructed in Example 3, §9.2.

Thus with Π-probability 1 we have a disc contacting the point O on the x axis, as shown in fig. 10.7.2.

Explanation: assuming $Q_1 = 0$ we represent Ea_3 as

$$-E' \sum_i I_{B(M_i)}(M_i^{-1}\{b_s\}) I_C(g(M_i)) I_{(0,t)}(x(M_i)) + o(l^2)$$

where $g(M)$ is the line on which lies the directed segment representing $M \in \mathbb{M}_2$, $x(M)$ is the abscissa of the source of M, $B(M)$ is the event 'the interval $(-x(M), -x(M) + t)$ is white', E' denotes expectation with respect to $\{M_i, M_i^{-1}\{b_s\}\}$. It remains to apply (9.1.4) and integrate using (3.13.3). Further

$$Ea_4 = t \cdot c_1(t) \cdot l^2 [\tilde{\Pi}_{Ot}((O, t) \text{ is white} \cap \{d\}) - \tilde{\Pi}_{Ot}((O, t) \text{ is white} \cap \{s\})] + o(l^2)$$

where asymptotically $c_1(t) \, l^2$ is the probability of the event.

There is exactly one pair of discs b_i, b_j in realization such that

$$\partial b_i \text{ hits } v_1 \quad \text{and} \quad \partial b_j \text{ hits } v_2$$

and $\tilde{\Pi}_{Ot}$ is the corresponding limiting conditional probability (see the Remark in §10.3, IV)). $\tilde{\Pi}_{Ot}$ is a probability distribution on the space

$$(0, 2\pi) \times (0, 2\pi) \times \mathcal{M}_{\mathbb{R}^2},$$

with $\tilde{\Pi}_{Ot}$-probability 1 the boundaries of two different discs pass through the points O and t on the x axis. The events $\{d\}$ and $\{s\}$ can be defined in terms of the angles ψ_1 and ψ_2 as shown in fig. 10.7.3.

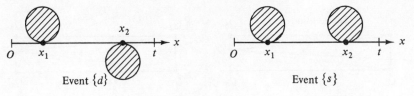

Event $\{d\}$ Event $\{s\}$

Figure 10.7.4

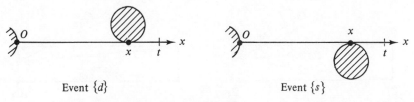

Event $\{d\}$ Event $\{s\}$

Figure 10.7.5 If the interval (O, x) is white then the events $\{d\}$ and $\{s\}$ are well defined

Also

$$E\alpha_1 = c_2(t)\cdot l^2 E^{\mathscr{F}\mathscr{F}}I_{(O,t)\text{ is white}}(I_{\{d\}} - I_{\{s\}})|x_1 - x_2| + o(l^2)$$

where asymptotically $c_2(t)\, l^2$ is the probability of the event that in the expression of α_1 in (10.7.3) there is exactly one non-zero summand, $E^{\mathscr{F}\mathscr{F}}$ stands for the expectation with respect to $\Pi_{Ot}^{\mathscr{F}\mathscr{F}}$, the corresponding limiting conditional distribution. This is a probability distribution in the union of two copies of the space

$$(O, t) \times (O, t) \times \mathscr{M}_{\mathbb{R}^2};$$

the copies correspond to the events $\{s\}$ and $\{d\}$. With $\Pi_{Ot}^{\mathscr{F}\mathscr{F}}$-probability 1 we have two discs contacting the x axis at some points x_1, $x_2 \in (O, t)$; see fig. 10.7.4.

Lastly

$$E\alpha_3 = c_3(t)l^2 E^{\mathscr{P}\mathscr{F}}I_{(O,t)\text{ is white}}(I_{\{d\}} - I_{\{s\}})x + o(l^2),$$

where asymptotically $c_3(t)l^2$ is the probability of the event 'in the expressions of α_3 in (10.7.3) there is exactly one non-zero summand'. $E^{\mathscr{P}\mathscr{F}}$ stands for the expectation with respect to $\Pi_{Ot}^{\mathscr{P}\mathscr{F}}$, the corresponding limiting conditional distribution. It is a probability distribution in the union of two copies of the space

$$(0, 2\pi) \times (O, t) \times \mathscr{M}_{\mathbb{R}^2}.$$

With $\Pi_{Ot}^{\mathscr{P}\mathscr{F}}$-probability 1 we have a disc b_i with $O \in \partial b_i$ and a disc b_j contacting the x axis at a point $x \in (O, t)$ (see fig. 10.7.5).

By symmetry we have

$$E\alpha_2 = 0.$$

The case of

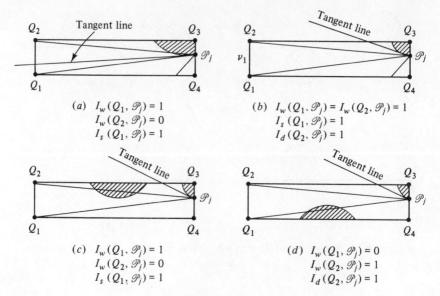

Figure 10.7.6

$$Ea_5 = E \sum_{Q_i, \mathscr{P}_j} |(Q_i, \mathscr{P}_j)| I_w(Q_i, \mathscr{P}_j)(I_s - I_d)$$

is more complicated since the probability to have at least one pair Q_i, $\mathscr{P}_j$ for which $I_w = 1$ is $O(l)$. Therefore it is necessary to perform subtraction of the terms corresponding to I_s and I_d.

First we consider the event (a) in which the tangent line at $\mathscr{P}_j$ hits the opposite vertical window (fig. 10.7.6(a) shows one of the four components of the event (a)). The contribution of the event (a) in Ea_5 is

$$2\lambda l^2 [\Pi((O, t) \text{ is white}) + \Pi((-t, O) \text{ is white})] + o(l^2).$$

The contribution of the part of the complement of (a) which is defined by the relations

$$\mathscr{P}_j \in v_2 \quad \text{and} \quad I_w(Q_1, \mathscr{P}_j) = I_w(Q_2, \mathscr{P}_j) = 1$$

or

$$\mathscr{P}_j \in v_1 \quad \text{and} \quad I_w(Q_3, \mathscr{P}_j) = I_w(Q_4, \mathscr{P}_j) = 1$$

(see fig. 10.7.6 (b)) is $o(l^2)$.

The union of four components of the type (c) and of four components of the type (d) (see fig. 10.7.6) coincides with the event 'there is exactly one non-zero summand in the expression of α_3 in (10.7.3)'. The corresponding contribution to Ea_5 equals

$$-c_3(t) l^2 t E^{\mathscr{P}^{\mathscr{F}}} I_{(O,t) \text{ is white}} (I_{\{d\}} - I_{\{s\}}) + o(l^2)$$

Putting all this together yields

$$\frac{\mathrm{d}}{\mathrm{d}t} P(\text{a segment of length } t \text{ is white})$$

$$= \frac{2\lambda}{t} \int_{-t}^{0} \Pi((x, x + t) \text{ is white}) \, \mathrm{d}x$$

$$- 2\lambda\Pi((-t, O) \text{ is white}) - 2\lambda\Pi((O, t) \text{ is white})$$

$$- tc_1(t)\{\tilde{\Pi}_{Ot}((O, t) \text{ is white} \cap \{d\}) - \tilde{\Pi}_{Ot}((O, t) \text{ is white} \cap \{s\})\}$$

$$- c_2(t) \cdot E^{\mathscr{F}\mathscr{F}} I_{(O, t) \text{is white}}(m)(I_{\{d\}}(m) - I_{\{s\}}(m))|x_1 - x_2|$$

$$- c_3(t) \cdot E^{\mathscr{P}\mathscr{F}} I_{(O, t) \text{is white}}(m)(I_{\{d\}}(m) - I_{\{s\}}(m))(t - x). \tag{10.7.4}$$

10.8 Exponential distribution of typical white intervals

If $\{b_i\}$ happens to be Poisson then each of the last three lines in (10.7.4) reduces to zero. We explain that annihilation of the two terms involving $\tilde{\Pi}_{Ot}$ follows from the remarks at the end of §8.9 and of §10.5, III. The $E^{\mathscr{F}\mathscr{F}}$ vanishes by virtue of the following nature of the probability distribution $\Pi_{Ot}^{\mathscr{F}\mathscr{F}}$ in the case of Poisson $\{b_i\}$: the event $\{d\}$ (or $\{s\}$), the variables x_1, x_2 and $m \in \mathscr{M}_{\mathbb{R}^2}$ are independent; the projection of $\Pi_{Ot}^{\mathscr{F}\mathscr{F}}$ on $(0, t) \times (0, t)$ is uniform while its projection on $\mathscr{M}_{\mathbb{R}^2}$ is Poisson (identical with the distribution of $\{b_i\}$);

$$\Pi_{Ot}^{\mathscr{F}\mathscr{F}}(\{s\}) = \Pi_{Ot}^{\mathscr{F}\mathscr{F}}(\{d\}) = \tfrac{1}{2}.$$

The term $E^{\mathscr{P}\mathscr{F}}$ vanishes for similar reasons.

Correctly speaking, all terms in (10.7.4) depend solely on the distribution of the marked black interval process $\{\delta_i, \psi_i', \psi_i''\}$ (see the end of §8.9) induced by the Boolean model $\bigcup b_i$ on a test line (say the x axis). Therefore it is natural to look for *classes* of processes $\{\delta_i, \psi_i', \psi_i''\}$ for which annihilations similar to the above occur. It was shown in [3] that if

(a) the triads $(|\delta_i|, \psi_i', \psi_i'')$ for different values of i are independent (and certain moment exist)

then the last three lines in (10.7.4) cancel.

Now let us also consider the sequence $\{v_i\}$ of white intervals (spaces between the black intervals δ_i) induced by our Boolean model on the test line. We will assume in what remains that

(b) the sequences $\{|v_i|\}$ and $\{(|\delta_i|, \psi_i', \psi_i'')\}$ are independent;
(c) $\{|v_i|\}$ is a sequence of independent lengths.

We stress that (a), (b) and (c) together yield a version of the renewal property as interpreted in §8.9.

If (a), (b) and (c) hold then the distribution of the typical white interval length is necessarily exponential.

Here is the derivation of this result whose prototype can be found in [3], chapter 10.

We use the notation

$$F(x) = P(\text{typical } |v_i| \text{ exceeds } x).$$

By virtue of the renewal properties of the process on the x axis

$$\Pi((O, t) \text{ is white}) = \Pi((-t, O) \text{ is white}) = qF(t)$$

and

$$\int_{-t}^{0} \Pi((x, x + t) \text{ is white}) \, dx = q \int_{0}^{t} F(t - x)F(x) \, dx,$$

where

$$q = \Pi(\text{the point } O \text{ remains on the boundary of the black set}).$$

Then by a version of Palm formula (8.7.4)

$$\frac{d}{dt} P((O, t) \text{ is white}) = -2\lambda qF(t).$$

Therefore (10.7.4) becomes

$$F(t) = t^{-1} \int_{0}^{t} F(t - x)F(x) \, dx. \qquad (10.8.1)$$

In terms of the Laplace transform

$$L(u) = \int_{0}^{\infty} \exp(-ut)F(t) \, dt$$

(10.8.1) is equivalent to

$$L'(u) = -L^2(u),$$

i.e.

$$L(u) = (u + C)^{-1} \quad \text{or} \quad F(x) = \exp(-Cx), \qquad C > 0.$$

This result is yet another illustration of the general principle according to which M_2-invariance often imposes rather heavy restrictions on the distributions of intersection processes (compare with §10.3, III).

REFERENCES

[1] G. Matheron, *Random Sets and Integral Geometry*, Wiley, New York, 1975.

[2] L. A. Santalo, *Integral Geometry and Geometric Probability*. Addison-Wesley, Reading, Mass., 1976.

[3] R. V. Ambartzumian, *Combinatorial Integral Geometry*. John Wiley and Sons, Chichester, 1982.

[4] E. Hewitt and K. A. Ross, *Abstract Harmonic Analysis*. Springer Verlag, 1963.

[5] P. Funk, 'Uber Flächen mit lauter geschlossenen geodätishen Linien', *Math. Annalen*, **B.74**, 1913, pp. 278–300.

[6] W. Blaschke, *Kreis und Kugel*, 2nd edn. W. de Gruyter, Berlin, 1956.

[7] R. Schneider and W. Weil, 'Zonoids and related topics', in *Convexity and its Applications*, P. Gruber and J. M. Wills eds. Birkhäuser, Basel, 1983, pp. 296–317.

[8] K. Krickeberg, 'Invariance properties of the correlation measure of line processes', in *Stochastic Geometry* (E. F. Harding and D. G. Kendall, eds.), pp. 76–88, Wiley, New York, 1974.

[9] O. Kallenberg, 'A counterexample to R. Davidson's conjecture on line processes', *Math. Proc. Camb. Phil. Soc.*, **82**, 301–7, 1977.

[10] F. Papangelou, 'On the Palm probabilities of processes of points and processes of lines', in *Stochastic Geometry* (E. F. Harding and D. G. Kendall, eds.), pp. 114–47, Wiley, New York, 1974.

[11] R. V. Ambartzumian, 'Random fields of segments and random mosaics on a plane', *Proc. Sixth Berkeley Symp. Math. Stat. Prob*, **III**, 369–81, 1972.

[12] M. G. Kendall and P. A. P. Moran, *Geometrical Probability*, Griffin, London, 1963.

[13] R. E. Miles, 'On random rotations of R^3', *Biometrica*, 1965 v. 52, pp. 636–9.

[14] K. Leichtweiss, *Konvexe Mengen*, VEB Deutscher Verlag der Wissenschaften, Berlin 1980.

[15] W. Blaschke, *Vorlesungen über Integralgeometrie*, 3rd edn., Deutsch. Verlag Wiss., Berlin, 1955.

[16] J. Favard, 'Définition de la longueur et de l'aire', *C.R. Acad. Sci. Paris*, 194, (1932), 344–6.

[17] D. Stoyan and J. Mecke, *Stochastische Geometrie: Eine Einführung*. Academic Verlag, Berlin, 1983.

[18] K. Matthes, J. Kerstan and J. Mecke, *Infinitely Divisible Point Processes.* Wiley Chichester (English edition), 1978.

[19] R. V. Ambartzumian, 'Stereology of random planar segment processes', *Rend. Sem. Mat. Torino*, **39**, 20, 1981.

[20] R. V. Ambartzumian, 'On Sylvester type problems for homogenous Poisson processes' (in Russian). *Izv. Akad. Nauk Armjan. SSR, Mathematics* 20, 4, 284–8.

[21] R. Aramian, 'Calculation of probabilities of affine shapes' (in Russian), *Izv. Acad. Nauk Armjan. SSR, Mathematics*, **XX**, 4, 1985, pp. 289–98.

[22] R. Deltheil, *Probabilités Géométriques.* Gauthier-Villars, Paris, 1926.

[23] J. J. Sylvester, 'On a funicular solution of Buffon's problem of the needle in its most general form', *Acta Math.*, **14**, 185–205, 1890.

[24] E. F. Harding, 'The number of partitions of a set of N points in k dimensions induced by hyperplanes', *Proc. Edinburgh Math. Soc.*, *II*, **15**, 285–9, 1967.

[25] L. Schläfli, *Gesammelte mathematische Abhandlungen*, vol. I, Birkhauser, Basel, 1950.

[26] D. Watson, 'On partitions of N points', *Proc. Edinburgh Math. Soc.*, *II*, **16**, 263–4, 1969.

[27] R. V. Ambartzumian, 'On combinatorial foundations of integral geometry', *Izv. Acad. Nauk Armjan. SSR, Mathematics*, **XVI**, 4, 1981, pp. 285–92.

[28] R. V. Ambartzumian, 'The solution to the Buffon–Sylvester problem in R^3', *Z. Wahrsch. Verw. Geb.*, **27**, 53–74, 1973.

[29] R. V. Ambartzumian, 'A synopsis of combinatorial integral geometry', *Adv. Math.* **37**, (1), July, 1980, pp. 7–15.

[30] H. Minkowski, 'Volumen und Oberfläche', *Mathematische Annalen*, **57**, 447–95, 1903.

[31] A. Ya. Khintchine, *Mathematical Methods in the Theory of Queueing*, Griffin, London, 1960.

[32] R. V. Ambartzumian, 'Palm distributions and superpositions of independent point processes', in *Stochastic Point Processes* (P. Lewis, eds.), Wiley-Interscience, New York, 1972.

[33] R. E. Miles, 'The various aggregates of random polygons determined by random lines in a plane', *Adv. Math.*, **10**, 256–90, 1973.

[34] R. V. Ambartzumian, 'Factorization in integral and stochastic geometry' in *Stochastic Geometry, Geometric Statistics, Stereology* (R. Ambartzumian and W. Weil eds.) Teubner-Texte zur Mathematic, B.65 (1984), pp. 14–35.

[35] J. J. Sylvester, *Collected Works.*

[36] G. L. L. Buffon, 'Essai d'arithmétique morale; supplement à *l'Histoire Naturelle*', **4**, Paris, 1977.

[37] A. Pleijel, 'Zwei kurze Beweise der isoperimetrischen Ungleichung', *Archiv Math.*, **7**, 317–19, 1956.

[38] A. Pleijel, 'Zwei kennzeichende Kreiseigenschaften der Kreis', *Archiv Math.*, **7**, 420–4, 1956.

[39] H. S. Sukiasian, 'Two results on triangular shapes', in *Stochastic Geometry, Geometrical Statistics, Stereology* (R. V. Ambartzumian and W. Weil, eds.), Teubner-Texte zur Math. B.65, 1984, pp. 210–21.

[40] V. K. Oganian, 'On triangular shapes in a planar Poisson point process', *Dokl. Akad. Nauk Armjan. SSR*, **81**, (2), 1985, pp. 59–63.

[41] D. G. Kendall, 'Shape manifolds, Procrustean metrics and complex projective spaces,' *Bull. London Math. Soc.* **16**, (2), 81–121, 1984.

[42] R. V. Ambartzumian, 'Stochastic geometry from the standpoint of integral geometry', *Adv. Appl. Prob.*, **9**, 792–823, 1977.

[43] R. Davidson, 'Construction of line processes: second order properties', in *Stochastic Geometry* (E. F. Harding and D. G. Kendall, eds.), pp. 148–61, Wiley, New York, 1974.

[44] V. K. Oganian, 'On Palm distributions of processes of lines in the plane', in *Stoch. Geometry, Geom. Statistics, Stereology* (R. V. Ambartzumian and W. Weil eds.) Teubner-Texte zur Mathematik B.65, 1984, pp. 124–32.

[45] G. Panina, *Translation-Invariant Measures and Convex Bodies in* $\mathbb{R}^3$. Zapiski Nauchnik Seminarov LOMI 157, 1987. (In Russian.)

[46] P. R. Halmos, *Measure Theory*, van Nostrand, New York, 1950.

[47] H. S. Sukiasian, 'Randomizable point systems' in *Stochastic and Integral Geometry*, *Acta Applicandae Mathematicae*, vol. 9, no 1–2, 1987.

[48] J. Mecke, 'An explicit description of Kallenberg's lattice type point process', *Math. Nachr.* **89**, 185–95, 1979.

[49] R. V. Ambartzumian, 'On random fiberfields in $\mathbb{R}^n$', *Dokl. Akad. Nauk SSSR*, **214**, (2), 245–8, 1974. (In Russian.)

[50] R. V. Ambartzumian, 'Homogeneous and isotropic point fields on the plane', *Math. Nachrichten*, **70**, 365–85, 1976. (In Russian.)

[51] B. Thiele, 'Eine Umkehrformel für einfache stationäre Punktprozesse endlicher Intensität im R^s, *Math. Nachrichten*, **9**, 171–9, 1971.

[52] R. V. Ambartzumian and B. S. Nahapetian, 'Palm distributions and limit theorems for point processes', *Dokl. Akad. Nauk Armjan SSR*, **LXXI**, (2), 1980.

[53] J. Mecke, 'Inequalities for intersection densities of superpositions of stationary Poisson hyperplane processes', *Proc. 2nd Internat. Workshop on Stereology and Stochastic Geometry*, 1983, memoirs No 6, Aarhus, 115–25.

[54] D. R. Cox, *Renewal Theory*, Methuen, London, 1962.

[55] H. Solomon, *Geometric Probability*, Society for Industrial and Applied Mathematics, Philadelphia, 1978.

[56] D. Husemoller *Fiber Bundles*, McGraw-Hill, New York, 1966.

[57] R. E. Miles, 'Random points, sets and tessellations on the surface of a sphere', *Sankhya, the Indian Journal of Statistics*, **33**, 145–71, 1971.

[58] R. V. Ambartzumian, 'Combinatorial integral geometry, metrics, and zonoids', *Acta Applicandae Mathematicae*, **9**, 1987, pp. 3–28.

[59] R. Davidson, 'Line processes, roads and fibers', in *Stochastic Geometry* (E. F. Harding and D. G. Kendall, eds.), Wiley, New York, 1974, pp. 248–51.

[60] S. Janson and O. Kallenberg, 'Maximizing the intersection density of fiber processes', *J. Appl. Prob.* **18**, 820–8, 1981.

[61] J. Mecke and C. Thomas, 'On an extreme value problem for flat processes', *Commun. Statist.-Stochastic Models*, **2** (2), 273–80, 1986.

[62] R. Aramian, 'Flag representations and curvature measures of convex bodies', *Izvestia Acad. Nauk Arm. SSR, Mathemat.* XXIII, no 1, 1988, pp. 97–101. (In Russian.)

[63] D. Stoyan, W. S. Kendall and J. Mecke, *Stochastic Geometry and Its Applications*, John Wiley, Chichester, 1987.

[64] R. V. Ambartzumian and H. S. Sukiasian, *Inclusion–Exclusion and Point Processes*. In press.

[65] E. Pinney. *Ordinary Difference-Differential Equations.* University of California Press, Berkeley, 1958.

[66] R. V. Ambartzumian and H. Sukiasian. 'On inner description of nonintersecting noninteracting balls', *Izvestia Akad. Nauk Armjan. SSR Mathem.* XVIII, No 3, 1983 pp. 206–15. (In Russian.)

[67] R. V. Ambartzumian. *Random Graph Approach to Gibbs Processes with Pair Interaction.* In press.

[68] D. G. Kendall. 'A survey of the statistical theory of shape', *Stat. Sci.*, **4**, no 2, 87–120, 1989.

INDEX OF KEY WORDS